Chemistry of Trace
Elements in Fly Ash

Chemistry of Trace Elements in Fly Ash

Edited by

Kenneth S. Sajwan
Savannah State University
Savannah, Georgia

Ashok K. Alva
USDA-ARS-PWA
Prosser, Washington

and

Robert F. Keefer
Formerly of West Virginia University
Morgantown, West Virginia

Kluwer Academic/Plenum Publishers
New York, Boston, Dordrecht, London, Moscow

Library of Congress Cataloging-in-Publication Data

Chemistry of trace elements in fly ash/edited by Kenneth S. Sajwan, Ashok K. Alva, and
  Robert F. Keefer.
        p.   cm.
    Papers presented as a part of the Sixth International Conference on Biogeochemistry
    of Trace Elements, held at the University of Guelph, Ontario, Canada, from July 29–August
  2, 2001.
    Includes bibliographical references and index.
    ISBN 0-306-47742-4
      1. Fly ash—Environmental aspects—Congresses.   2. Fly ash—Analysis—Congresses.   3.
  Trace elements—Analysis—Congresses.   I. Sajwan, Kenneth S.   II. Alva, Ashok K.   III.
  Keefer, Robert F.   IV. International Conference on Biogeochemistry of Trace Elements
  (6th: 2001: University of Guelph, Ontario)

    TD884.5.C46 2003
    621.31′2132′0287—dc21

                                                                          2003044719

Proceedings of the symposium Chemistry of Trace Elements in Fly Ash, which took place at the 6th
International Conference on the Biogeochemistry of Trace Elements, held July 29–August 2, 2001, at the
University of Guelph, Canada

ISBN: 0-306-47742-4

© 2003 Kluwer Academic/Plenum Publishers, New York
233 Spring Street, New York, New York 10013

http://www.wkap.nl/

10   9   8   7   6   5   4   3   2   1

A C.I.P. record for this book is available from the Library of Congress.

Permissions for books published in Europe: *permissions@wkap.nl*
Permissions for books published in the United States of America: *permissions@wkap.com*

Printed in the United States of America

IN GRATITUDE TO

My family, Maria, Mia, and Joseph
Who've been through it all before,

and

Dr. Les Evans
University of Guelph
For his major contribution in organizing the Sixth International Conference on the
Biogeochemistry of Trace Elements, of which the present work is a part.

CONTRIBUTORS

S.A. Aburime, Department of Engineering, Clark Atlanta University, 223 James P. Brawley Dr., Atlanta, GA 30314, USA.

Domy C. Adriano, Savannah River Ecology Laboratory, University of Georgia, Drawer E, Aiken SC 29802, USA.

A.K. Alva, USDA-ARS, Pacific West Area, Vegetable and Forage Crops Research Unit, 24106 N. Bunn Rd. Prosser, WA 99350, USA.

Christopher Barton, USDA Forest Service, Center for Forested Wetlands Research, c/o Savannah River ecology Laboratory, Drawer E, Aiken, SC 29802, USA.

D.K. Bhumbla, Division of Plant and Soil Sciences, West Virginia University, Morgantown, WV 26506-6108, USA.

Herbert Bryan, Tropical Research and Education Center, IFAS, University of Florida, Homestead, FL 33031, USA.

D. Chaudhuri, Department of Geology & Geophysics, Indian Institute of Technology, Kharagpur, West Bengal, India 721302.

C.S. Chetty, Department of Natural Sciences and Mathematics, Savannah Stte University, Savannah, GA 32404, USA.

Don Cowthon, Texas Agricultural Experiment Station and Tarlton State University, Stephenville, TX, 76402, USA.

R.M. Dankar, Advanced Analytical Center for Environmental Sciences, The University of Georgia, Savannah River Ecology Laboratory, Drawer E, Aiken, SC 29802, USA.

S.S. Dhaliwal, Division of Plant and Soil Sciences, West Virginia University, Morgantown, WV 26505-6108, USA.

Stanislaw Dudka, University of Georgia, Department of Crop and Soil Sciences, Athens, GA 30602-7272, USA.

Gian S. Ghuman, Department of Natural Sciences and Mathematics, Savannah State University, Savannah, GA 31404, USA.

B.R. Hart, Department of Earth Sciences, University of Western Ontario, London, Canada, N6A 5B7.

M.R. Hajarnavis, National Environmental Engineering Institute, Nagpur, MS 440020, India.

Zhenli He, Indian River Research and Education Center, IFAS, University of Florida, Ft. Piears, FL 34954, USA.

M. Hope-Simpson, Agrosystem Atlantic, P.O. Box 1674, Truro, Nova Scotia, Canada B2N 5Z5.

William Hopkins, Savannah River Ecology Laboratory, University of Georgia, P.O. Drawer E, Aiken SC 29802, USA.

K.H. Hostler, Citrus Research and Education Center, University of Florida, Lake Alfred, FL 33850, USA.

Brian P. Jackson, Savannah River Ecology Laboratory, University of Georgia, P.O. Drawer E, Aiken, SC 29802, USA.

Ursala Kaantee, Finnsementti Oy, FIN-21600 Parainen, Finland.

A.D. Karathanasis, Agronomy Department, University of Kentucky, Lexington KY 40546, USA.

Bon-Jun Koo, Advanced Analytical Center for Environmental Sciences, The University of Georgia, Savannah River Ecology Laboratory, Drawer E, Aiken, SC 29802, USA.

Anna S. Knox, Savannah River Ecology Laboratory, University of Georgia, Drawer E, Aiken SC 29802, USA.

Joanna Kyziol, Polish Academy of Sciences, Institute of Environmental Engineering, 34 M. Sklodowska-Curie St., 34, 41-819 Zabrze, Poland.

Yuncong Li, Tropical Research and Education Center, IFAS, University of Florida, Homestead, FL 33031, USA.

Shas V. Mattigod, Pacific Northwest National Laboratories, Richland, WA 99352, USA.

William P. Miller, University of Georgia, Department of Crop and Soil Sciences, Athens, GA 30602-7272, USA.

G.L. Mills, Savannah River Ecology Laboratory, University of Georgia, Aiken, SC 29802, USA.

Arun B. Mukherjee, Department of Limnology and Environmental Protection, P.O. Box 62, FIN-00014, University of Helsinki, Finland.

Linda Paddock, University of Georgia, Savannah River Ecology Laboratory, Drawer E, Aiken, SC 29802, USA.

S. Paramasivam, Department of Natural Sciences and Mathematics, Savannah State University, Savannah, GA 31404 USA.

M.A. Powell, Department of Earth Sciences, University of Western Ontario, London, Canada, N6A 5B7.

Tracy. Punshon, Savannah River Ecology Laboratory, University of Georgia, Drawer E, Aiken SC 29802, USA/Rutgers University, Division of Life Sciences, 604 Allison Road, Piscataway, NJ 08854, USA.

Thomas R. Quinn, Pacific North West National Laboratory, Richland, WA 99352, USA.

G.R. Reddy, Department of Natural Sciences and Mathematics, Savannah State University, Savannah, GA 31404, USA.

W. Richards, Nova Scotia Power Inc., Generation Services, Point Aconi Generating Station, P.O. Box 1609, Bras d'Or, Nova Scotia, Canada B1Y 3Y6.

Christopher Romanek, Department of Geology, University of Georgia, Athens, GA 30602, USA.

Kenneth S. Sajwan, Department of Natural Sciences and Mathematics, Savannah State University, Savannah, GA 31404, USA.

Maxim J. Schlossberg, University of Georgia, Department of Crop and Soil Sciences, Athens, GA 30602-7272, USA.

John C. Seaman, Savannah River ecology Laboratory, University of Georgia, Drawer E, Aiken SC 29802, USA.

B.S. Sekhon, Division of Plant and Soil Sciences, West Virginia University, Morgantown, WV 26506-6108, USA.

Jennifer S. Simmons, West Virginia Water Research Institute, National Mine Land Reclamation Center, West Virginia University, Morgantown, WV 26506-6064, USA.

Gulab Singh, Central Fuel Research Institute, CSIR-Council for Scientific and Industrial Research, Dhanbad, 828108, India.

John Sloan, Texas Agricultural Experiment Station, Dallas, TX 75252, USA.

Sesbastian Stefaniak, Polish Academy of Sciences, Institute of Environmental Engineering, 34 M. Sklodowska-Curie St., 34, 41-819 Zabrze, Poland

Peter Stoffella, Indian River Research and Education Center, IFAS, University of Florida, Ft. Piears, FL 34954, USA.

Jadwiga Szczepanska, University of Mining and Metallurgy, Department of Hydrogeology and Water Protection, 30 Mickiewicza Av., 30-059 Krakow, Poland.

Prem S.M. Tripathi, Central Fuel Research Institute, CSIR-Council for Scientific and Industrial Research, Dhanbad 828108, India.

S. Tripathy, Department of Geology & Geophysics, Indian Institute of Technology, Kharagpur, West Bengal, India 721302.

Irena Twardowska, Polish Academy of Sciences, Institute of Environmental Engineering, 34 M. Sklodowska-Curie St., 34, 41-819 Zabrze, Poland.

D. van Cliff, Citrus Research and Education Center, University of Florida, Lake Alfred, FL 33850, USA.

H. Veeresh, Department of Geology & Geophysics, Indian Institute of Technology, Kharagpur, West Bengal. India 721302.

Min Zhang, Tropical Research and Education Center, IFAS, University of Florida, Homestead, FL 33031, USA.

Ron Zevenhoven, Energy Engineering and Environmental Protection, Helsinki University of Technology, P.O. Box 4400, FIN-02015 Espoo, Finland.

Paul F. Ziemkiewicz, West Virginia Water Research Institute, National Mine Land Reclamation Center, West Virginia University, Morgantown, WV 26506-6064, USA.

PREFACE

The accumulation of large amounts of ash from fossil fuel combustion for electric power plant generation is becoming a major environmental concern in the United States. Furthermore, stringent environmental regulations mandated by the Environmental Protection Agency through the Clean Air Act, Clean Water Act, Resource Conservation and Recovery Act, as well as state and local environmental regulations may result in even more ash production with subsequent contact with the environment. The concentrations of trace elements in coal residues are extremely variable and depend on the composition of the original coal, conditions during combustion, the efficiency of emission control devices, storage and handling of byproducts, and climate.

The research papers in this book were presented as a part of the Sixth International Conference on the Biogeochemistry of Trace Elements held at the University of Guelph, Ontario, Canada, from July 29-August 2, 2001. The purpose of this conference was to present current knowledge on the source, pathways, behavior and effects of trace elements in soils, waters, plants and animals. In addition, the book also includes invited research papers from scientists who have done significant research in the area of coal and coal combustion byproducts. All the research papers presented herein have been subjected to peer review. The editors have arranged the articles systematically by topic, beginning with the introductory chapter entitled "The Production and Use of Coal Combustion Products" followed by the sections on Environmental Impacts of Coal Combustion Residues, Trace Elements in Fly Ashes, Transport and Leachability of Metals from Coal and Coal Ash Piles, and the Use of Coal Ash as Agricultural Soil Amendment.

This book has been published for a variety of readers, including public health and environmental professionals, industrial hygienists, environmental consultants, waste management professionals, and academicians. It may also prove valuable to scientists conducting research on coal and coal combustion byproducts.

The editors wish to thank the contributing authors for their diligence in providing the changes requested by the reviewers and for their patience in waiting so long for those to go into print.

Kenneth S. Sajwan
Ashok K. Alva
Robert F. Keefer

CONTENTS

INTRODUCTION

ENVIRONMENTAL IMPACT OF COAL COMBUSTION RESIDUES

TRECE ELEMENTS IN FLY ASHES

THE PRODUCTION AND USE OF COAL COMBUSTION PRODUCTS

Tracy Punshon, John C. Seaman[1], and Kenneth S. Sajwan[2]

[1] Savannah River Ecology Laboratory, University of Georgia, Drawer E, Aiken SC 29802 USA
[2] Department of Natural Sciences and Mathematics, Savannah State University Savannah GA 31404 USA

ABSTRACT

Coal combustion byproducts (CCBs) arising from energy generation are the most abundant waste streams worldwide. Legislation aimed at reducing environmental pollution associated with coal combustion will continue to add to this waste stream into the future, increasing the need to develop pertinent and safe end uses for these materials. While production of CCBs continues to rise so also do the costs associated with their disposal and landfilling. This chapter presents updated information about the production of the main categories of CCB in the U.S., outlining their individual characteristics and describing their various end uses. Further, it introduces the reader to current research on potential novel end uses of CCBs, and their effect on the environment.

INTRODUCTION

During 2000, 860 million metric tons (Mt) of coal were burned in the U.S.[1], producing 98 million Mt of coal combustion products (CCPs). This term is used in reference to the various residues arising from the combustion of coal for electrical energy, and has evolved over recent years from 'coal combustion waste' and 'byproduct'. The changing nomenclature indicates the increasing recognition of its standing as a potentially beneficial commodity[1, 2]. CCPs consist of fly ash (FA), bottom ash (BA), boiler slag (BS) and flue gas desulfurization residue (FGD or synthetic gypsum), the latter being the most recent addition to the product group following the Clean Air Act of 1990. Currently, FA is the most abundant of the CCPs; worldwide, the coal-fired power plants produce in excess of 500 million Mt FA every year[3]. Disposal of these products through insufficient reuse is an immense burden on the environment, and the development of economically and environmentally sound uses is a recognized global need[4-6]. The majority of unused CCPs are disposed of in lagoons[7], disposal mounds and landfill sites, although due to diminishing space and increasing expense associated with landfilling the latter option is becoming increasingly scarce and, research shows, environmentally unsafe in the long-term[8, 9].

The main hazards associated with CCPs arise from the potentially toxic trace metals, metalloids and excessive concentrations of soluble salts that leach from them[10], provoking various toxicity responses in biota[11]. Worldwide, CCPs are most commonly employed in various civil engineering applications; such as concrete production and road construction[12], although there is a great need to expand potential uses for these abundant materials. Overall, the total amount of CCPs used in the U.S. in 2000 amounted to only 29% of the total produced[1], leaving 69 million Mt for disposal. Finding disposal and recycling solutions congruent with good environmental stewardship represents an enormous challenge to science and engineering. Recent scientific research carried out on these materials – included in the following chapters of this volume – continues to form a vital information base for safe CCP reuse, disposal, and risk assessment.

FLY ASH

Fly ash is a fine powdered ferroaluminosilicate material made up of hollow, glassy particles enriched with Ca, K and Na[13, 14], and with various trace elements (such as As, B, Mo, Se and Sr) condensed upon the surface[13, 15, 16]. Various types of particles can be found within composite FA[17]; *cenospheres* – true hollow particles, *plerospheres* – filled will smaller aggregations known as microspheres, and opaque magnetite spheres, which are related to the pyritic content of the source coal[15]. Due to its fine composition, FA is collected by mechanical filters or electrostatic precipitators from the flue gas. It is also pozzolanic in nature; a siliceous (or combination siliceous and aluminous) material that, when in finely divided form and in the presence of moisture, chemically reacts with calcium hydroxide at normal temperatures to form cementitious compounds[18].

In the United States, classification of FA by American Standards of Testing and Materials (ASTM C618) separates the ashes into either class F or C. Specifications used in the rest of the world have been based on this classification[18]. In the European Union (EU) a common specification known as EN 450 is being developed, although classifications tend to be specific to each country of origin. In the U.S. class F coal is produced from burning anthracite or bituminous coal, and class C from subbituminous coal characteristic of the western U.S.[19]. Class C ashes are pozzolanic, and can be used for the production of cement, whereas Class F are not. This distinction greatly influences potential re-use options. In general, fly ashes from subbituminous and lignite coals contain greater quantities of CaO, MgO and SO_3[18]. A minimum limit of 50% is set for the content of SiO_2, Al_2O_3, and Fe_2O_3 in class C ash, and 70% for class F.

BOTTOM ASH AND BOILER SLAG

Bottom ash is defined as an uncombusted material that settles to the bottom of the boiler; boiler slag is formed when operating temperatures exceed ash fusion temperature and the slag remains in a molten state until it is drained from the bottom of the combustion chamber[20]. Bottom ash is granular and is similar to concrete sand[4]. Boiler slag is a shiny, black granular material that has abrasive properties, and is often used as grit for snow and ice control, structural embankments, aggregate and as road base material (Table 1). The utilization potential of BA is influenced by its physical characteristics, for example the grain-size distribution, staining potential and colour[21]. In the scientific literature, the BA derived from coal and municipal solid waste (MSW) are frequently confused; and some workers suggest that these materials have considerable similarities[22], or are similar in nature to FA. However, in the present volume, which focuses on the chemical properties of the CCPs, distinctions are made between those arising from coal, and those from the combustion of other solid materials.

FLUE GAS DESULFURIZATION RESIDUES

The Clean Air Act Amendments of 1990 (CAAA '90 Public Law 101-549) placed stringent restrictions on the production of sulfur oxide (SO_x) from coal combustion, with a two phase implementation plan, which required electric utility companies to reduce SO_2

Table 1. Production and use of category I (Dry) and II (ponded) coal combustion product (CCP) in the US during the year 2000 (million metric tons)[35].

	Fly Ash	Bottom ash	Boiler slag	FGD residue
Produced	**57.14**	**15.38**	**2.43**	**23.28**
Used				
Cement, concrete or grout materials	9.610	0.381	*	0.319
Raw feed for cement clinker	1.029	0.158	*	*
Flowable fill	0.632	0.010	0.016	0.030
Structural fill	2.370	1.227	0.032	0.497
Road base/sub base	1.096	0.759	*	0.085
Soil modification	0.102	0.025	*	0.000
Mineral filler	0.108	0.093	0.011	0.001
Snow and ice control	0.003	0.755	0.053	*
Blasting grit/roofing granules	*	0.133	1.905	*
Mining applications	1.045	0.333	*	0.166
Wallboard	*	*	*	3.022
Waste stabilization or solidification	1.803	0.032	*	0.019
Agricultural application	0.013	0.004	*	0.069
Miscellaneous	0.414	0.572	0.089	0.173
Total	**18.407**	**4.483**	**2.108**	**4.380**
% Use	**31.9%**	**29.2%**	**86.5%**	**18.8%**

Total Produced	**98.22**
Total Used	**29.19**
Total % Use	**29.7%**

emissions thereby reducing atmospheric pollution and the incidence of acid rain[23]. The majority of utility companies previously used high-sulfur bituminous coal, which was thought to have significantly contributed to incidences of acid rain in North America. Following the instatement of the act, many companies switched to low-sulfur coal or fuel oil for partial and rapid compliance with the new regulations, although retrofitting power plants with flue-gas scrubbing systems was ultimately necessary to fully comply with the act. This change effectively resulted in the creation of a new waste stream, termed flue-gas desulfurization residue (FGD).

Flue gas desulfurization residue is an alkaline material produced when SO_x is extracted from the flue gas of coal-fired power plants using a range of sorbents[24, 25]. There are several technologies currently in use, differentiated by the sorbent used and the method of SO_x extraction. They are: (1) lime or limestone-based, (2) magnesium-based, (3) ammonium sulfate based (used in Europe[23]) and (4) dry injection techniques. Respectively they generate: (1) calcium sulfite ($CaSO_3$) and calcium sulfate ($CaSO_4$); (2) magnesium sulfite ($MgSO_3$) and magnesium sulfate ($MgSO_4$); (3) ammonium sulfite [$(NH_4)SO_3$], ammonium sulfate [$(NH_4)_2SO_4$], ammonium carbonate [$(NH_4)_2CO_3$] and (4) sodium carbonate (Na_2CO_3) and sodium sulfate (Na_2SO_4). Dry FGD systems have a number of advantages over wet-production systems; they are less costly to maintain (because processes such as thickening, centrifugation, mixing and vacuum filtration are not required), they save energy by removing the need for reheating and pumping steps, and they can be handled by existing systems ordinarily used for FA. The only disadvantage, and one which is by far outweighed by the advantages, is that relatively more sorbent is needed to efficiently scrub flue gases in a dry system, because the solid-gas reaction proceeds at a slower rate than the liquid-gas reaction[23].

FGD residues are a rapidly changing group of CCPs; research is continually underway that aims to increase scrubbing efficiency[26] and this results in higher sulfur content of the final product. In common with other CCPs, the quality of the product also varies, based on the type of coal burned, the type of scrubbing system used (i.e., wet or dry[25]) and the handling and stabilization procedures. Stabilization usually takes the form of mixing the FGD with FA[27], and this often changes the re-use options of the stabilized material. Fly ash and additional quicklime are usually added to stabilize FGD filter cake prior to landfilling[28]. Due to their alkaline nature, FGD residues have potential value as neutralizing agents[2, 28] for agricultural soils which suffer from excessive acidity[29], or for the alleviation of excessive sodicity[30, 31]. Beneficial reuse of FGD and other CCPs is generally based on bulk chemical analysis, although recent investigative analyses of typical FGD and FA material demonstrate heterogeneous distribution of chemical species within stabilized FGD residues[32], which typically consist of mixtures of CCPs, excess sorbents and reaction products. Further, the stability of these species over time cannot be presumed; this is of particular importance when the compressive strength of CCP-based construction materials is a feature of its use. For example, ettringite [$Ca_6Al_2(SO_4)_3(OH)_{12} \cdot 26H_2O$] minerals, which degrade the strength of cementitious minerals, can form over time and may also influence the movement and sorption reactions of potentially toxic trace elements[33, 34].

PRODUCTION

The extent of CCP production and use in 2000 are shown in Table 1, expressed as millions of Mt[35]. Currently, the most widely employed applications of CCPs remain in civil engineering[35]. The greatest re-use of FA and BA are in structural fill materials, with 2.3 and 1.3 million Mt used respectively. For BS, the majority is applied as blasting grit, or used in the manufacture of roofing materials – representing a particularly efficient reuse rate for this material (Table 1); whereas FGD residues are primarily used as wallboard. Only 0.01 million Mt FA is currently applied to agricultural soils. There is an

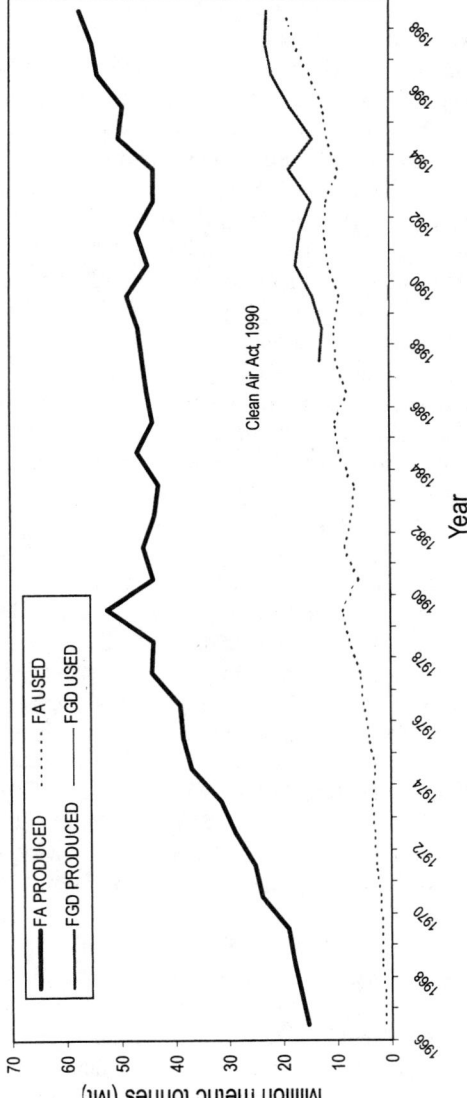

Figure 1. Production and use (million metric tons) of coal fly ash (FA) and flue gas desulfurization (FDG) residue between 1966 and 1998[64]

abundance of experimental work that has shown enrichment of potentially toxic trace elements in plants grown on FA treated soils[25, 36-49], including (but not limited to) B, Mo, As and Se. The leaching characteristics of the various products are the main limiting factor to their widespread and safe use.

USE OF CCPs

Analysis of production and reuse data for abundant products such as FA in previous years (Figure 1) shows that the percentage of total FA that is recycled has increased concurrently with increasing FA production; from 7% in 1966 to 29% in 2000; an increase in use of approximately 0.5 million Mt per year. On average the production of FA has increased approximately 4.5% per year between 1966 and the present, or approximately 1.3 million Mt a year. The amount of CCPs produced and our reliance on coal as an energy source maintains the need for advances in both science and engineering to increase reuse rates considerably. FGD is a relatively new material, and as such the rate of use is much lower, as research strives to find safe applications for the material. Rates of reuse of this material remain low; and are still below 10%. This is a concern because the production of FGD residues is quickly increasing, with only 12.9 million Mt produced in 1987 and approximately 23 million Mt produced in 2000 (Table 1).

Novel uses of CCPs have been found in the fields of restoration, remediation and stabilization. In terms of restoration, the combination of FA with other abundant industrial by-products such as sewage sludge[50] and biosolids, can result in a high quality restorative material[2, 51, 52], and thus applied to eroding soils that require physical stabilization in addition to chemical improvement. Application of FA to soils has been shown to improve the physical quality of the soil by introducing different sized particles into the profile, enhancing the water-holding capacity of sandy soils, and improving drainage in clay soils[51, 53]. Enrichment of various potentially toxic trace elements is a frequently undesirable effect, and in terms of limiting potential health risks, moderate rates of CCP are recommended for application to the majority of soils. However, an important stage in the effective recycling of any abundant waste material lies in determining what it primarily contains, and where these constituents are currently in short supply. One novel and safe application of CCPs may lie in using the trace elements they contain to amend nutrient deficiencies in depleted, disturbed or eroded soils. Punshon et al.,[51] applied FA and poultry biosolids to the site of previous soil-borrow area, in an attempt to replace lost topsoil and revegetate the site. The study was conducted in the southeastern U.S., an area with a typically leached and nutrient deficient soil profile. The eroded soils benefited from co-application of FA and biosolids, showing slower water infiltration rates, and a lower bulk density. Further, application of poultry biosolid in the mixture amended Cu and Zn deficiency. Sajwan et al[50] combined FA with sewage sludge (SS) and applied various mixture ratios (SS:FA of 4:1, 4:2, 4:3 and 4:4) to a crop of *Sorghum vulgaris* var. Sudanese Hitche. ("sorgrass") and found stimulation in biomass at rates of 50-100 tons acre[-1] of all ratios of SS:FA mixtures. Again, safe application rates have to be determined prior to use. The combination of by products such as CCPs and biosolid materials (both of which pose environmental health concern when stockpiled individually in large quantities) can produce a material which contains both trace elements, nutrients and organic matter.

When monitoring studies are extended for greater periods of time, deleterious environmental effects such as the increase in electrical conductivity of the soil solution – a common side-effect of FA application[54] – can be seen to subside after several years[25]. Further research indicates that there are many more uses for FA. FA is similar to volcanic ash, also an aluminosilicate-based material, which forms zeolite when alkalized[3]. Zeolite materials have many uses as drying agents, deodorants, water-softening agents, soil conditioners and as *in situ* soil amendments[55-58], and regions of the

6

world in which these substances occur naturally are limited[3]. Furthermore, processing FA into potassium silicate fertilizers is an additional proposed end use[3]. Potassium silicate fertilizers produced from FA are soluble in weak citric acid solutions, which prevents rainwater leaching from the soil profile, and retains the nutrients for plant uptake[59]. This would also recycle nutrients such as Ca, Mg and Fe back into the soil. Finally there are even studies which indicate that FA can be used in the manufacture of sorbents for use in dry injection flue gas desulfurization processes[60].

Reuse of FGD materials, although slower at present, may have considerably more potential, especially if technologies continue to produce purer materials. Clark et al.[30] list the beneficial effects of applying FGD material to soil for remedying various chemical and physical soil problems. Amendment can mitigate soil acidity, provide a source of nutrients to plants and animals (e.g. Mg, K, Zn, Cu and B), reduce Al toxicity[61], reduce surface crusting, soil compaction as well as improving water infiltration and holding capacity to prevent erosion and reduce sodicity[30, 31]. However, research into increasing the safety of civil engineering applications of FGD materials is also ongoing [62, 63], so that these materials can be used without the hazards associated with metals or salts leaching out.

CONCLUSION

There is great impetus for continued research into the beneficial reuse and recycling of CCPs. The information presented here indicates both the enormity of the tasks facing science and engineering in finding suitable end uses for CCP, and the immense potential that exists for making coal combustion for energy a more sustainable process.

ACKNOWLEDGEMENTS

This work was supported by Financial Assistance Award Number DE-FC09-96SR18546 from the Department of Energy to the University of Georgia Research Foundation.

REFERENCES

1. Kalyoncu, R. Coal combustion products http://minerals.usgs.gov/minerals/pubs/commodity/coal/874400.pdf, 2000.
2. Schlossberg, M.J., Sumner, M., Miller, W.P. and Dudka, S. Utilization of coal combustion by-products (CBP) in horticulture and turfgrass industries: technical and environmental feasibility studies. Proceedings of *6th International Conference on the Biogeochemistry of Trace Elements*. 2001. Guelph, Canada.
3. Kikuchi, R. Application of coal ash to environmental improvement. Transformation into zeolite, potassium fertilizer, and FGD absorbent. *Resources Conservation & Recycling* 27, 333-346. 1999.
4. Keefer, R.F. Coal ashes - Industrial wastes or beneficial byproducts?, in *Trace Elements in Coal and Coal Combustion Residues.*, R.F. Keefer and K.S. Sajwan, Editors. 1993, Lewis Publishers: Ann Arbor. 3-9.
5. Horiuchi, S., Kawaguchi, M. and Yasuhara, K. Effective use of fly ash slurry as fill material. *Journal of Hazardous Materials* 76, 301-337. 2000.
6. Praharaj, T., Powell, M.A., Hart, B.R. and Tripathy, S. Leachability of elements from sub-bituminous coal fly ash from India. *Environment International* 27, 609-6154. 2002.

7. Twardowska, I., Tripathi, P.S.M. and Das, R.P. Trace elements and their mobility in coal ash/fly ash from Indian power plants in view of its disposal and bulk use in agriculture. Proceedings of *6th International Conference on the Biogeochemistry of Trace Elements*. 2001. Guelph, Canada.

8. Suter II, G.W., Luxmoore, R.J. and Smith, E.D. Compacted soil barriers at abandoned landfill sites are likely to fail in the long term *Journal of Environmental Quality* 22, 217-226. 1993.

9. Danker, R., Adriano, D.C., Barton, C. and Punshon, T. Revegetation of a coal fly ash - reject landfill. Proceedings of *6th International Conference on the Biogeochemistry of Trace Elements (ICOBTE)*. 2001. Guelph, Canada.

10. Adriano, D.C., Page, A.L., Elseewi, A.A., Chang, A.C. and Straughan, I. Utilization and disposal of fly ash and other coal residues in terrestrial ecosystems. A review. *Journal of Environmental Quality* 9, 333-334. 1980.

11. Jackson, B., Shaw-Allen, P., Mills, G., Hopkins, W. and Jagoe, C. Trace element-protein interactions in fish from a fly ash settling basin. A study using size exclusion chromatography coupled to ICP-MS. Proceedings of *6th International Conference on the Biogeochemistry of Trace Elements*. 2001. Guelph, Canada.

12. ACAA *Fly Ash Facts for Highway Engineers*. 1995, American Coal Ash Association Inc., 2760 Einsenhower Avenue, Suite 304, Alexandria, Virginia, 22314: Washington D.C.

13. Page, A.L., Elseewi, A.A. and Straughan, I. Physical and chemical properties of fly ash from coal-fired power plants with references to environmental impacts. *Residue Rev.* 71, 83-120. 1979.

14. El-Mogazi, D.D., Lisk, D.J. and Weinstein, L.H. A review of physical, chemical and biological properties of fly ash and effects on environmental ecosystems. *Science of the Total Environment* 74, 1-37. 1988.

15. Mattigod, S.V., Rai, D., Eary, L.E. and Ainsworth, C.C. Geochemical Factors Controlling the Mobilization of Inorganic Constituents from Fossil Fuel Combustion Residues: I. Review of the major elements. *J. Environ Qual.* 19, 188-201. 1990.

16. USEPA *Wastes from the combustion of coal by electricity power plants*. 1988, U.S. Environmental Protection Agency: Washington DC.

17. Fisher, G.L. and Natusch, D.F.S. Size dependence of the physical and chemical properties of coal fly ash., in *Analytical methods for coal and coal products.*, C. Karr Jr., Editor. 1979, Academic Press: New York. 489-541.

18. Manz, O.E. Coal fly ash: a retrospective and future look. *Fuel* 78, 133-136. 1999.

19. Halstead, W.J. *Use of fly ash in concrete*. 1986, Transportation Research Board,NRC: Washington.

20. Bluedorn II, D.C. Recent environmental regulation of coal combustion wastes - revised. Proceedings of *2001 Conference on Unburned Carbon (UBC) on Utility Fly Ash*. 2001: National Energy Technology Laboratory.

21. Kula, I., Olgun, A., Sevine, V. and Erdogan, Y. An investigation on the use of tincal ore waste, fly ash and coal bottom ash as Portland cement replacement materials. *Cement and Concrete Research* 32, 227-232. 2002.

22. Goodwin, R.W. and Schuetzenduebel, W.G. Residues from mass burn systems: testing, disposal and Utilization issues. Proceedings of *New York State Legislative Commission of Solid Waste Management and Materials Policy Conference*. 1988. New York,.

23. Kalyoncu, R. Coal Combustion Products http://minerals.usgs.gov/minerals/pubs/commodity/coal/874498.pdf, 1998.

24. Punshon, T., Knox, A.S., Adriano, D.C., Seaman, J.C. and Weber, T.J. Flue Gas Desulfurization residue (FGD): Potential Applications and Environmental Issues., in *Biochemistry of Trace Elements in Coal and Coal Combustion Byproducts.*, K.S. Sajwan and R.F. Keefer, Editors. 1999, Lewis Publishers: Boca Raton, FL. 7-28.

25. Punshon, T., Seaman, J.C. and Adriano, D.C. The effect of flue gas desulfurization residue on corn (*Zea mays* L.) growth and leachate salinity: Multiple season data from amended mesocosms., in *Chemistry of Trace Elements in Fly Ash*, K.S. Sajwan, A.K. Alva, and R.F. Keefer, Editors. 2002, Kluwer Academic/Plenum Press: New York, NY. In press.

26. Baege, R. and Sauer, H. Recent developments in CFB-FGD technology. *VGB Powertech* 80(2), 57-60. 2000.

27. Punshon, T., Adriano, D.C. and Weber, J.T. Effect of Flue Gas Desulfurization Residue on Plant Establishment and Soil and Leachate Quality. *J. Environ. Qual.* 30(3), 1071-1080. 2001.

28. Chen, L., Dick, W.A. and Nelson, S. Flue gas desulfurization addition to acid soil: alfalfa productivity and environmental quality. *Environmental Pollution* 114(2), 161-168. 2001.

29. Stehouwer, R.C., Sutton, P. and Dick, W.A. Transport and plant uptake of soil-applied dry flue gas desulfurization by products. *Soil Science* 161(9), 562-574. 1996.

30. Clark, R.B., Ritchey, K.D. and Baligar, V.C. Benefits and constraints for use of FGD products on agricultural land. *Fuel* 80(6), 821-828. 2001.

31. Chun, S., Nishiyama, M. and Matsumoto, S. Sodic soils reclaimed with by-product from flue-gas desulfurization: corn production and soil quality. *Environmental Pollution* 114(3), 453-459. 2001.

32. Laperche, V. and Bigham, J.M. Quantitative, chemical and mineralogical characterization of flue gas desulfurization by-products. *Journal of Environmental Quality* 31(3), 979-988. 2002.

33. Day, R.L. *The effect of secondary ettringite formation on the durability of concrete: A literature analysis.* 1992, Department of Civil Engineering, University of Calgary: Alberta, Canada.

34. McCarthy, G.J., Hasset, D.G. and Bender, J.A. Synthesis, crystal chemistry and stability of ettringite, a material with potential in hazardous waste immobilization., in *Advanced Cementitious Systems: Mechanisms and Properties. Materials Research Society Symposium Proceedings 245.* 1992, Elsevier, NY.

35. ACAA 2000 Coal Combustion Product (CCP) Production and Use. http://www.acaa-usa.org/CCP%20Survey/PDF/00SurveyComplete.PDF, 2000.

36. Dosskey, M.G. and Adriano, D.C. Trace element toxicity in VA mycorrhizal cucumber grown on weathered coal fly ash. *Soil Biol. Biochem.* 25(11), 1547-1552. 1993.

37. Arthur, M.A., Rubin, G. and Woodbury, P.B. Uptake and accumulation of selenium by terrestrial plants growing on a coal fly ash landfill. 2. Forage and root caps. *Environ. Toxicol. Chem.* 11(9), 1289-1299. 1992.

38. Aitken, R.L. and Bell, L.C. Plant uptake and phytotoxicity of boron in Australian fly ashes. *Plant and Soil* 84, 245-257. 1985.

39. Cary, E.E., Gilbert, M., Bache, C.A., Gutenmann, W.H. and Lisk, D.J. Elemental composition of potted vegetables and millet grown on hard coal bottom ash-amended soil. *Bulletin of Environmental Contamination & Toxicology* 31, 418-423. 1983.

40. Furr, A.K., Parkinson, T.F., Gutenmann, W.H., Pakkala, I.S. and Lisk, D.I. Elemental content of Vegetables, Grains and Forages Field-Grown on Fly-Ash Amended Soil. *J. Agric. Food Chem.* 26(2), 357-359. 1978.

41. Furr, A.K., Parkinson, T.F., Elfving, D.C., Gutenmann, W.H., Pakkala, I.S. and Lisk, D.J. Elemental Content of Apple, Millet, and Vegetables grown in pots of neutral soil amended with fly ash. *J. Agric. Food Chem.* 27(1), 135-138. 1979.

42. Furr, A.K., Kelly, W.C., Bache, C.A., Gutenmann, W.H. and Lisk, D.J. Multielement Uptake by Vegetables and Millet Grown in Pots on Fly Ash Amended Soil. *J. Agric. Food Chem.* 24(4), 885-888. 1976.

43. Sale, L.Y., Naeth, M.A. and Chanasyk, D.S. Growth response of Barley on Unweathered Fly Ash Amended soil. *J. Environ Qual.* 25, 684-691. 1996.

44. Elseewi, A.A. and Page, A.L. Molybdenum enrichment of plants grown on fly-ash amended soils. *J. Environ Qual.* 13, 394-398. 1984.

45. Fail, J.L. and Wochok, Z.S. Soybean growth on fly ash-amended strip mine spoils. *Plant & Soil* 48, 473-484. 1977.

46. Gissel-Nielsen, G. and Bertelsen, F. Inorganic element uptake by barley from soil supplemented with flue gas desulfurization waste and fly ash. *Environmental Geochemistry and Health* 10, 21-25. 1988.

47. Kukier, U. and Sumner, M.E. Boron availability to plants from coal combustion by-products. *Water, Air & Soil Pollution* 87(1-4), 93-110. 1996.

48. Gollany, H.T., Bloom, P.R., Thomes, M.R., Gustin, F., Hassett, D. and Roffman, H. Mobilization of B, Hg, Mo, Na, S and V from a Coal Fly Ash Stabilized Soil. Proceedings of *6th International Conference on the Biogeochemistry of Trace Elements (ICOBTE)*. 2001. Guelph, Canada.: ICOBTE.

49. Yuncong, L., ZHnag, M., Stoffella, P., Bryan, H. and He, Z. Influence of fly ash compost application on distribution of metals in soil, water and plants. Proceedings of *6th International Conference on the biogeochemistry of Trace Elements*. 2001. Guelph, Canada.

50. Sajwan, K.S., Ornes, W.H. and Youngblood, T. The effect of fly ash/sewage sludge mixtures and application rates on biomass production. *Journal of Environmental Science and Health* A30(6), 1327-1337. 1995.

51. Punshon, T., Adriano, D.C. and Weber, J.T. Restoration of drastically eroded land using coal fly ash and poultry biosolid. *The Science of the Total Environment* In press. 2002.

52. Sloan, J.J. and Cawthorn, D. Mine soil remediation using coal ash and compost mixtures. Proceedings of *6th International Conference on the Biogeochemistry of Trace Elements*. 2001. Guelph, Canada.

53. Chang, A.C., Lund, L.J., Page, A.L. and Warneke, J.E. Physical properties of fly-ash amended soils. *Journal of Environmental Quality* 6, 267-270. 1977.

54. Carlson, C.L. and Adriano, D.C. Environmental impacts of coal combustion residues *Journal of Environmental Quality* 22(2), 227-247. 1993.

55. Chen, X., Wright, J.V., Conca, J.L. and Peurrung, L.M. Evaluation of heavy metal remediation using mineral apatite. *Water, Air and Soil Pollution* 98, 57-78. 1997.

56. Chlopecka, A. and Adriano, D.C. Influence of zeolite, apatite and Fe-oxide on Cd and Pb uptake by crops. *The Science of the Total Environment* 207, 195-206. 1997.

57. Ma, Q.Y., Traina, S.J., Logan, T.J. and Ryan, J.A. In situ lead immobilization by apatite *Environ. Sci. Technol.* 27(9), 1803-1810. 1993.

58. Edwards, R., Rebedea, I., Lepp, N.W. and Lovell, A.J. An Investigation into the Mechanism by which Synthetic Zeolites Reduce Labile Metal Concentrations in Soil. *Environmental Geochemistry and Health* 21, 157-173. 1999.

59.   Yorita, G. Utilization of coal ash from the power industry. *Bulletin of the Electricity Association* 3, 26-28. 1993.

60.   Nagashima, N., Arashi, N. and Kanda, O. Operation results of the first commercial dry desulfurization plant in Hokkaido Electric Power Co. Proceedings of *SO$_2$ Control Symposium*. 1993. Boston: Electrical Research Institute.

61.   Stout, W.L., Sharpley, A.N. and Pionke, H.B. Reducing soil phosphorus solubility with coal combustion by products. *J. Environ Qual.* 27, 111-118. 1998.

62.   Schlieper, H., Duda, A., Jager, R., Kanig, M. and Kwasny-Echterhagen, R. FGD Gypsum - A raw material for new binder systems. *ACAA Compendium* 1973-1997, Paper 97-48. 1997.

63.   Taulbee, D.N., Graham, U.M., Rathbone, R., Robl, T.L. and Schram, W.H. Leaching characteristics of light-weight pellets prepared from both wet and dry FGD wastes. *ACAA Compendium, 1973-1997* (Paper 97-40). 1997.

64.   Kelly, T.D. and Kalyoncu, R. Coal combustion products statistics. http://minerals.usgs.gov/minerals/pubs/of01-006/coalcombustionproducts.html, 2001.

OCCURRENCE AND MOBILIZATION POTENTIAL OF TRACE ELEMENTS FROM DISPOSED COAL COMBUSTION FLY ASH

Irena Twardowska[1], Jadwiga Szczepanska[2], and Sebastian Stefaniak[1]

[1]Polish Academy of Sciences
Institute of Environmental Engineering
34 M.Sklodowska-Curie St., 41-819 Zabrze, Poland
[2]University of Mining and Metallurgy
Department of Hydrogeology and Water Protection
30 Mickiewicza Av., 30-059 Krakow, Poland

ABSTRACT

Despite many beneficial properties, high amounts of fly ash (FA) from coal-fired power plants have been disposed in landfills. These amounts are continuously growing in consequence of the limited possibility to utilize all generated FA due to the lack of market for it, or too high shipping/handling costs compared to the natural competing materials. Besides landfilling, the increasing utilization of FA as agricultural soil amendment that give an opportunity to dispose this material in a big scale also leads to its spreading on the vast land surface where it is exposed to the atmospheric conditions. To assess the effect of weathering processes on the mobilization potential of trace elements in disposed FA, the unique studies were carried out that comprised sampling FA along the vertical profile of 12 years' old FA pond in the post-closure period, extracting pore solution from the material by a pressure method and direct analysis of its chemical composition using ICP-OES technique. Studied FA represented alkaline aluminum silicate material of a composition and trace element enrichment within the range typical for FA from the majority of other hard coal-fired power plants. The chemical composition of pore solutions along the three vertical profiles in the fly ash disposal pond in the post-closure period after 12 years' operation was found to reflect both the altered water flow (vertical downward redistribution of ions) and the changed equilibria conditions. While the pore solution in FA in H-6 profile reflects the Dissolution stage (II) (pH 7-10), in the looser FA profiles H-2 and H-3 it indicates alteration of buffering properties of the system that can be defined as Delayed Release (III) stage. The character of pore solutions along these profiles suggests that the major buffering mechanisms controlling pH after depletion of carbonates comprise reactions involving hydrolysis of aluminum ions from amorphous phase exposed to the direct contact with percolating water due to the devitrification of glaze, with further formation of the secondary minerals. These processes in simplified form can be described as reactions between dissolved silica, water, as well as kaolinite and gibbsite at the stage of their formation. This caused high non-linear release of trace elements from FA and significant qualitative/quantitative increase of its contamination potential with respect to the ground water and soils in adjacent area (decrease of pH to min. 4.3-4.5, and delayed extensive release of Zn, Fe, Mn, Mo, Cd, Cr, Be, B, V in high concentrations). In conformity with pH-Eh-stability fields metals can be grouped according to the similar release-dissolution response to controlling parameters, e.g.: I: (Zn-Cd-W-Be); II:(Fe-Mn) (reverse pH-dependent increase). Several metals (mainly oxyanions with broad fields of aqueous species) show weak influence of pH-Eh parameters (Li, Mo, Se, Sr, B), while Al, Cu and V are immobilized at pH 4.3-5.0. The screening study proved (i) possibility of FA acidification and discontinuous non-linear time-delayed increase of its pollution potential to the hazardous level due to weathering transformations (ii) necessity of life-cycle screening/monitoring of FA disposal sites for trace element release as a function of controlling factors along the vertical profile of anthropogenic or natural vadose zone. The results suggest also caution in use of FA as acidic soils improver.

1. INTRODUCTION

Continuous efforts to find sufficient and environmentally friendly alternative sources of energy up to now have not resulted in substantial changes: coal remains the major fuel used for generating electricity worldwide and both its production and consumption shows general increasing trend. Over past 25 years almost 50 % growth in global hard coal production occurred – from 2,400 million tons (Mt) in 1976 to 3,639 Mt in 2000, i.e. over 1,200 Mt, and further increase in coal use for electric power production is anticipated. Currently, coal generates about 38 % of the world's electricity; according to 2000 data, countries heavily dependent on coal for electric power generation include e.g. China (80%, estimate), India (66%, estimate) and USA (56%). The European Union (EU-15) generates as a whole 25% coal-based power (1999 data), though in several its Member States over 50% of electricity is generated from coal (Greece, Denmark, Germany). The highest dependence on coal exhibits Poland which electric power is almost thoroughly coal-based (96%).[1]

Coal burning results in generation of huge amount of coal combustion waste (CCW) generated by coal fired electric utilities. This waste comprises fly ash, bottom ash, boiler slag and FGD products (if flue-gas desulfurization is applied). Of these, fly ash (FA) is a predominant kind of CCW that determines the total amount of waste used or disposed of. The beneficial chemical and physical properties of FA make it suitable for a number of applications. This induces many proponents of FA utilization to back up the term of "coal combustion products" or "by-products" that suggests its full use in technically sound, commercially competitive and environmentally safe way with no considerable adverse environmental impact during storage.[2] The current status of FA utilization, along with increase in coal use for electric power production, though, does not support this optimistic view that can be exemplified in FA market in the United States. The recent data by US Department of Energy (U.S. DOE) and American Coal Ash Association (ACAA), report increase the U.S. coal production for 1% per year, to 1,268 Mt in 2015, while the trends in FA use suggest prospective rate of FA utilization at the present level close to 25%, up to 30% of the total amount generated. The largest volume of FA is used in cement, concrete and grout, and this application shows a permanent upward growth trend, while other traditional applications such as road base/sub-base or structural fill are of minor importance in the total balance and do not exhibit increase. Of the applications developed in the last decade (waste stabilization and flowable fill) only the first one shows some promise for higher utilization of FA, though still inadequate to the increase rates of this waste generation.[3] The limiting factors for the rapid growth of traditional and emerging markets for FA and other CCW are requirements to meet the standards on the one hand, and the unfavorable competing with natural materials (e.g. due to high shipping costs) on the other hand. In view of the omnipresence of inexpensive and easily available natural materials, the faster increase of FA market needs serious legislative support that would either strongly favor use FA as engineering materials compared to the natural ones, or put adequately stringent financial restrictions on FA disposal. The lack of such protectionist regulations is a barrier to greater increase of FA use. The ACAA's goal "to gain recognition and acceptance of CCW as engineering material **on par** with (unrestricted) competing virgin, processed and manufactured materials" does not seem realistic unless the aforementioned regulatory handicap is applied that would add higher cost-efficiency to other benefits of FA compared to natural materials. The results of deep-laid protectionistic legislation in Poland resulted in 75.7% use of 18,100 Mt CCW generated in 2000 that places this country at the top position among large CCW generating states with respect to its utilization. Nevertheless, a huge amount of 249,359 Mt of CCW generated in previous years is still stored in disposal sites throughout the country, and this amount continuously increases, in 2000 for 3,865 Mt due to still 24.3 % of CCW generated that was not utilized.[4]

To summarize, despite potentially vast markets of coal combustion waste (CCW) use, a large its portion, in particular fly ash (FA), is disposed of annually in surface ponds or landfills, and the total stored amount grows continuously in the countries heavily dependent on coal for electric power generation. FA exposed to the atmospheric conditions may pose an adverse environmental impact that requires adequate life cycle assessment. We still do not know much about long-term weathering transformations of bulk waste and quite often overlook the potential for "time-bomb effect", i.e. for non-linear time-delayed release of chemicals from the disposed material in harmful concentrations. At the same time, many CCW ponds and landfills of other waste that are not hazardous under RCRA have been located on land that is considered of low value for other use, such as abandoned sand and gravel quarries or disused strip mines, which are susceptible to groundwater contamination. Most of such landfills are unlined. Significant number of that impoundments is located in areas with thin, permeable vadose zone over aquifers used as a source of drinking water. Old CCW ponds/landfills can be particularly problematic as these utilities were sited when ground water protection was rarely considered due to the inadequate general knowledge.

Here, the potential risk from FA is exemplified in a case study of time-delayed post-closure transformations of pore solution quality in the Przechlebie disposal site (Upper Silesia coal basin USCB, Poland).

2. MATERIAL AND SITE CHARACTERISTICS

In this surface pond, "pure" FA from Rybnik power plant (without FGD products) was disposed hydraulically from 1979 to 1991 in the disused sand quarry of the maximum depth of 20 m. Due to construction of embankments of coal mining waste, the disposal of FA: water slurry was continued above the surface level, up to the final surface area of 53 ha, FA thickness up to 25-30 m and total volume 23 Mm^3 (Fig. 1). After closure, FA pond surface was stabilized with carbamide resin spread from helicopter to prevent dusting and in following years after advanced dewatering was reclaimed biologically with grass cover. The disposed FA represents alkaline aluminum silicate material [5]: $CaO+MgO/SO_3 + 0.04$ $Al_2O_3 =$ 1.7-3.6 (mean 2.6) that is within the range for FA from other European power plants. The detailed discussion of this FA phase, petrographical and chemical composition in the background of coal and "pure" FA characteristics from different European power plants has been presented elsewhere and in general shows high similarity.[6] Combustion processes result in enriching the majority of trace elements in FA (except Hg, I and F) in about an order of magnitude compared to that in coal. Contents of trace elements (in mg/kg) show declining order typical for FA from other European power plants [7]: $[10^3]$ (Ba>Sr>Mn>V) $>>$ [10^2] (Rb, Cr, Zr, Ce, Zn, Ni, Cu) > [>10] (Co, Pb, La,Y, Nd, Sc, Th, Cs, As) > [$\cong$ 10] (Sm, Be, U, Mo, Br, Sb) > [<10] (Yb, Hf, Bi, W, Se) > [10^{-1}] (Eu, Ta, Tb, Lu, Hg, Cd, Ag) $>>$ [10^{-2}] (Au, Ir).

The characteristic feature of "pure" FA (without FGDS admixture) is domination of alkali cations in leachate that results in high pH (over 10-12) and lack of equilibria constraints. In petrographic composition prevail superficially vitrified spherical and irregular particles filled with amorphous aluminum silicate relics of clay minerals. These properties make FA potentially attractive for use in agriculture as soil improver/amendment, in particular

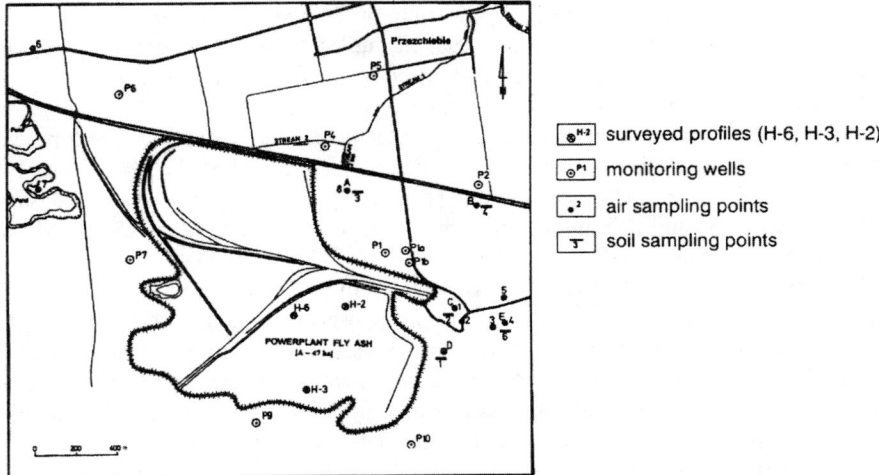

Figure 1. Przezchlebie fly ash (FA) pond of the Rybnik power plant (USCB, Poland) in the post-closure period. H-6, H-3, H-2: screening boreholes for pore solution sampling along the vertical profiles of FA pond; P1 – P10 – disused wells of the local ground water monitoring network.

for application to the acidic soils for their neutralization, micronutrient supply and improvement of physical properties. The FA pond described here represents a routine way of FA disposal worldwide - in the form of water: FA mixture to enable its hydraulic transport from the power plant. Due to continuous pulp supply, the hydrogeological conditions within FA pond during the operating stage were specific for the saturation zone.

In the post-closure period, gradual dewatering of the pond and the transformation of the hydrogeological conditions within the FA layer occurred, from the saturated zone typical for the operating stage into the vadose zone conditions, when vertical percolation of atmospheric precipitation had started. The material disposed in the pond displayed high to moderate hydraulic conductivity ($k = 10^{-3}$ to 10^{-4} cm/s) dependent of the compaction, and a high porosity ($n = 0.58 - 0.50$) adequate to the hydraulic conductivity of mean and fine sands. This assures free percolation of atmospheric precipitation through the waste layer.

The newly formed anthropogenic zone was sampled along the vertical profiles in the three drilled boreholes 5 to 11 m deep (H-6, H-2, H-3) in 1993 after 12 years' operation and 2 years after closure. The age of sampled FA along the profiles ranged from 3 to 9 years. Pore solution from FA samples was extracted by the pressure method under nitrogen and analyzed for the metal content using ICP-OES technique (ICP Perkin Elmer Plasma 40). The elemental speciation in pore solution and QA/QC testing was performed with use of the geochemical computer programs WATEQ 4F and MINTEQ Visual 2.1 that give similar results and are complementary to each other.

3. RESULTS AND DISCUSSION

The chemical composition of pore solutions along the three vertical profiles in the fly ash disposal pond in the post-closure period after 12 years' operation reflects both the altered water flow compared to the saturated zone under operational stage, and the changed macro- and trace component release and migration due to changed equilibria conditions. Percolating precipitation water flow results in the well-known phenomenon of a vertical downward redistribution of leached macroconstituents' concentrations/loads, which is particularly strong with respect to ions that are not bound by equilibria limitations, such as chlorides and sulfides balanced by alkali cations.[8] Due to it, the highest concentrations of leached constituents will occur at the bottom of the landfill where the percolating stream enters the natural vadose zone and further reaches the unprotected Quaternary aquifer.

The leaching behavior of trace elements is controlled by several basic variables, of these Liquid to Solid ratio L/S (reflecting the time factor), pH, redox Eh and complexation have been considered the most important ones. The natural moisture content m_n in studied FA profiles ranged from 23.2 to 40.7 % wt., mean values 26.7% (H6), 30.8% (H2) and 37.70% (H3); mean liquid/solid ratio L/S accounts for these profiles 1: 3.7, 1: 3.2 and 1: 2.7, i.e. considerably lower than in the pulp and the water-saturated layer under the operational conditions, while cumulative L/S ratio is continuously growing in time.

The long-term leaching behavior of constituents from the disposed waste in principle may follow the III-stage generic leach pattern that comprise Wash-out (I), solubility controlled Dissolution (II) and Delayed Release (III) stages.[9] This pattern reflects a situation when due to change of solubility-controlling factor, the massive release of constituents at a high rate may occur at some point delayed in time. In general, the correct prediction of the possibility of occurrence and intensity of the Delayed Release (III) phase appears to be the particularly problematic task. In the most frequent case, the development of this phase is determined by two kinetically defined processes of acid generation and depletion of available buffering constituents. The availability of buffering constituents quite often is not adequate to their depletion: heterogeneity of waste cause that acidifying and buffering agents are not occurring in the waste matrix in close proximity required for the direct interaction. The complexity of real systems makes the correct prediction of the Delayed Release (III) stage development extremely difficult and requires full and detailed information that besides the chemical composition, comprise also phase composition of a waste material, including the forms, dispersion and specific surface of the phases in the matrix, which influence their reactivity and availability, as well as the trends in weathering transformations that they may undergo in the waste layer during its exposure to the atmospheric conditions.

Due to the changing location of the pulp outlet in the operational stage, the hydrogeochemical conditions within the fly ash pond were not uniform and reflected different stages of waste transformations that permit to illustrate these stages.

While the pore solution in FA along H-6 profile still reflects the Dissolution stage under operating conditions (pH 7-10), in the looser FA profiles H-2 and H-3 where vertical percolation of precipitation and exchange of pore solution is considerably more advanced, it indicates alteration of buffering properties of the system that can be defined as Delayed Release stage. The major process that induce dramatic change of FA properties and results in deep transformation of chemical composition of pore solution seems to be the weathering process that causes the devitrification of surface glaze covering FA particulates and the exposure of amorphous intrinsic content of the particle to atmospheric conditions. The character of pore solutions along these profiles suggests that the major buffering mechanisms controlling pH are: (i) depletion of calcium carbonates and sulfates in the form of calcite, anhydrite and gypsum, along with formation of new secondary minerals such as ettringite $Ca_6Al_2(SO_4)_3 (OH)_{12} 26H_2O$ (c – crystalline phase) and other hydrated sulfaluminates; (ii) reactions involving II-stage hydrolysis of aluminum ions from amorphous phase exposed to the direct contact with percolating water due to the devitrification of glaze, with further formation of the secondary minerals, at this stage mainly gibbsite $Al_2O_3 H_2O$ (c); (iii) dissolution of silica and kaolinite $H_4Al_2Si_2O_9$ (c) formation. These processes in simplified form can be described as reactions between kaolinite and gibbsite at the stage of their formation, dissolved silica and water. Detailed discussion of weathering transformations of FA at these stages has been presented elsewhere.[10] In short, these processes result in shifting from highly alkaline pH value of the material and pore solutions at the initial Wash-out (I) stage (pH 10- >12) through moderately alkaline to close to neutral pH 7- 8 at the Dissolution (II) stage, up to acidic pH value at the level of pH 4 – 5 at the Delayed Release (III) stage.

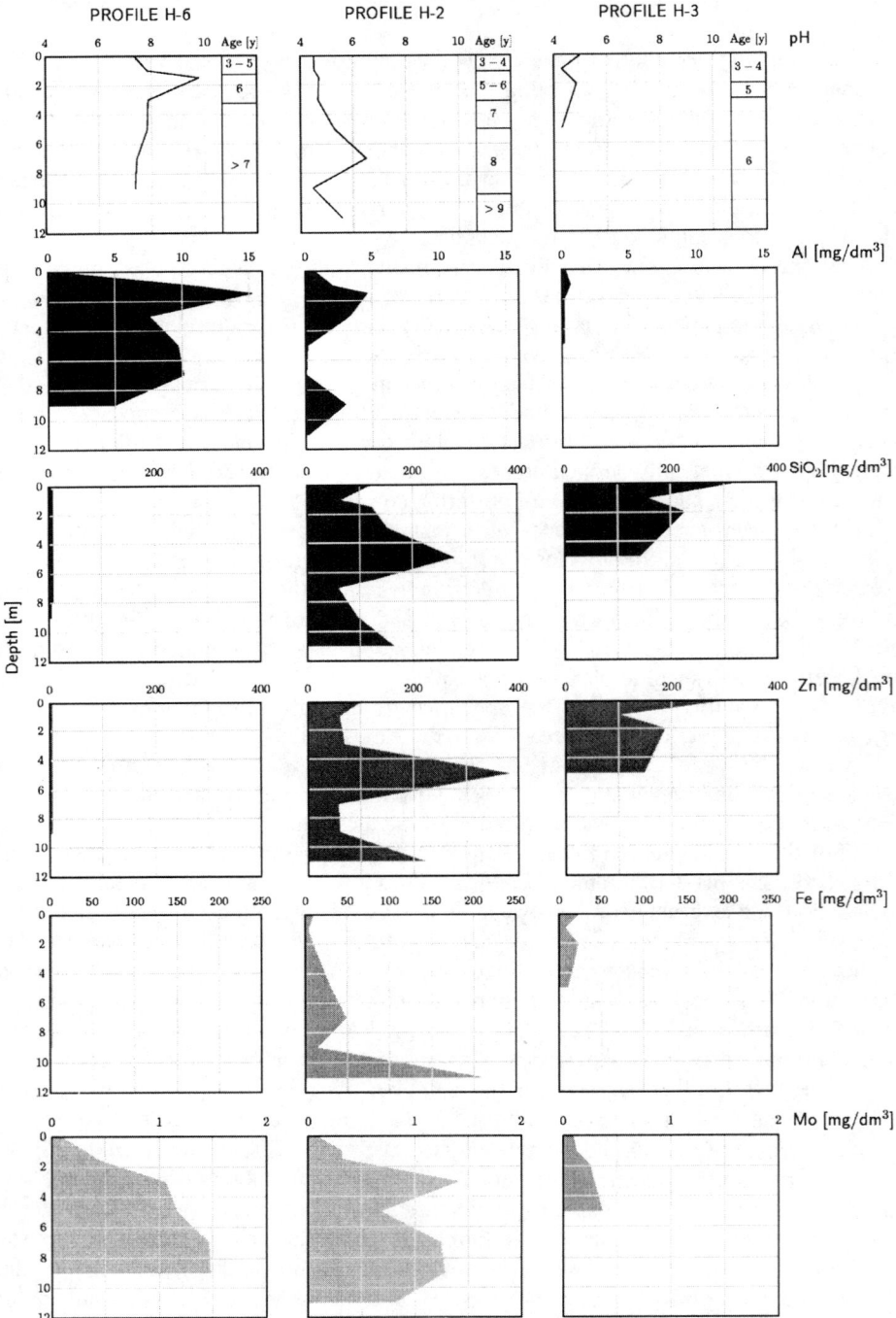

Figure 2. Example patterns of pore solution chemical transformations along the vertical profile of FA pond: selected elements (Al, SiO₂, Zn, Fe, Mo) concentrations in the Dissolution (II) (profile H-6) and Delayed Release (III) stages (profiles H-3 and H-2).

Table 1. Concentration range of selected constituents and trace elements in pore solutions along the vertical profiles of FA pond of the Rybnik power plant (USCB, Poland) in the post-closure period: Dissolution (II) (profile H-6) and Delayed Release (III) stages (profiles H-2 and H-3).

Fly ash age: 3-9 years

| Parameter, Constituent | Concentration range in pore solutions, mg/L (mg dm^{-3}) | | | | | | | | | | |
| | Borehole H 6 (0.1-9.0 m) | | | Borehole H 2 (0.1-11.0 m) | | | Borehole H 1 (0.1-5.0 m) | | | Drinking water | |
	min	max	mean	min	max	mean	min	max	mean	MCL[2], Poland	RBS[3], US EPA
Conductivity[1]	961	2150	1696	1194	1913	1536	1119	1756	1508		
pH	7.35	9.8	7.93	*4.46*	*6.5*	*5.02*	*4.29*	*5.04*	*4.62*	6.5-8.5	
Ca	23.26	88.41	40.26	2.29	87.29	45.31	3.67	73.63	27.98		
Cl	19.25	294.6	205.73	108.15	1116.06	566.14	108.6	589.62	337.71	300	
SO$_4$	227.53	828.89	542.8	393.01	1296.35	786.89	164.50	1745.25	731.64	200	
HCO$_3$	38.44	147.05	96.407	4.27	4.27	4.27	nd	nd	nd		
SiO$_2$	4.83	17.26	11.05	63.46	281.18	154.40	135.19	313.66	207.81		
F	*1.62*	*7.85*	*4.17*	*8.03*	*33.76*	*17.15*	*12.98*	*30.17*	*20.33*	1.5	2.2
TDS	681.40	1976.40	1501.40	1198.40	5079.80	2880.16	1564.90	3712.6	2292.25	800	
Al	*0.744*	*15.26*	*8.51*	<0.06	*4.622*	*1.767*	<0.06	*0.663*	0.257	0.3	110
B	< 0.01	*3.064*	*1.39*	*2.868*	**7.987**	**5.590**	**3.343**	**10.46**	**6.546**	[1.0]	3.30
Ba	0.069	0.204	0.117	0.096	0.341	0.203	<0.002	0.488	0.210		2.6
Be	< 0.002	< 0.002	< 0.002	0.002	**0.012**	**0.0044**	0.002	**0.007**	**0.005**		0.0016
Cd	< 0.005	< 0.005	< 0.005	0.013	**0.175**	**0.049**	0.045	**0.167**	**0.092**	0.005	0.018
Cr$_1$	0.059	0.170	0.105	<0.010	0.940	0.289	0.121	0.310	0.219	0.01[4]	0.18[4]
Cu	*0.069*	*0.200*	*0.140*	<0.005	0.036	0.010	<0.005	0.042	0.018	0.05	1.4
Fe	0.043	0.191	0.080	*3.981*	*209.3*	*42.180*	*7.222*	*23.59*	*15.999*	0.5	
Li	0.537	**3.512**	**2.160**	0.430	**4.000**	**2.822**	0.345	**2.548**	**1.334**		0.73
Mn	<0.002	0.012	0.007	0.081	**4.061**	**0.932**	**0.222**	**0.399**	**0.302**	0.1	0.18
Mo	0.088	**1.48**	**0.862**	0.086	**1.412**	**0.667**	0.090	**0.361**	**0.206**		0.18
Sr	0.503	1.304	0.755	0.229	1.828	1.166	0.152	1.565	0.876		22.0
Ti	<0.005	<0.005	<0.005	0.006	0.050	0.035	0.002	0.040	0.020		
V	0.076	**1.120**	**0.743**	<0.010	**0.900**	0.241	0.020	0.115	0.082		0.26
W	<0.010	0.400	0.099	*1.044*	*7.800*	*2.887*	*1.910*	0.700	*4.510*	[1.0]	
Zn	0.004	0.052	0.018	**38.84**	**279.20**	**88.75**	**73.17**	**272.57**	**141.63**	5.0	11.0

[1] µS cm^{-1}
[2] MCL for drinking water (Polish regulations) [3] RBC (Risk-Based Concentrations) by US EPA, 1994[12]
Values that exceed RBC by US EPA are bold; *Values that exceed Polish MCL are italic*
nd – not detected; [4] Cr^{6+}

There is relatively abundant information concerning leaching behavior of "pure" FA at the Wash-out (I) stage that in general has been considered the most problematic due to high pH and alkalinity, unconstrained release of macro-component ions, particularly sulfates balanced by alkali ions, release of aqueous aluminum as AlO_2^-, as well as oxyanions such as As, Mo, V or B species. The reported field studies were conducted in the period when this stage already ceased.

The extensive leaching of highly soluble components in the dewatering stage and also during the transformation of the water flow mode in the FA disposal site into vertical percolation of atmospheric precipitation under vadose zone conditions resulted in the development of the Dissolution (II) stage that generally lead to the conclusion that the contaminant mobility, in particular problematic trace element concentrations reached the safe level below MCL values due to decrease of pH within the range from moderately alkaline to close to neutral, when the most of metals are stably immobile in the solid phase.[11] This stage is illustrated by the composition of pore solution along the H-6 profile (Table 1, Fig. 2). In these solutions, entirely maximum and mean cons of Mo and V still exceeded risk-based level (RBC),[12] while in the uppermost surface layer due to washout reduction of these mobile elements in FA matrix their concentrations reached permissible values. In this after-closure period, monitoring of the ground water in the vicinity of the site terminated. Due to it, the development of the Delayed Release (III) stage caused by the aforementioned processes of the further FA weathering transformations that involve reactions between gibbsite and kaolinite and their formation as secondary minerals has been overlooked .

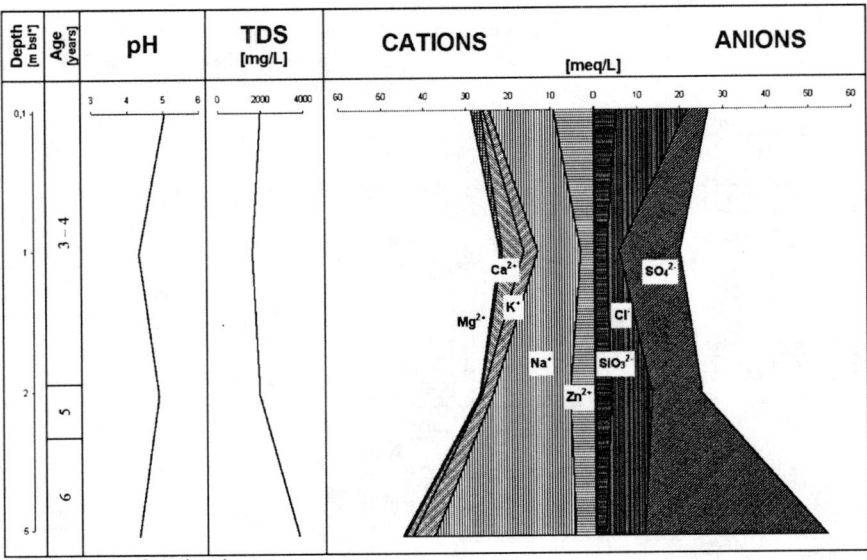

*bsl - below surface level

Figure 3. Hydrogeochemical profile of pore solutions in the FA layer at the Delayed Release (III) stage (H-2 borehole). Acidification, massive Zn and SiO$_2$ release and vertical redistribution of major constituents.

At the Delayed Release (III) stage high non-linear release of trace elements from FA and significant qualitative/quantitative increase of its contamination potential with respect to the ground water was observed (profile H-2 and H-1). The composition of pore solution along these profiles displayed decrease of pH to min. 4.3-4.5, and extensive release of Zn, Fe, Mn, Mo, Cd, Be, Cr, B, V in high concentrations showing characteristic patterns that reflect also construction of FA pond in layers depending upon the heightening of embankments (Table 1, Fig. 2). In conformity with pH-Eh-stability fields, some metal concentration patterns along the profiles display the similar release-immobilization response to controlling parameters, e.g.: I (Zn-Cd-W-Be); II (Fe-Mn) (reverse pH-dependent increase). Several metals (mainly oxyanions) show either weak influence of pH-Eh parameters (B, Cr, Ba) or almost none (Li, Mo, Sr), while concentrations of Al, V and Cu in pore solutions at pH 4.3-5.0 considerably decreased for different reasons.

Particularly high mobilization showed Zn that reached concentrations up to 300-400 mg/L that changed its status from the trace element into a macro-component (Fig. 3). Another abundant element was Fe; in concentrations much above natural level occurred also B and W. Zn and Mo appeared to exceed RBC for more than an order of magnitude; also B, Be, Cd, Mn and V exceeded RBC (As was not analyzed in this study).

Metal speciation by MINTEQ Visual 2.1 and WATEQ4F at the Dissolution (II) stage (profile H-6) shows abundant occurrence of aluminum in solution predominantly in the form of hydroxides, mainly Al(OH)$_4$; formation of gibbsite Al(OH)$_3$ is also observed at activity of dissolved Al below 10^{-4} that determines the narrower pH range for Al(OH)$_3$ solid; accessory amounts of fluoride compounds were also calculated, mainly as AlF$_{3\,aq}$ and AlF$_2$. At this stage, trace elements in cationic form occur in very low concentrations, mostly as carbonate, hydroxide and sulfate species and free ions, e.g. Zn occurs as Zn^{2+}, Zn(SO$_4$)$_{2\,aq}$, ZnCO$_3$ and ZnHCO$_3$ species. In higher amounts are present oxyanions that show very large fields of water soluble or metastable forms such as B, which is present mostly as boric acid H$_3$BO$_{3\,aq}$ or Mo that occurs in solution mainly as HMoO$_{4\,aq}$ species.

At the Delayed Release (III) stage that is determined by kaolinite formation, dissolution of silica in the H$_4$SiO$_4$ form and massive release to pore solution of metal cations, in particular

Table 2. Example offprints of the selected element speciation in pore solution along the vertical profiles of disused FA pond at the Dissolution (II) and Delayed Release (III) stages (computed by WATEQ 4F)

Species	Valence	Profile H – 6 Dissolution (II) Stage 1,0 m below surface level (pH - 7,85; TDS - 1113 mg/L; COND. – 1361µS/cm)				Profile H – 2 Delayed Release (III) Stage 5,0 m below surface level (pH – 5,28; TDS - 3984 mg/L; COND. - 1870µS/cm)			
		Anal. ppm	Calc. ppm	Anal. Molal	Calc. Molal	Anal. ppm	Calc. Ppm	Anal. Molal	Calc. Molal
SiO$_2$ tot.	0	14,480		2,413 E-04		281,180		4,698 E-03	
H$_4$SiO$_4$ (aq)	0		22,942		2,390 E-04		449,689		4,698 E-03
H$_3$SiO$_4$	-1		0,218		2,294 E-06		0,013		1,330 E-07
H$_2$SiO$_4$	-2		0,00047		5,017 E-10		0,000000		9,954 E-14
SiF$_6$	-2		0,000000		5,428 E-28		0,000012		8,529 E-11
Al	3	11,180000	0,000000	4,148 E-04	4,368 E-12	0,171000	0,000000	6,363 E-06	8,871 E-14
AlOH	2		0,000070		1,600 E-09		0,000000		5,850 E-14
Al(OH)$_2$	1		0,36		5,839 E-07		0,000000		4,512 E-14
Al(OH)$_3$	0		3,521		4,519 E-05		0,000000		8,571 E-15
Al(OH)$_4$	-1		34,936		3,681 E-04		0,000000		2,055 E-16
AlF	2		0,000173		3,777 E-09		0,000019		4,202 E-10
AlF$_2$	1		0,015		2,303 E-07		0,011		1,648 E-07
AlF$_3$ (aq)	0		0,052		6,224 E-07		0,278		3,326 E-06
AlF$_4$	-1		0,06169		5,998 E-08		0,294		2,871 E-06
AlSO$_4$	1		0,000000		3,842 E-12		0,000000		5,348 E-14
Al(SO$_4$)$_2$	-1		0,000000		7,630 E-13		0,000000		1,382 E-14
AlHSO$_4$	2		0,000000		2,184 E-20		0,000000		1,436 E-19
Zn	2	0;006000	0,00,891	9,189 E-08	4,427 E-08	382,310	277,347	5,872 E-03	4,261 E-03
ZnHCO$_3$	1		0,000559		4,432 E-09				
ZnCO$_3$	0		0,002538		2,026 E-08				
Zn(CO$_3$)$_2$	-2		0,000552		2,984 E-09				
ZnCl	1		0,000027		2,654 E-10		13,331		1,328 E-04
ZnCl$_2$ (aq)	0		0,000000		8,094 E-13		0,332		2,445 E-05
ZnCl$_3$	-1		0,000000		3,505 E-15		0,013		7,666 E-08
ZnCl$_4$	-2		0,000000		8,864 E-18		0,000336		1,630 E-09
ZnF	1		0,000006		6,940 E-11		3,608		4,295 E-05
ZnOH	1		0,000188		2,282 E-09		0,038		4,642 E-07
Zn(OH)$_2$	0		0,000160		1,611 E-09		0,000080		8,041 E-10
Zn(OH)$_3$	-1		0,000000		4,150 E-13		0,000000		6,098 E-16
Zn(OH)$_4$	-2		0,000000		7,005 E-18		0,000000		3,519 E-23
ZnOHCl (aq)	0		0,000024		2,007 E-10		0,29		2,465 E-07
ZnSO$_4$ (aq)	0		0,002414		1,497 E-08		216,417		1,346 E-03
Zn(SO)$_2$	-2		0,000136		5,271 E-10		22,032		8,592 E-05
Cd	2					0,175	0,067	1,563 E-06	5,967 E-07
CdCl	1						0,097		6,598 E-07
CdCl$_2$ (aq)	0						0,008827		4,836 E-08
CdCl$_3$	-1						0,000186		8,527 E-10
CdF	1						0,000701		5,360 E-09
CdF$_2$ (aq)	0						0,000002		1,459 E-11
CdOH	1						0,000000		4,931 E-12
Cd(OH)$_2$	0						0,000000		3,996 E-17
Cd(OH)$_3$	-1						0,000000		1,075 E-24
Cd(OH)$_4$	-2						0,000000		3,489 E-33
Cd$_2$OH	3						0,000000		3,409 E-17
CdOHCl (aq)	0						0,000007		4,112 E-11
CdNO$_3$	1						0,000033		1,878 E-10
CdSO$_4$ (aq)	0						0,048		2,319 E-07
Cd(SO$_4$)$_2$	-2						0,004145		1,997 E-08
B tot	0	0,957		8,862 E-05		7,343		6,819 E-04	
H$_3$BO$_3$ (aq)	0		5,228		8,464 E-05		41,933		6,810 E-04
H$_2$BO$_3$	-1		0,241		3,974 E-06		0,005709		9,426 E-08
BF(OH)$_3$	-1		0,000523		6,479 E-09		0,038		4,669 E-07
BF$_2$(OH)$_2$	-1		0,000000		1,634 E-12		0,030		3,584 E-07
BF$_3$OH	-1		0,000000		4,218 E-18		0,000238		2,815 E-09
BF$_4$	-1		0,000000		4,045 E23		0,000007		8,216 E-11

Zn, mostly as a free ion and as sulfate species $ZnSO_4$ $_{aq}$, and in minor amounts as $ZnCl_2$ $_{aq}$, $Zn(SO_4)_2$, and ZnF; occurrence of other Zn species is negligible; the prevailing species of other metal cations in solution are also free ions and sulfate species. Oxyanions in this phase are present in the same forms as at the Dissolution (II) stage. The selected element speciations in pore solution for these two stages are given in Table 2 where the example offprints of metal speciation computed by MINTEQ Visual 2.1 and WATEQ4F has been presented.

The reported results prove that FA besides beneficial properties display also environmentally detrimental qualities that should be taken into consideration and adequately controlled at either siting or construction of FA disposal facilities, or at using this waste material for the purposes that involve land spreading, interaction with other materials (e.g. soil, biosolids in mixtures) and exposure to atmospheric conditions. The typical application of this kind is FA use as soil improver/amendment.

Despite continuous attention paid to FA properties and environmental implications of its disposal and use, there is still no sufficient information, nor univocal assessment of the potential long-term effect of FA on waters (in particular on ground water) and soil. This study is unique with respect to the methodology and the direct analysis of the pore solution in the FA layer of the defined age. In other known studies the research was focused either on the receiving water, or on the leachate from the FA dump, or on the lysimetric studies with disturbed or undisturbed core that were limited to the relatively thin layer. Some studies underestimate coal combustion waste, in that FA, as a source of the environmental contamination and do not consider any potential hazard from this source to the environment at any stage of its disposal.[12] These opinions are somewhat alarming in particular that originate from India, the country where electric power generation is predominantly coal-based and which is going to double its coal production and combustion for power generation in a short time. Other evidences, though, do not support these optimistic conclusions clearly showing the possibility of the adverse environmental impact of coal combustion waste storage on surface and ground water quality.[13, 14] These sources point out enrichment of major and trace elements in FA compared to coal and easy leaching of soluble compounds from this material. Monitoring observations of ground water quality, soil/subsoil system and surface vegetation in the vicinity of the waste material disposal originating from two different sources: coal pile and FA of different leaching behavior does not enable to define the particular detrimental effect of each source on ground water and soil. Nevertheless, the authors confirmed contamination of the shallow aquifer in the vicinity of the dump and emphasized the importance of use weathered FA of high buffering capacity i.e. low B and high Ca and Mg fly ash as amendment of agricultural soils and acidic mining waste dumps[13]. Another recent study reports more dramatic effects of coal FA – Coal Reject landfill, where pH of leachate was found to drop down to pH 1.5.[14] This level of acidity confirms the authors' assumption that (fly ash generation and disposal) "may pose more serious ecological problems than previously believed", as at this level of acidity even more massive release and leaching of toxic trace elements than has been reported in the presented study can be anticipated, not only from the FA or coal reject, but also from soil and subsoil of the vadose zone. The authors[14] refer to high level of highly toxic metals, e.g. excess of Se found in animals exposed to coal FA and coal reject of the investigated site.

These data confirm the results reported in this study and are also support the warning against FA application as amendment of the acidic soils or for the joint disposal as a mixture with the material susceptible to acidification due to weathering transformations that cause Delayed Release of contaminants from FA. The FA properties presented by the authors of this chapter[9] and also by other authors in the monographic review related to biogeochemistry of trace elements in CCW[15] show clearly that trace elements are significantly (about 10-fold) enriched in this material compared to coal and mean content in soils; this material has high hydraulic conductivity and is fairly leachable; it also does not have permanently high buffering capacity. As a result, this material is susceptible to acidification and massive trace element release.

CONCLUSIONS

The reported results emphasize that FA of a typical composition exposed to the atmospheric conditions has a potential for deep weathering transformations and related change of leaching behavior, which should be strongly taken into consideration. As can be derived from the approach of many soil scientists[13] they treat FA as a rather stable material with a permanent highly alkaline reaction, therefore suitable as the acidic soils amendment. The composition of pore solutions in the material from 3-4 to over 9 years' old shows though that FA in the post-closure period of a pond undergo changes that cause gradual shift from highly alkaline to moderately alkaline/close to neutral pH values, with possible further acidification to pH 4.5 in the process of devitrification and formation of crystalline phases (gibbsite and kaolinite). In this phase, the massive release of trace elements occurs, up to the macro-amounts adequate to the Delayed Release stage. This can adversely affect ground water quality, and also increase availability of trace metals to the receptors in FA-amended soil. These observations, confirmed by the deep acidification (<pH 1.5) of FA-Coal Reject landfill along with the mobilization of toxic metals and their high levels in exposed animals observed by other authors,[12,13] suggest the need to be very careful with application of FA to the acidic soils or with disposal it in admixture with the material susceptible to acidification, to avoid the development of environmentally detrimental consequences in a long run. The screening study proved (i) possibility of discontinuous non-linear time delayed increase of pollution potential of FA in the disposal sites to the hazardous level due to externally or intrinsically induced transformations of FA properties resulted from the environmental exposure and alteration of controlling factor values; (ii) necessity of life-cycle screening/monitoring of FA disposal sites for trace element release as a function of the primary (pH-Eh, ionic strength, ionic composition of solute) and secondary controlling factors (L/S-liquid to solid ratio, water flow conditions) along the vertical profile of an anthropogenic or natural vadose zone.

ACKNOWLEDGEMENT

This study was supported financially by PPH UTEX Ltd. in Rybnik, Silesia, Poland.

REFERENCES

1. WCI – World Coal Institute, Key coal statistics for 2000, *Ecoal*, 40, 8, December 2001.
2. Collins, S., Managing powerplant wastes, Special Report, *Power*, 8, 15, 1992.
3. Stewart, B. R., Coal combustion product (CCP) production and use. Survey results, in *Biogeochemistry of Trace Elements in Coal and Coal Combustion Byproducts*. Sajwan, K.S., Alva A.K. and Keefer, R.F., Eds., Kluwer Academic/Plenum Publishers, New York, 1999, chap. 1.
4. GUS – Central Statistical Office, *Environment 2001. Information and Statistical Papers*. GUS, Warsaw, 2001, chap. 6 (in Polish).
5. van der Sloot, H. A., Piepers, O., and Kok, A., *A Standard Leaching Test for Combustion Residues*, BEOP-31, Netherlands Energy Research Foundation ECN, Petten, 1984, 64.
6. Twardowska, I., Environmental aspects of power plants fly ash utilization in deep coal mine workings, in *Biogeochemistry of Trace Elements in Coal and Coal Combustion Byproducts*. Sajwan, K.S., Alva A.K. and Keefer, R.F., Eds., Kluwer Academic/Plenum Publishers, New York, 1999, chap.3.
7. Meij, R, and Schaftenaar, H. P. C., Hydrology and chemistry of pulverized fuel ash in a lysimeter or the translation of the results of the Dutch column leaching test into field conditions, in *Environmental Aspects of Construction with Waste Materials, WASCON'94 Int. Conf., Maastricht, the Netherlands*. Goumans J. J. J. M, van der Sloot, H. A., and Aalbers, Th. G., Eds., Elsevier, Amsterdam, 1994, 491.
8. Twardowska, I., *Mechasnism and Dynamics of Coal Mining Waste Leaching at the Dumping Sites.*, Ossolinski National Foundation, Polish Acad. of Sci. Publishers, Wroclaw, 1981, chap. 5 (in Polish).

9. van der Sloot, H. A., Hjelmar, O, Aalbers, Th. G., Wahlström, M., and Fällman, A.-F., *Proposed Leaching Test for Granular Solid Wastes*, Rep. ECN-C-93-012, Netherlands Energy Research Foundation ECN, Petten, 1993, 75.

10. Twardowska, I., and Szczepanska, J., Solid waste: terminological and long-term environmental risk assessment problems exemplified in a power plant fly ash study. *Sci.Total Environ.*, 285, 29, 2002.

11. Brookins, D.G., *Eh-pH Diagrams for Geochemistry.* Springer-Verlag, Berlin Heidelberg New York, 1987, 176.

12. Smith, R. L.: *Risk-Based Concentrations: A Method to Prioritize Environmental Problems Using Limited Data,* US EPA, Region 3, Philadelphia, 1994, 21.

13. Singh, G:, Environmental evaluation of coal combustion residues utilization in mining areas, in *Clean Coal. Proceedings of the International Symposium on Clean Coal Initiatives*, New Delhi, India, T. N. Singh, and M. L. Gupta, Eds., Oxford & IBH Publ.Co.Pvt.Ltd., New Delhi –Calcutta, 1999, 463.

14. Ghuman, G.S., Sajwan, K.S., and Denham, M.E.,:Impact of coal pile leachate and fly ash on soiland droundwater, in *Biogeochemistry of Trace Elements in Coal and Coal Combustion Byproducts.* Sajwan, K.S., Alva A.K. and Keefer, R.F., Eds., Kluwer Academic/Plenum Publishers, New York, 1999, chap. 14.

15. Danker, R., Adriano, D.C., Barton, C., and Punshon, T., Revegetation of a coal fly ash-reject landfill, in *Sixth Intern. Conf. Biogeochemistry of Trace Elements, ICOBTE 2001 Conf. Proc.,* Guelph, Ontario, Canada, University of Guelph, 2001, 381.

16. Sajwan, K.S., Alva A.K. and Keefer, R.F., Eds., *Biogeochemistry of Trace Elements in Coal and Coal Combustion Byproducts.* Kluwer Academic/Plenum Publishers, New York, 1999, 359.

TRACE ELEMENTS AND THEIR MOBILITY IN COAL ASH/FLY ASH FROM INDIAN POWER PLANTS IN VIEW OF ITS DISPOSAL AND BULK USE IN AGRICULTURE

Irena TWARDOWSKA[1], Prem S.M.TRIPATHI [2], Gulab SINGH,[2]
and Joanna KYZIOL[1]

[1]Polish Academy of Science
Institute of Environmental Engineering
34 M.Sklodowska-Curie St. 34, 41-819 Zabrze, Poland
[2]Central Fuel Research Institute
CSIR – Council for Scientific and Industrial Research
Dhanbad 828108, India

ABSTRACT

The dynamic growth of coal consumption in India in order to meet the demand in power, along with high ash content in coal (~30-40% wt.) will result in generation of about 140 Mt/yr of fly ash (FA) by 2020. This creates a problem of its environmentally safe utilization and disposal. The use of FA in agriculture as soil amendment seems to be particularly attractive to India as high-volume low-technology application, and potential sink for almost unlimited amounts of coal combustion waste. In the chapter, the macro- and trace element mobility in FA has been analyzed and exemplified in field studies: (i) on impact of FA slurry pond on ground water quality; (ii) on effect of FA-amendment on a crop yield along with trace element contents and uptake from FA-amended soil. With respect to major chemical and phase characteristics, FA from Indian power plants does not differ from that generated in other countries of the world. Trace element content in FA is of about an order of magnitude higher than in coal and markedly exceeds the average concentrations in soil. FA can be classified as alkaline aluminum silicate with predominantly low CaO content, and low buffering potential ratio BPR ranging from 0.64 to 4.25, mean <2. This suggests the susceptibility of FA to acidification in time due to weathering transformations and developing the extensive trace constituents release from the disposed FA in the Delayed Release (III) stage of leaching. The analysis of 1996 survey showed deep adverse alteration of ground water quality in the vicinity of FA slurry pond under operation that reflected the Washout (I) stage of leaching. Release of soluble macro-constituents was a major process, while the most of trace elements was within their stability fields in solid phase. Due to the weathering transformations, the change of leaching pattern in time according to Dissolution (II) and Delayed release (III) stages is anticipated. The field trials with use as soil amendment of different doses of FA (from 25 to 500 t/ha) in acidic red and alkaline alluvial soils conducted by CFRI (CSIR) India in 1994-2000, showed increase of different crop yield up to 45-75% at FA dose of 200 t/ha, and from 16 to 33% at FA dose of 500 t/ha. The simultaneous increase of total and DTPA-extractable trace element concentrations in both types of soil was also noticed. Crops displayed diverse susceptibility to metal uptake: soybean > linseed > jowar > wheat. In the light of presented data, FA disposal in unprotected surface ponds is not environmentally safe. Application in agriculture seems to be a prospective sink for FA, though its large-area uncontrolled agricultural use may cause an irreversible soil or water contamination in the long-range period. The caution and pollution prevention principles suggest avoidance of FA application in acidic soils, and use it entirely in the well-buffered alkaline/neutral soils, with careful selection of FA doses and cultivated crops.

1. INTRODUCTION

Currently, with its 310 Mt of hard coal production in 2000, India holds the third place in the list of major producers of hard coal after PR China and the U.S.A. It belongs also to the major coal importers: according to the key coal statistics, in 2000 India imported 9.1 Mt of steam coal and 15.4 Mt of coking coal. Coal is the major fuel used for electricity generation: estimated dependence of India on coal for this purpose accounts for 66%.[1] At power production of 70,000 MW in 2000, coal consumption for electricity accounted for 250 Mt that did not cover the needs. To meet the growing demand in power, further almost double increase of the installed thermal power generation capacity by 2020 is planned that will result in coal consumption of about 380 Mt. The specifics of Indian hard coal deposits are high ash content, around 30-40 % wt., which is adequate to ash generation of 50-60 Mt/h for 200 MW unit. In total, it means estimated ash generation 90 Mt in 2000, and its increase by 2020 to about 140 Mt.[2] According to the same source, for each MW of installed capacity, about 1 acre of land is required for FA disposal in a layer 8-10 m.

The most widespread FA disposal facilities in India are FA-slurry surface ponds with open water circuit that collect 85-90% of the solid waste generated by power plants, while around 10% constitute dry or wet deposits on the adjacent land as emission from the stacks. In general, all new power plants in India have their own FA ponds, but quite a few due to the shortage of space, partly discharge FA slurry directly to the receiving stream.[2] According to Indian Management and Handling Rule of 1989, FA is not hazardous waste. The existing Environmental Protection Act in India does not set any regulations concerning FA disposal or limitations on contaminant discharge from FA pond into the environment. The only standards to be followed comprise pH values (pH 6.5-5.5), suspended solids (100 mg/L), and oil and grease (20 mg/L) in the effluent from FA pond. The results of leach studies carried out by the Indian researchers suggest effluent from the FA ponds to be "environmentally benign".[3,4,5] A literature review of studies on FA impacts on the basis of only three sources supports this conclusion.[2]

Despite of extremely liberal environmental regulations in India, huge amounts of FA generated and disposed currently and anticipated to be generated in the nearest future make this waste highly problematic and induce some activity of Indian central and state governmental institutions such as Central Pollution Control Board and the Ministry of Environment and Forest directed to reduction of FA disposal. Taking into consideration widely known beneficial properties of FA that can be used in a multitude of applications,[5,6,7] their recommendations suggested FA utilization by the year 2001 to achieve 50%, with the emphasis on use in production of construction materials (bricks, cement, tiles), in agriculture as soil improver/ amendment, for abandoned mines backfilling (stowing) and for lining of irrigation canals. Unfortunately, these recommendations have not been supported by any enforcement or incentive mechanisms that resulted in the lack of significant progress in this area. The current coal ash utilization level is negligible and according to a rough estimation ranges between 2-3 % and 6 %.[8,9] Though numerous reports have showed successful use of FA in different, up to pilot or small full scale applications in construction material production, they also pointed out the failure of FA commercial viability in competition with natural materials.[10]

The use of FA in agriculture as soil amendment and fertilizer seems to be particularly attractive to India as high-volume low-technology application, and potential sink for almost unlimited amounts of coal combustion waste. Therefore, the content and mobility of trace elements in FA from Indian power plants in the actual conditions of its disposal and use are of a vital importance.

In this chapter, the trace element mobility has been analyzed and exemplified in field studies on ground water quality in wells in the vicinity of the FA slurry pond under operation, and on trace element contents and uptake from FA-amended soils, on the background of effect of FA application on crop yield.

2. CHARACTERISTICS OF COAL ASH FROM INDIAN POWER PLANTS

The available data on Indian hard coal burned in different power plants and coal ash/fly ash from these power plants[5,11,12,13,14] show that with respect to its major chemical and phase characteristics, Indian FA does not differ considerably from each other and from FA originating from the different parts of the world, among them from the European power plants[15,16], though the specific features of this material can be also distinguished. In particular, Indian coal ash/fly ash displays wide range of CaO content, predominantly low, and also low alkali concentrations (Table 1).

Table 1. Chemical composition of fly ash from several Indian thermal power stations under NTPC, A.P.Genco, NALCO and other corporations (after [5,11,12,13])

Major constituents (% wt.)	NTPC, 8 Thermal Power Stations (TPS)[11]			A.P.Genco , 4 Thermal Power Stations (TPS)[12]			Average FA, a few TPS[**5]	Average FA, a few TPS[13]	NALCO[11]
	from	to	mean	from	to	mean	mean	mean	mean
SiO$_2$	55.50	64.76	62.04	62.50	65.60	63.39	56.80	49-67	62.72
Al$_2$O$_3$	23.45	36.40		24.0	27.5	26.34	24.10	20-30	29.93
Fe$_2$O$_3$	4.40	7.73	5.23	3.17	5.32	3.90	4.10	5-22	2.13
TiO$_2$				0.50	1.55	1.23	1.20	0.1-2.0	
CaO	0.10	4.80	1.19	1.00	2.20	1.56	2.00	0.1-2.0	2.33
MgO	0.46	1.20	0.63	0.05	0.60	0.38	1.40	0.1-1.0	0.72
Na$_2$O	0.15	0.60	0.31	0.04	0.65	0.39		0.1-0.2	0.19
K$_2$O	0.70	1.13	0.88	0.85	1.72	1.30		0.1-1.0	0.25
SO$_3$	0.07	0.35	0.18	0.12		(0.12)		0.1-1.0	0.18
BPR*	0.64	4.25	1.47	1.22	2.15	1.68			2.19

* BPR – Buffering Potential Ratio = CaO+MgO / SO$_2$ + 0.04 Al$_2$O$_3$.(dimensionless)[17]
** Power Plants from East India[5]

The Buffering Potential Ratio BPR = CaO+MgO/SO$_3$ + 0.04 Al$_2$O$_3$ for FA from European power plants fired by hard coal ranges from 1.5 to 3.6, which classifies this material as alkaline aluminum silicate.[17] For FA considered typical for Indian power plants this ratio varies widely from 0.64 to 4.25 (Table 1) with a domination of low-buffered material with low CaO content ranging from 0.10 to 2.20.[5,11,12,13] For 8 power plants of NTPC, the average BPR is 1.47, for NALCO power plant it accounts for 2.19, though for other power plants higher values may also occur, as some reported CaO contents in FA appeared to be as high as 5-14 %[14]. Nevertheless, for average coal ash/FA collected from a few thermal power stations in eastern India[5] and the majority of other Indian power plants, the BPR = CaO+MgO/SO$_3$ + 0.04 Al$_2$O$_3$ and alkali content is markedly lower than e.g. for FA from disposal pond of Rybnik power plant in Poland, which showed acidification and intensive trace element release due to weathering transformations in the post-closure period (see Chapter XX). This suggests the possibility of the similar weathering transformations that could cause acidification of the material in time and the similar leaching behavior with the development of the time delayed release of trace elements in hazardous concentrations. Thus, both storage and application of FA requires assessment of its buffering capacity and taking adequate preventive measures with respect to the material susceptible to acidification and extensive leaching of trace constituents in the Delayed Release (III) stage. Unfortunately, there is no reliable data on the "safe" $CaO+MgO/SO_3 + 0.04 Al_2O_3$ ratio, and probably sharp border between the permanently alkaline material and that susceptible to acidification in time does not exist. Besides buffering capacity of the material, also the intensity of the weathering processes such as devitrification, formation of gibbsite and subsequent kaolinitization of the amorphous phase significantly affect the development of the Delayed Release (III) stage.[16] The information on the preconditions and kinetics of transformation processes of FA are still scarce and need further extensive studies. On the other hand, the principle of precaution and prevention suggests avoidance of any FA management option that poses potential risk to the environment and health at any stage of its life cycle.

Comparison of the elemental composition of coal and FA from European coal-fired power plants with that from India shows similarity of concentrations and trace element enrichment of about an order of magnitude higher than in coal; in coal ash (mixture 80/20 of FA and bottom ash) trace element content is somewhat lower than in FA [14, 18,19,20] (Table 2). In general, trace element concentrations in FA are also significantly higher than in natural soils under normal geochemical conditions[21] (Table 3). Of these elements, particularly environmentally problematic are oxyanions of a high toxicity such as As, B, CrVI, Mo, Se that have a broad stability field in solution and are mobile at any pH range, as well as toxic heavy metal cations that are able to accumulate in the tissue of living organisms (Pb, Cd, Ni). Many metals are useful macro- and microelements but exert detrimental effect in higher concentrations, e.g. Ca, Mg, Fe, Zn, Mn, Cu. Most of metal cations are mobilizable at low pH, but several of them form mobile species at extreme pH, both acidic and alkaline (e.g. Pb, Ni, Zn). In general, metal cations show high mobility in the acidic pH range. Besides pH-Eh that are the major controlling factors, also availability of trace elements for contact with pore solution, conditions of water circulation and presence of complexing agents determine leaching behavior of trace elements from FA under different conditions and stages of the environmental exposure determined by both FA characteristics and its applied management option in a definite area.

3. ENVIRONMENTAL EFFECTS OF FLY ASH DISPOSAL AND USE

3.1. Impact of FA Pond on Ground Water Quality

The reports on "no adverse effect" of FA disposal ponds on the water quality[3,4,5] induced the authors to analyze the results of a survey[22] carried out in response to the World Bank enquiry for evaluating the impact of the 17-years' old coal ash pond under operation (Maharastra State Electricity Board MSEB, Maharastra, Chandrapur) on the ground water quality in the vicinity of the disposal site. Surface pond for disposal of coal ash is sited in the submerged depression of the total area 27 km^2 in the Erai River basin near Chargaon village (Fig. 1). The ash pond that has a storage capacity of about 166 Mm^3 is expected to be filled up in 30 years. The excess of water is discharged as the overflow directly into the Kankalya Nalla, the tributary of the Erai River.

Within the ash pond area, the shallowest unprotected aquifer occurs 3-10 m below ground level in alluvial sediments. The general direction of ground water flow is from NW-NE to SE towards the main drainage watercourse of the Erai River. The alluvial aquifer has been used as a major source of water supply from the shallow dug wells for the numerous villages in the area sited predominantly within or down-gradient of the ash pond area.

Table 2. Concentrations of major constituents and trace elements in Indian ccal and FA/coal ash from thermal power stations (TPS) (TPS) compared to European coal/FA

Constituents	The Netherlands[18]		Coal[14]	Fly ash[19]	Coal ash[19]	Fly ash		Tamil Nadu TPS[20]	
	Coal	Fly ash		Mean (From-to)	From-to	Chandrapur STPS*	Bhusawal TPP*	Mettur	Ennore
Major constituents, wt %									
Al.	1.65	15.0	0.85+0.01	14.94 (12.43-19.29)	12.76-13.90				
C	73.2	4.3	78.11+0.37					0.08	.0.26
Ca	0.14	1.2	0.20+0.006	0.94 (0.07-3.43)	1.43-1.55				
Fe	0.51	4.7	0.75+0.045	3.42 (1.49-5.41)	2.79-2.80	3.15-3.17	2.98	1.7	2.5
K	0.17	1.5	0.074+0.0028	0.52 (0.08-0.94)	0.81			0.32	0.10
Mg	0.08	0.7	0.038+0.0008	0.43 (0.28-0.72)	0.84-3.06			1.3	2.6
N	1.6	0.3	1.56+0.07					0.01	0.04
Na	0.04	0.4	0.051+0.0011	0.19 (0.07-0.28)	0.36				
P	0.01	0.10			0.24			0.19	0.21
S	0.7	0.1	1.89+0.06	0.14 (0.03—0.87)				0.3	0.6
Si	2.82	25.7		29.01 (25.92-30.24)	26.51-27.22			58	51.
Ti	0.08	0.8	0.045+ 0.0017		0.72				
Trace elements (mg kg-1)									
As	3.7	34	3.72+0.09			3.4-3.7	1.9		
B	43	163							
Ba	158	1438	67.50+2.1	400	120-350				
Cd	0.10	0.9	0.06+0.0027			11.8-12.1	9.7		
Co	5.8	52	2.29+0.17	14	5-25	54-57	49	3.8	4.8
Cr	14.4	131		145	40-100	79-82	68		
Cu	16.6	151	6.28+0.30	72	20-60	60-61	54	30	20
Ga	2.0	18		50	15-25			44	35
Hg	0.16	0.23				<0.008	<0.005		
La	7.6	69		108	15-20				
Mn	46	415	12.40+1.0			213-216	194	30	25
Mo	3.0	27							
Ni	11	98	6.10+ 0,27			141-142.5	136		
Pb	8.5	77	3.67+0.26	95	10-30	10-41	36	2.8	3.1
Rb	9.2	84	5.05 +0.11						
Se	2.2	13	1.29+0.11	10	1-8				
Sr	107	971		164	40-350				
Th	2.9	26	1.342+0.036	112	39-85				
U	1.5	13	0.436+0.012	21	8-27				
V	29	262		321	55-150				
Y				47	20-50				
Zn	24	218	11.89+0.78	295	180-460	106 -118	90	40	190

* Data obtained by the authors within the current research project of CFRI (CSIR) Dhanbad, India

The quality of the ground water within the ash pond and in its vicinity was assessed on the basis of sampling in triplicate (with two weeks' intervals) and analyzing water quality from 11 dug wells and the effluent from the ash pond. As a control, the dug well up-gradient of the ash pond was selected, which was unaffected by the leachate from the pond (Fig. 1). More details on ash pond location and hydrogeological conditions in the area, as well as on the scope of the survey have been presented elsewhere[23].

The survey of 1996 showed significant adverse alteration of well water quality within and down-gradient of the pond (Table 4). The major changes consisted in the multiple increase of macro-constituent concentrations typical for the leachate from a power plant FA (TDS, chloride, sulfate, hardness, Ca, Mg, alkalinity). Analyzed trace element contents (B, Cd, Cr^{6+}, Hg, Pb) though were distinctly higher than the background concentrations, only occasionally exceeded MCL (Cd, Pb) due to pH values within the stability field for the most of these elements (pH 7.4-8,3). Only Hg and of other constituents also phenols were permanently above MCL, while NO_3 exceeded background concentrations. The strongest deterioration of ground water quality occurred within the ash pond (wells 11, 3, 10) that made

Table 3. Natural background concentrations of metals in agricultural soils (after [21]).

Countries	Soil	Cd From to	Cd Mean	Cr From to	Cr Mean	Cu From to	Cu Mean	Hg From to	Hg Mean	Ni From to	Ni Mean	Pb From to	Pb Mean	Zn From to	Zn Mean	As From to	As Mean	Mo From to	Mo Mean	Se From to	Se Mean
Australia	Sand			1.4-3.5									57	39-86				2.6-3.7			
	Loess			13–30	16																
	Clay				24																
Canada	Sand	0.10-1.8	0.43	2.6-34.0				0.01-0.7	0.06			2.5-47.5	10			1.1-28.9	5.8	0.4-2.5	1.5	0.1-0.32	0.27
	Loess	0.12-1.6	0.64													1.3-16.7	4.8				
	Clay							0.02-0.78	0.13	3-98	23	1.5-50	16.5	15-20				0.93-4.7	1.7	0.13-1.67	0.43
	Different															<1-30	5.8				
Europe	Sand	0.01-0.24	0.07	2-360	27	1-26	8.0		0.04	1-110	14	4-81	16	7-150	30	0.5-15	2	0.2-3.0	1.5	0.06-0.38	0.14
	Loess	0.24-0.36	0.30	21-35	29	8-54	19			7-70	19	14-96	44	20-180	88			0.6-3	3	0.17-0.34	0.23
	Clay	0.04-0.80	0.27	14-80	38	16-70	37	0.45-1.1		10-104	25	13-52	25	40-50		1.4-10	4.5	0.1-6.0		0.18-0.6	0.3
	Different			4-307	53	11-323	23	0.004-0.99	0.088	1.3-68	16	4-81	27			0.1-95	9	0.4-5	1.05	0.09-2.32	0.41
Japan	Sand															1.2-6.8	4.0				
	Different			3.5-810	50	5-176	34	0.08-0.49	0.28		28	5-189	35			0.4-70	11	0.2-11.3	2.6		
Russia	Sand		0.32			1.5-29	11							4-57	30			0.3-2.9	1.5	0.05-0.32	0.18
	Loess					11-36*	25							40-55	50			1.8-3.3	2.2		
	Clay				51	4-21	12							9-77	35			0.6-4	2.0		
	Different							0.06-0.29	0.1									0.8-3.6	2.2		
USA	Sand	0.08-0.47	0.21	3-200	40	1-70	14	0.01-0.54	0.08	<5-70	13	<10-70	17	5-165	40	<0.1-30	5.1	0.75-6.4	2.5	0.02-0.7	0.5
	Loess	0.13-0.55	0.27	10-1000	55	7-100	25	0.02-0.32	0.06	5-30	17	10-30	19	20-110	60	1.7-27	7.7	1.2-7.2	4.1	0.1-1.9	0.26
	Clay			20-150	65	7-70	29	0.0-0.90	0.13	5-50	21	10-70	22	20-220	70					0.1-1.9	0.5
	Different					3-300	26									<1-93	7	0.8-3.3	2.0		

*Alluvial soils

30

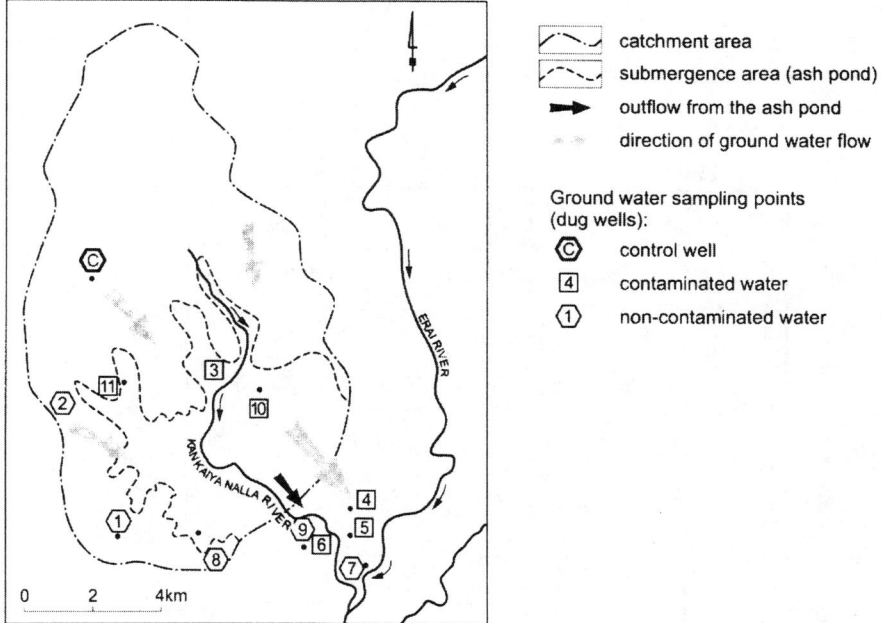

Figure 1.   Location and impact of the 17-years' old coal ash pond under operation
(Maharastra State Electricity Board MSEB, Maharastra, Chandrapur, India) on the
ground water quality.

water unfit for any use. Down-gradient of the ash pond (wells 4, 6, 5) groundwater quality got improved due to dilution, but still did not meet the standards. To summarize, the general pattern of adverse changes of groundwater quality caused by coal ash disposal reflects the release of soluble constituents in the Washout (I) stage of leaching under the saturated zone conditions. Owing to weakly alkaline pH values, most of trace elements remain within their stability fields as solid phases, while leaching of soluble macro-constituents is the major process, which strongly affect ground water quality. Hydrogeological conditions of FA pond represent both horizontal flow typical for the saturated zone in the surface ash layer, and vertical percolation of water through the underlying ash layers, with downward concentration/loads redistribution and occurrence of maximum concentrations at the bottom that is typical for the vadose zone. Elevated concentrations of Hg and phenols also in waters not affected by the FA pond (wells 2, 1, 8, 7) suggest dry and wet deposition from the power plant stack emission as a main source of these anthropogenic contaminants, in conformity with their environmental behavior.

Low concentrations of contaminants in the excess water overflow (point 9 - Fig, 1, Table 4) that are often used as the proof of a lack of adverse impact of ash disposed in the pond on water quality[2,3,4,5] are misleading. They result from the method of ash: water slurry preparation from fresh water, high water: ash ratio and short contact time of water with FA in the open circuit. In this particular case, further extensive deterioration of the unprotected shallow aquifer during the 30 years long operational stage of the ash pond is anticipated. Large disposal area and volume, despite of a lack of sectional construction, may result in the development of the different mode of soluble constituents release, including trace elements, through Dissolution (II) stage of a lower dynamics of both macro- and trace constituents release, up to the Delayed Release (III) stage with intensive leaching of trace elements following the acidification of a material due to weathering transformations that is reported in XX chapter of this book. Similar to that, increase of acidification and further long-term

Table 4. Ground water quality in sampling points located in the vicinity of coal ash pond site and in the direction of ground water flow in Washout (I) and Dissolution (II) stages (MSEB Chandrapur, Maharastra, India); concentrations in mg/L (mg dm^{-3}) (after [22])

Parameter	Control well up-gradient	Wells along the W border (up-gradient)			Well shielded by riverbed	Discharge (Overflow weir)	Wells within the ash pond			Wells down-gradient			DWS*) (MCL)	DL
Sampling points	C	2	1	8	7	9	11	3	10	4	5	6		
pH	8.35	7.66	7.75	7.81	7.79	7.96	7.51	7.45	7.37	7.41	7.67	7.36	6.5-8.5	0.1
Alkalinity CaCO$_3$	268.5	266.7	319.3	236	194.7	82	560	360	512	320.7	717.33	322	200	-
Total Dissolved Solids (TDS)	330	438.7	810	394.7	359.3	290	3513	2716	2122.7	1824.7	1332	1173	500	-
Total Hardness CaCO$_3$	194	286.7	460	264	216	168	1172	1010	720.7	884	147.33	660	300	-
Calcium Ca	48	78.13	90.4	72.3	69.9	57.6	174.13	190.4	93.87	156.1	24	178.4	75	-
Magnesium Mg	36	22.2	56.9	20.3	9.4	5.83	179.65	129.8	118.1	109.8	21.22	52	30	-
Chloride Cl	16.56	44.16	82.84	47.4	21.28	28.4	817.4	733.4	377.4	384.3	75.09	183.9	250	0.1
Sulphate SO$_4$	150	1	88.3	1.73	81.3	97.5	658	376.7	435.8	210	317.5	83	200	0.1
Fluoride F	1.41	0.94	0.66	0.5	0.041	1.68	1.24	0.428	1.107	0.78	1.55	0.263	1	0.018
Nitrate NO$_3$	0.24	0.52	2.47	0.69	0.21	0.109	5.98	4.58	3.36	4.95	0.3	3.62	45	0.01
Arsenic As	0.001	0.0065	0.011	0.016	0.0071	0.037	0.0072	0.011	0.03	0.011	0.006	0.0064	0.05	0.001
Boron B	0.07	0.097	0.126	0.054	0.054	0.45	0.318	0.211	0.383	0.179	0.43	0.102	1	0.01
Cadmium Cd	<0.001	0.0013	0.0027	0.003	0.005	0.001	0.008	0.007	0.02	0.006	0.0057	0.0063	0.01	0.001
Chromium Cr^{6+}	0.006	0.0063	0.0057	0.004	0.004	0.01	0.008	0.012	0.008	0.008	0.0157	0.0063	0.05	0.001
Copper Cu	0.011	0.0061	0.0073	0.004	0.005	0.003	0.016	0.013	0.009	0.011	0.0157	0.014	0.05	0.001
Iron Fe	0.31	0.388	0.325	0.247	0.173	0.55	0.213	0.516	0.152	0.205	0.2027	0.174	0.3	0.001
Mercury Hg	<0.001	0.0025	0.0037	0.004	0.002	0.006	0.002	0.004	0.005	0.005	0.002	0.001	0.001	0.001
Lead Pb	0.016	0.004	0.012	0.005	0.005	0.0007	0.06	0.054	0.023	0.051	0.0047	0.037	0.05	0.001
Selenium Se	0.004	0.008	0.013	0.005	0.006	0.0096	0.0058	0.007	0.006	0.008	0.005	0.004	0.01	0.001
Zinc Zn	0.037	0.107	0.068	0.035	0.052	0.012	0.0073	0.053	0.061	0.096	0.067	0.086	5	0.001
Phenols	0.001	0.015	0.022	0.022	0.025	0.028	0.28	0.04	0.037	0.039	0.026	0.083	0.001	0.001
α radiation Bq/l	<0.04	<0.04	<0.04	<0.04	<0.04	<0.04	<0.04	<0.04	<0.04	<0.04	<0.04	<0.04	Absent	0.04
β radiation Bq/l	<0.08	<0.08	<0.08	<0.08	<0.08	<0.08	<0.08	<0.08	<0.08	<0.08	<0.08	<0.08	Absent	0.08

DWS – Drinking water standard established by Bureau of Indian Standards
*) – According to IS 10500, 1991= SDWA/EPA
C– Control ground water from Chalbardi village
DL – Detection limit
Value – exceeds MCL
Value – exceeds the background level

Sampling points (Fig. 1):
1 – Tadoli village
2 – Ghodpeth village
3 – Kachrala village
4 – Tirwanja village
5 – Kawathi village
6 – Chhota Nagpur village
7 – Wichota village
8 – Chargaon village
9 – Discharge from FA pond
10 – Awanda village
11 – Gunjala village

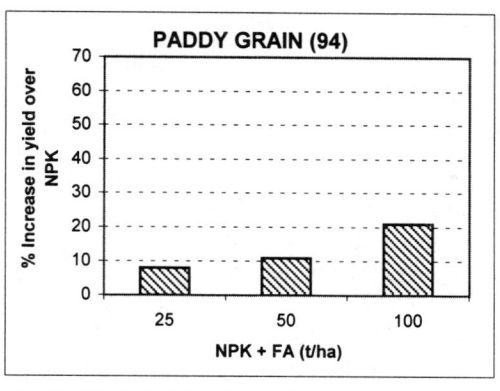

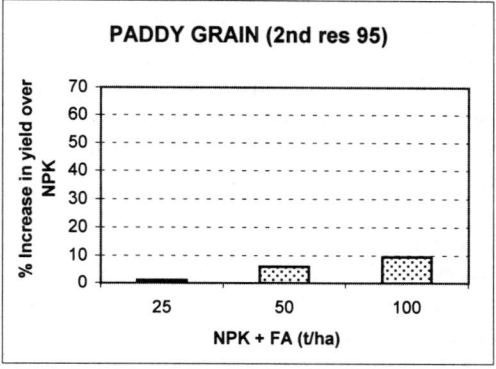

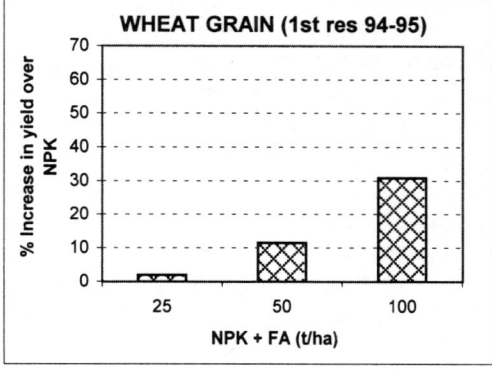

Figure 2.   Effect on agricultural crops of FA application in red and lateritic soil (pH <5.5) at BKTPP (field trials 1994-95).

intensification of the Delayed Release (III) adverse impact of ash pond on ground water in the post-closure period may develop as a result of devitrification and crystalline minerals formation (gibbsite, kaolinite, other clay minerals and aluminosilicates) from amorphous phase.[16]

Besides opinions about FA as a negligible source of the aquatic environment contamination cited above, the evidences of strong long-term adverse environmental impact of FA disposal sites are constantly increasing.[14, 24, 25] This leads to conclusion that much more attention and efforts to understand the short- and long-term processes resulting in health hazard and deterioration of the environment should be paid to avoid or attenuate the hazards and utilize beneficial properties of waste from coal-based power generation.

3.2. Studies on Agricultural Use of FA in India

The interest in FA use as soil amendment/ improver is growing steadily in many countries, among them in India and in the USA, though the enthusiastic approach to this form of this waste material utilization is not common. In particular, in many countries of the European Union (e.g. in the Netherlands) and other European countries (e.g. in Poland) the use of FA in agriculture is not permitted, and therefore the research in this area are scarce, if any. The limit values of heavy metal concentrations in soil according to the regulations of the European Community (EC), national regulations of the EU Member States and other countries (Poland, Canada, USA and New Zealand) display significant differences that reflect also different viewpoints and protocols adopted with respect to soil protection and risk assessment (Table 5).[26,27,28,29,30,31.] EPA 503[30] proposes definitely the highest limit values, much beyond the maximum concentration levels of trace elements in soil regulated by the EC and national legislation of European countries in force for application of sewage sludge. In turn, the new EC directive that is currently under development[28] considers increasing stringency of regulations compared to the current EC regulations in force. It should be underlined that these regulations were developed for the specific purpose of sewage sludge application as soil amendment. There are no regulatory limit values of trace element concentrations in agricultural soils in India.

In India, several research centers, among them research institutions under Council of Scientific and Industrial Research (CSIR) are involved in long-term research projects on FA utilization for agricultural purposes. In this chapter, the experiments conducted at the Central Fuel Research Institute (CFRI-CSIR) in Dhanbad since 1994 will be discussed, on the background of other similar research projects.

The field trials using different doses of FA ranging from 25 t/ha to 500 t/ha (that is 0.8 – 16% wt. or 1.3 – 25% v/v for soil layer 20 cm thick) were conducted in acidic red and lateritic soils (pH 5.5) and in alkaline alluvial soils (pH 8.0) in 1994-1995. FA used in field trials to amend the acidic red and lateritic soils originated from Bakreswar Thermal Power Project (BKTPP) under the management of West Bengal Power Development Corporation (WBPDC) and from Farakka Super Thermal Power Project (FSTPP) under National Thermal Power Corporation (NTPC). FA for amending alkaline alluvial soils came from FSTPP. The experiments and field trials in the same soil types have been continued also in years 1997-2000 using FA from Chandrapur STPS and Bhusawal TPP. Basic parameters and some trace metal content (total and DTPA extractable) in the soils used in the field trials (alkaline alluvial A, B, C) and acidic red soils (D), as well as in corresponding FA added to these soils are presented in Table 6 (DTPA extractable trace elements are considered bioavailable to plants). The comparison of these parameters for soil and FA shows FA superiority over soils with respect to Water Holding Capacity (WHC), porosity and bulk density. Concentrations of trace elements (Cu, Zn and Pb) in FA were distinctly higher than in amended soils; DTPA extractable Cu and Zn were lower, while DTPA extractable Pb was higher in FA. Both total and DTPA extractable contents of Fe and Mn appeared to be higher in soils than in FA. The data on bioavailability based on DTPA extractable metals, though, do not characterize bioavailability of metals in FA-amended soils that results from FA interaction with soil, and may also change in time.

The crops grown in these soils included paddy, wheat, maize, soybean, mustard, till, potato, radish and carrot. The major purpose of FA use was (i) improving some of the important physico-chemical properties of the soil such as hydraulic conductivity, water holding capacity WHC, bulk density, particle size and pozzolanic properties; (ii) neutralizing the acidic soil of red and lateritic type (pH 5.0-5.5). Crops grown in FA-affected soil (where FA was not added in purpose) were also investigated.

From the field trials 1994-95 it was concluded that FA application in doses from 25 to 100 t/ha to the acidic soils made them 10-30% more productive over control with respect to different crops (example – Fig. 2). Application of FA to alkaline soils in amounts from 25 to

Table 5. Limit values for heavy metal concentrations in the soil comparing the EU, Canada, USA and New Zealand (mg/kg d.m.)[26,27,28,29,30,31]

Country	Regulations	Cd	Cr	Cu	Hg	Ni	Pb	Zn	As	Mo	Se
EC/soil amended with sludge											
(86/278/EC)[27]	lower limit	1.0	100[7]	50	1.0	30	50	150	–	–	–
	upper limit	3.0	150	140	1.5	75	300	300	–	–	–
DG ENV.E.3 /LM draft 2000[28]	5 pH <6	0.5	30	20	0.1	15	70	60	–	–	–
	6 pH <7	1.0	60	50	0.5	50	70	150	–	–	–
	pH 7	1.5	100	100	1.0	70	100	200	–	–	–
EU-15 (national regulations)[1]	lower limit	0.4	30	36	0.2	15	40	60	–	–	–
	upper limit	3.0	400[2]	135[2]	2.0	112	300	450[2]	50[3]	4[3]	3[3]
Germany[3]	clay	1.5	100	60	1	70	100	200		–	–
	loam	1	60	40	0.5	50	70	150		–	–
	sand	0.4	30	20	0.1	15	40	60		–	–
Poland, 1999[29] (national regulations)	agricultural use soils: -heavy	3	100	75	1.5	50	80	180	–	–	–
	-mean	2	75	50	1.2	35	60	120	–	–	–
	-llight	1	50	25	0.8	20	40	80	–	–	–
	non-agricultural use soils::: -heavy	5	200	100	2	60	100	300	–	–	–
	-mean	4	150	75	1.5	45	75	220	–	–	–
	-llight	3	100	50	1	30	50	150	–	–	–
Canada		2.0	–	–	0.5	18	50	185	7.5	2.0	1.4
USA[4]											
EPA 503 proposed reg.[30]	soils – upper limit	19.5	1500	750	8.5	210	150	1400	20.5	9	50
EPA: Risk-Based Concentrations (RBC)[31]	soils – residential area	39 n	Cr(III) 78000 n Cr(VI) 390 n	2900 n	23 n	1600 n		23000 n	23 n	390 n	390 n
	soils – industrial area	510 n	Cr(III) 100000 n Cr(VI) 5100 n	38000 n	310 n	20000 n		310000 n	310 n	5100 n	5100 n
New Zealand		3.0	600	140	1.0	35	300	300	–	–	–

1) Limit Values for application of sewage sludge; 2) Soil pH 6.0.–7.0; 3) German Compost Ordinance
4) EPA 503 proposed regulations[30] and EPA Region III Risk-Based Concentrations (n - noncarcinogenuc effect)[31]

35

Table 6. Physico-chemical characteristics of Indian soil and fly ash (FA).

Parameters	Units	SOIL — Alkaline Alluvial Soil A (Field A)	SOIL B (Field B)	SOIL C (Field C)	SOIL — Aciddic Red Soil D (Red Soil)	FLY ASH A (Chandrapur STPS)	FLY ASH B (Bhusawal TPP)	FLY ASH C (Chandrapur STPS)	D (A and B added)
pH		8.71	7.95-8.03	8.5	5.06	7.3-8.35	8.2-8.64	8.49	
Electro Conductivity	dS/m	0.173	0.241-0.247	0.35	0.054	0.13-0.34	0.109-0.520	0.30	
WHC	%	39.71	35.73-36.70	55.2	31.2	43.90-45.35	44.96-50.11	47.80	
Porosity	%	47.17	43.63-44.35	49.3	34.1	52.46-58.7	57.9-59.46	53.18	
Bulk Density	g/cc	1.56	1.45-1.48	1.58	1.61	0.89-1.02	0.94-1.03	1.08	

HEAVY METALS

Elements	Units	A Total	A DTPA-extractable	B Total	B DTPA-extractable	C Total	C DTPA-extractable	Red Soil Total	FA A Total	FA A DTPA-extractable	FA B Total	FA B DTPA-extractable	FA C Total	FA C DTPA-extractable
Cu	mg/kg	35.6	2.18	33.8	1.14-1.21	33.8	1.75	39.5	60.5-61.2	1.03	53.8	0.92		1.38
Zn	mg/kg	51.8	1.56	63.9	0.42-0.65	63.9	1.60	45.3	105.6-118.3	1.2	89.6	0.77		1.31
Mn	mg/kg	315.8	10.9	298.6	6.90-7.18	298.6	9.73	179.6	212.8-215.7	0.71	194.2	0.68		1.21
Fe	mg/kg	35000	13.6	32500	9.3-11.03	32500	10.9	23100	31500-31700	6.70	29800	5.90		7.30
Pb	mg/kg	5.70	1.60	19.9	1.95-2.10	19.9	2.00	19.7	9.6-40.8	3.10	35.6	2.74		2.80
Ni	mg/kg	135.7	0.33	125.8	0.35-0.38	125.8	0.33	175.3	141.2-142.5	0.61	135.7	0.43		0.56
Co	mg/kg	20.6	0.02	18.8	0.02-0.03	18.8	0.02	29.6	53.8-57.4	0.03	49.1	0.03		0.05
Cr	mg/kg	26.8	<0.005	27.2	<0.005	27.2	<0.005	57.3	78.6-81.8	<0.005	68.4	<0.005		<0.005
Cd	mg/kg	5.1	<0.005	4.9	<0.005	4.9	<0.005	6.3	11.8-12.1	<0.005	9.7	<0.005		<0.005
As	mg/kg	0.9	<0.005	1.0	<0.005	1.0	<0.005	0.61	3.4-3.7	<0.005	1.9	<0.005		<0.005
Hg	mg/kg	<0.005	<0.005	<0.005	<0.005	<0.005	<0.005	<0.005	<0.008	<0.005	0.006	<0.005		<0.005

RADIOACTIVITY (Bq/kg)

	Units	A	B	C	Red Soil	FA A	FA B	FA C
α-radiation	Bq/kg	240	200.4-201	320	275.3	541-551	521	473
β-radiation	Bq/kg	610	589.3-611	740	628.6	1573-1581	1479	1089

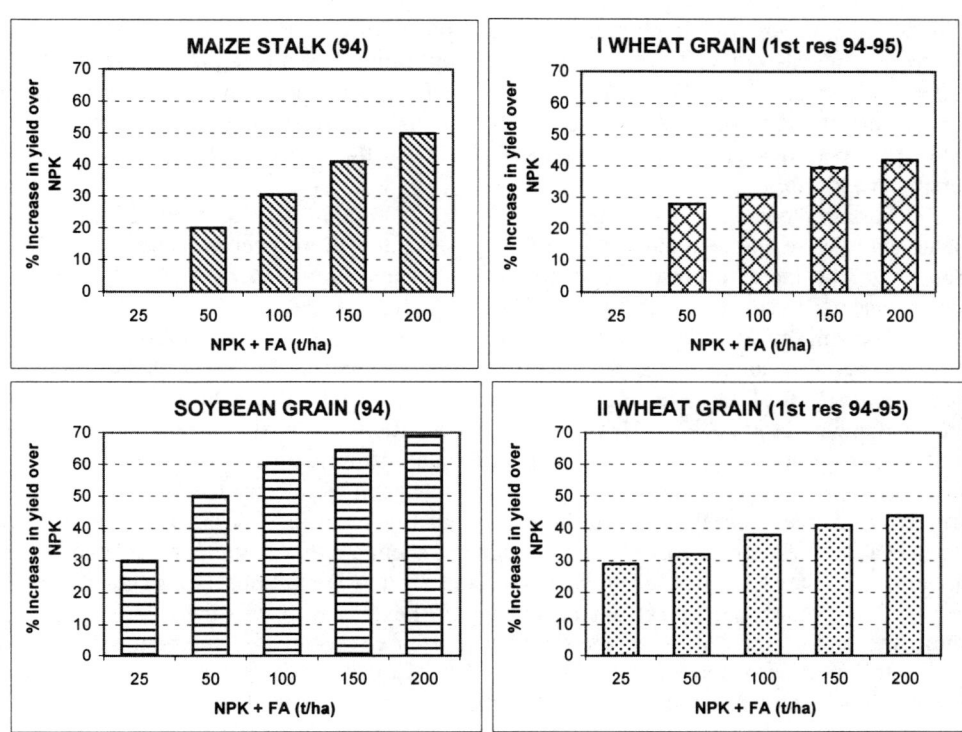

Figure 3.   Effect on agricultural crops of FA application in alluvial (pH> 8.0) soil at FSTPP
(field trials 1994-96).

Table 7. Effect of FA application on concentration of trace and toxic metals in crops
cultivated in acidic soils at BKTPP (field trials 1994-95).

Fertilizer + FA doze	Part of a plant	Trace elements (mg/kg, dry wt)								
		As	Cd	Co	Cr	Cu	Mn	Ni	Pb	Zn
Paddy crop										
NPK	Grain	nd	nd	3,6	nd	2,5	22,5	3,4	0,6	11,9
	Straw	nd	nd	5,5	nd	3,9	139,8	5,0	1,3	43,6
NPK + 25 t/ha	Grain	nd	nd	4,0	nd	2,4	20,9	3,5	0,7	12,0
	Straw	nd	nd	6,4	nd	4,0	138,2	4,9	1,0	43,7
NPK + 50 t/ha	Grain	nd	nd	3,8	nd	2,2	20,0	3,6	0,6	11,0
	Straw	nd	nd	6,3	nd	3,8	118,4	5,4	1,2	40,6
NPK + 100 t/ha	Grain	nd	nd	4,0	nd	2,0	19,8	3,8	0,5	10,9
	Straw	nd	nd	6,0	nd	3,6	113,0	6,0	1,1	38,7
Wheat grain										
NPK	Grain	<0,05	0,10	nd	0,55	nd	nd	nd	0,27	12,81
+ 200 t/ha		<0,05	0,11	nd	0,70	nd	nd	nd	0,30	10,80
Maize corn										
NPK	Grain	<0,05	<0,05	nd	0,67	nd	nd	nd	0,89	14,21
+ 200 t/ha		<0,05	<0,05	nd	0,60	nd	nd	nd	0,75	10,25

nd - not determined

200 t/ha was found to increase crop yield significantly, up to 45-70% at the FA dose 200 t/ha (Fig. 3), due to improving physico-chemical properties of the soil. No adverse effect of heavy/toxic metals uptake by crops due to FA application to acidic soils in doses up to 200 t/ha was observed (Table 7). The FA dozes over 300 t/ha applied either to acidic or alkaline soil did not increase the yield of crops, but also no adverse effect on crop growth during cultivation period was observed. Therefore, FA dose up to 200 t/ha was assumed to be beneficial and safe.

Further studies carried out by CFRI 1n 1997-2000 on FA effect from different power plants onto acidic and alkaline soils gave similar results with respect to increase of different crop yield; the increase of soil productivity in the range from 16 to 33% with respect to control was observed also at FA doses of 500 t/ha (Table 8, Fig. 4).

The simultaneous increase of total heavy metal concentrations in FA-amended acidic red soils was also noticed (Fig. 5). In FA-enriched alkaline soil, higher DTPA–extractability compared to control suggested higher availability of metals to plants, though pH of soil remained unchanged (Table 9). Heavy metal concentrations in FA-affected soil did not exceed limit values proposed by EPA 503[30], and were much below EPA-RBC[31], though Cd and Ni concentrations both in non-amended and FA-amended soil were considerably above the more stringent and diverse limit values of the EC [27,28] and national regulations of the EC Member States (Table 5). Other authors express the opinion about a rather low risk from trace element mobilization from FA-amended soils in Washout (I) and Dissolution (II) stages at alkaline pH[31,32]. The reduction of plant available trace elements due to alkaline FA application to acidic mine soils treated with sewage sludge was reported recently[33]. On the other hand, there is also an evidence of the elevated bioavailability of Zn, Fe, Mo in soils amended with FA compost (composted mixture of FA and biosolids, 1:1 ratio), as well as of the potential of Zn and Pb to be leached into groundwater from such soils. These observations lead to the conclusion about the need of developing guidelines that establish an upper limit for the total amount of FA application rate to the agricultural land[34].

Comparison of trace element concentrations in crops grown in FA-enriched soil showed diverse susceptibility of trace metal uptake by different crops: soybean>linseed>jowar>wheat. The increase of concentrations varied depending upon metals and crops. In soybean and linseed grain, all investigated metal concentrations were found to increase, in soybean in the range from 3.5 % (Fe) to 23.8 % (Ni) and in linseed from 2.8 (Fe) to 16.2% (Pb). In jowar grain, only Cu, Mn, Pb and Cr concentrations were higher than in control, in the range from 2.2% (Cu) to 7.6% (Cr) while contents of Zn, Fe, Ni and Co showed distinct decrease. In wheat grain from FA-affected soil, contents of all metals were lower than in control that confirmed well-known wheat resistance to metal uptake (Table 9).

Table 8. Yield pattern of crops grown in the fields affected by FA from Chandrapur STPS (~500 t FA/ha) in the period of 1997-2000.

Number of fields	Seson / year (field number)	Crop	Yield Qt/ha		% increase over control
			Control	FA-affected	
8	Kharif 97 (6) Rabi 99 (2)	Soybean	7.7 - 11.1	9.8 - 14.2	26.8 - 33.3
5	Rabi 97-98 (4) Rabi 99-2000 (1)	Wheat	8.1 - 10.4	10.5 - 13.4	19.4 - 29.5
4	Rabi 97-98 (1) Kharif 98-99 (1) Kharif 99-2000 (2)	Arhar	6.6 - 9.6	7.8 - 11.2	16.4 - 20.6
4	Kharif 98 (3) Kharif 98-99 (1)	Jowar	5.3 - 7.0	7.1 - 8.4	20.0 - 34.0
1	Rabi 98-99 (1)	Gram	6.8	8.6	26.5
1	Kharif 97 (1)	Moong	4.8	6.4	33.3
5	Rabi 97-98 (2) Rabi 98-99 (2) Rabi 99-2000 Rabi -2000	Linseed	4.3 -5.1	5.1 - 6.7	20.4 - 24

Table 9. Effect of FA on the concentration of trace elements in Chandrapur STPS FA-affected soil and grain of crops grown in this soil compared to control fields (total and DTPA extrablable).

Trace elements (mg/kg)	FA	Soil			Crop							
		Total	DTPA Extractable		Soybean		Wheat		Jowar		Linseed	
		Control	Control	FA-affected	Control	FA-affected	Control	FA-affected	Control	FA-affected	Control	FA-affected
pH	8.49	8.50	8.5	8.40								
Cu	55.8	33.8	1.75	2.20	5.16	6.00	4.50	3.60	2.65	2.71	6.59	7.22
Zn	126.6	63.9	1.60	1.74	12.79	13.52	16.30	14.90	17.7	17.5	17.0	17.8
Mn	210.9	298.6	9.73	10.6	16.58	18.11	17.10	16.60	23.8	25.0	45.8	47.5
Fe	30500	32500	10.9	13.3	40.87	42.33	89.2	87.9	80.3	79.8	49.5	50.9
Pb	29.1	19.9	2.00	2.29	0.79	0.83	0.41	0.33	0.54	0.57	0.37	0.43
Ni	145.6	125.8	0.33	0.42	0.42	0.52	0.39	0.25	0.46	0.44	0.34	0.36
Co	55.2	18.8	0.02	0.04	0.36	0.43	0.55	0.49	0.70	0.69	0.51	0.54
Cr	73.1	27.2	< 0.005	< 0.005	1.31	1.44	0.49	0.38	0.79	0.85	0.30	0.35
Cd	10.3	4.9	< 0.005	< 0.005	< 0.005	< 0.005	< 0.005	< 0.005	< 0.005	< 0.005	< 0.005	< 0.005
As	3.4	1.0	< 0.005	< 0.005	< 0.005	< 0.005	< 0.005	< 0.005	< 0.005	< 0.005	< 0.005	< 0.005
Hg	0.007	<0.005	< 0.005	< 0.005	< 0.005	< 0.005	< 0.005	< 0.005	< 0.005	< 0.005	< 0.005	< 0.005

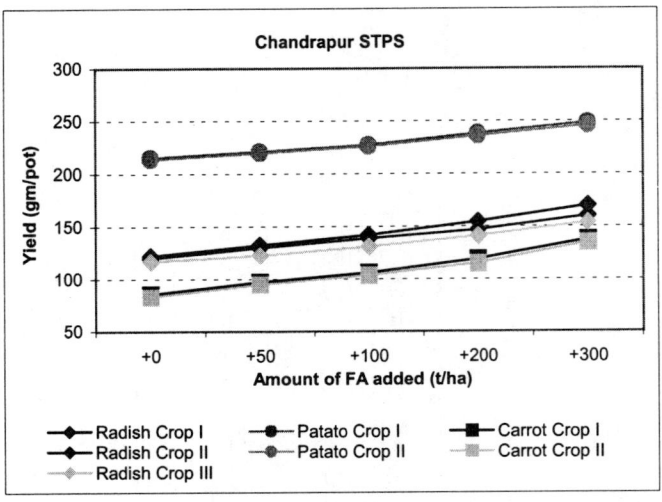

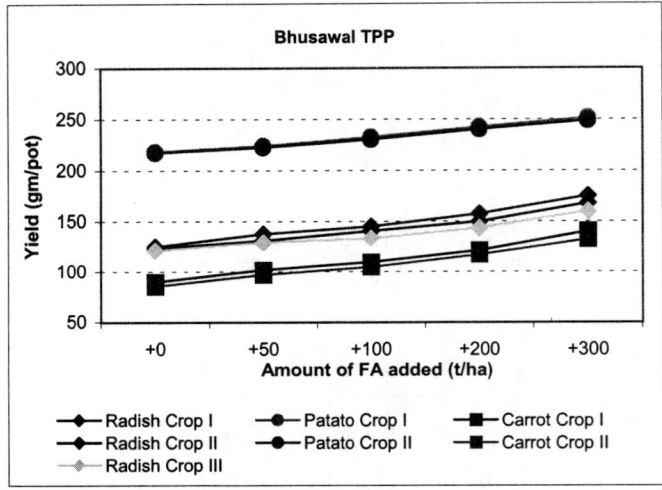

Figure 4.   Effect on crop yield of FA application from Chandrapur STPS and Bhusawal TPP
(pot experiments) ·

Other source[36] noted no significant influence of high loads of As added to soil (6-9 kg/ha year, 41 kg/ha of cumulative loading) from FA or FA/biosolids blends on element concentration in mature plants of corn grain or on its leaching from soil profile. A low risk from Ni, Pb, Cd, and Cr concentrations in shoots of maize grown in acidic soils amended with FA and other coal combustion residues was also reported[36]. Nevertheless, besides encouraging results and opinions concerning beneficial effect and environmental safety of FA use either alone or in FA/biosolids mixtures there is also an evidence of serious environmental implications of FA use and disposal that drive to the conclusion that "coal as a major energy source may pose more serious ecological problems than previously believed"[25]. This conclusion is based on the growing number of reports on acidification and massive heavy metal release from FA disposal sites discussed in the chapter XX and elsewhere[16, 23], as well as from FA/reject coal landfills[25], and on finding high levels of heavy metals in animals exposed to coal fly ash and coal rejects from burning coal.[37] These signals cannot be neglected and suggest that more attention should be paid to the time-dependent

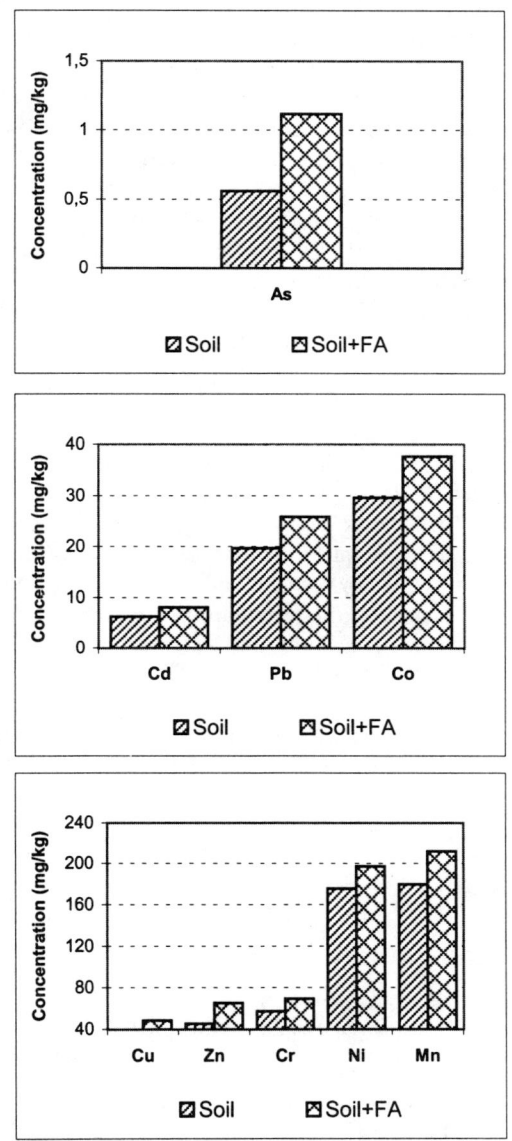

Figure 5.   Increase of heavy metal concentration in FA-amended acidic red soil (300 t/ha).

transformations of FA exposed to the atmospheric conditions due to weathering processes, in order to utilize safely beneficial properties of this material and adequately reduce potential and actual adverse effects.

4. CONCLUSIONS

In the light of presented data, FA disposal in surface ponds is not environmentally safe, while application in agriculture seems to be a prospective sink for FA, in particular when a significant increase of crop yield after application up to about 500 t/ha have been observed. Though, considering low lime content in Indian CCW (predominantly <1 – 2 % CaO) typical also for the majority of this waste in other countries, and hence its short-term neutralizing capacity, large-area uncontrolled disposal or agricultural use of FA may bring about an irreversible water or soil contamination in the long-range period (in the Delayed Release stage). High heavy metal enrichment including mobile oxyanions of proven toxicity (B, As, Mo, Se) and adverse weathering transformations of FA properties, as well as susceptibility to extensive trace metal release due to acidification caused by devitrification and mineralization of amorphous phase support this anxiety. These premises resulted in a ban on CCW use in agriculture as soil amendment in many countries, e.g. in the EU Member States and also in Poland, and limited use in other countries (e.g. in Japan) as presumable environmentally problematic application, while use of biosolids in agriculture is generally acceptable provided the regulations are strictly followed. The safe utilization of CCW in agriculture, particularly for the industrial dumping sites reclamation with use of technical species, e.g. aromatic plants growth[38] though is sound and prospective, does not solve the problem of bulk utilization of FA.

Currently, most of the studies on FA application in agriculture have been conducted with use of the freshly generated FA, when its environmental behavior and interactions are determined by the properties of strongly alkaline non-weathered vitrified material in the Washout (I) or Dissolution (II) stages of macro- and trace elements release, and all the conclusions and suggestions are relevant to these stages. Hence, further extensive studies on the long-term environmental behavior of this anthropogenic waste material should be conducted, in particular to elucidate the weathering transformations and interactions of weathered FA with soil and plants also in the Delayed Release (III) stage. Until these questions are not clear, the application of FA to the acidic soils should be avoided. If there are no regulatory restrictions, the caution and pollution prevention principle suggests FA application entirely to the well-buffered alkaline/neutral soils in the doses that increase the crop yields but do not cause significant increase of metal bioavailability to plants. Plant selection for cultivation in FA-amended fields should consider their susceptibility to metal uptake and accumulation in the edible parts.

REFERENCES

1. Anonymous, Key coal statistics for 2000, *Ecoal,* 40, 8, 2001.
2. Raghuveer, S., Coal ash and its impact on environment, *Journ. CAII,* III, 31, 1999.
3. Gambhir, S.K., Environmental management of coal ash in Chandrapura the thermal power plant – a case study, *Journ. of CAII,* III, 70, 1999.
4. Singh, G., and Gambhir, S. K.:, Environmental evaluation of flyash in its disposal environment, in *Coal Science, Technology, Industry, Business & Environment,* .Narasimhan, K. S., and .Sen, S, Eds., Allied Publishers Ltd, New Delhi-Mumbai-Calcutta-Lucknow-Madras-Nagpur-Bangalore-Hyderabad-Ahmedabad, 1996, 546.
5. Singh, G: Environmental evaluation of coal combustion residues utilization in mining areas, in *Clean Coal. Proceedings of the International Symposium on Clean Coal Initiatives,* New Delhi, India, (T.N.Singh, T. N., and Gupta, M. L., Eds., Oxford & IBH Publ.Co.Pvt.Ltd., New Delhi Calcutta, 1999, 463.
6. Collins, S., Managing powerplant wastes. Special report. *Power,* 8, 15, 1992.
7. Cabrera, J. G., and Woolley, G. R., Fly ash utilization in civil engineering, in *Environmental Aspects of Construction with Waste Materials, Proceedings of the International Conference WASCON'94,* Maastricht, The Netherlands, Goumans, J. J. J. M., van der Sloot, H. A., and Th.G.Aalbers, Th. G., Eds., Elsevier, Amsterdam-London-New York-Tokyo, 1994, 345.
8 Dalal, G. G., Generating "cash from ash". A strategy for ash utilisation, *Journ.f CAII,* III, 65, 1999.
9. Prasad, B., Bose, J. M., and Dubey, A. K., Present situation of fly ash disposal and utilization in India: an appraisal, in *Indo-Polish Workshop on Fly Ash Management,* Calcutta, R. P. .Das, Ed., RRL Bhubaneswar, CMRI and CFRI Dhanbad, CGCRI Calcutta, 2000, 7.1.
10. Maji, P., Commercial production of coal ash bricks – constraints, *Journ. CAII,* III, 84, 1999.
11. Mishra, C. R., and Seth, M. M., Status of utilization of coal ash at NALCO, *Journ. CAII,* III, 37, 1999.
12. Rao, D. V. V. P., and Reddy, M. S., Fly ash utilization in A.P. Genco – a status report, *Journ. CAII,* III, 52, 1999.

13. Prusty, B.K., Singh, T. B. and Tewary, B.K., Utilization of fly ash in reclamation of colliery waste – a green approach, in *Clean Coal,* Singh, T. N., and Gupta, M. L., Eds., Oxford & IBN Publ. Co., New Delhi Calcutta, 1999, 687.

14. Khandekar, M. P., Bhide, A.D., and Sajwan, K.S., Trace elements in Indian coal and coal fly ash, in *Biogeochemistry of Trace Elements in Coal and Coal Combustion Byproducts,* Sajwan, K.S., Alva, A. K., and Keefer, R. F., Eds., Kluwer Academic/Plenum Publishers, New York, 1999, chap.6.

15. Twardowska, I., Environmental aspects of power plants fly ash utilization in deep coal mine workings, in *Biogeochemistry of Trace Elements in Coal and Coal Combustion Byproducts,* Sajwan, K.S., Alva, A. K., and Keefer, R. F., Eds., Kluwer Academic/Plenum Publishers, New York, 1999, chap. 3.

16. Twardowska, I., Szczepanska, J., Solid waste: terminologial and long-term environmental risk assessment problems exemplified in a power plant fly ash study, *Sci. Total Environ.,* 285, 29, 2002.

17. van der Sloot, H.A., Piepers, O., and Kok, A.: *A Standard Leaching Test for Combustion Residues,* BEOP-31, Bureau for Energy Research Projects. Netherlands Energy Research Foundation, ECN, Petten, the Netherlands, 1984.

18. Meij, R., and Schaftenaar, H.P.C.,. Hydrology and chemistry of pulverized fuel ash in a lysimeter or the translation of the results of the Dutch column leaching test into field conditions, in *Environmental Aspects of Construction with Waste Materials, Proceedings of the International Conference WASCON'94.* Goumans, J.J.J.M., H.A.van der Sloot, Th.G.Aalbers, eds Maastricht, the Netherlands, June 1994.), Elsevier, Amsterdam-London-New York-Tokyo, 1994, 491.

19. R.P.Das, RRL(CSIR) Bhubaneswar – personal communications (unpublished data), 2000.

20. Selvakumari, G., and Baskar, M., Tnau's endeavour for effective utilization of flyash, *Journ. CAII,* III, 46, 1999

21. Kabata-Pendias, A., *Trace Elements in Soil and Plants,* 3rd Edition, CRC Press, Boca Raton, 2001, 432.

22. AIC Watson Consultants Ltd., Report on Ground water Strata of Ashbund / Aquifer at CTPS, Chandrapur Submitted to Maharastra State Electricity Board, Bombay, 1996.

23. Twardowska, I., Singh, G., and Tripathi P.S.M., Problems of monitoring and long-term risk assessment for ground water from high0volume solid waste sites in industrialized and developing countries, in *Environmental Monitoring and Remediation Technolpgies II,* Vo-Dinh, T., and Spellicy, R.L., Eds., *Proceedings of SPIE,* 3853, 344, 1999.

24. Ghuman, G.S., Sajwan, K.S., and Denham, M.E., Impact of coal pile leachate and fly ash on soil and groundwater, in *Biogeochemistry of Trace Elements in Coal and Coal Combustion Byproducts.* Sajwan, K.S., Alva A.K. and Keefer, R.F., Eds., Kluwer Academic/Plenum Publishers, New York, 1999, chap. 14.

25. Danker, R., Adriano, D.C., Barton, C., and Punshon, T., Revegetation of a coal fly ash-reject landfill, in *Sixth Intern. Conf. Biogeochemistry of Trace Elements, ICOBTE 2001 Conf. Proc.,* Guelph, Ontario, Canada, University of Guelph, 2001, 381.

26. Amlinger, F., A European survey on the legal basis for separate collection and composting of organic waste, in *Report: EU-Symposium "Compost – Quality Approach in the European Union",* Vienna, Austria, Fed. Ministry for the Environ., Youth and Family Affairs, 1998, 13.

27. Council Directive 86/278/EEC of 12 June 1986 on the protection of the environment, and in particular of the soil, when sewage sludge is used in the agriculture, *OJ L* 181 04. 07.1986, 6, with amendments *OJ L* 377 31.12. 1991, 48; 194 N; incorporated by *OJ L* 001 03.01.1994, 494.

28. Working Document on Sludge, 3rd draft, *DG ENV.E.3/LM,* EC, Brussels, 2000.

29. Minister of the Environ. Protection, Natural Resources and Forestry Regulation of 11 August 1999 on use of sewage sludge for non-industrial purposes, *Dz.U.* 813 31.08.1999.

30. US EPA, Part 503, Stanards for the use or disposal of sewage sludge, *Fed. Register.,* 58, 9248, 1993, 415.

31. Smith, R.L., *Risk-Based Concentrations: A Method to Prioritize Environmental Problems Using Limited Data,* US EPA, Reg. 3, Philadelphia, PA, 1994.

32. Alva, A.K., Bilski, J.J., Sajwan, K.S., and van Clief, D., Leaching of metals from soils amended with fly ash and organic byproducts, in *Biogeochemistry of Trace Elements in Coal and Coal Combustion Byproducts,* Sajwan, K.S., Alva, A. K., and Keefer, R. F., Eds., Kluwer Academic/Plenum Publishers, New York, 1999, chap. 11.

33. Alva, A.K., Paramasivam, S., Prakash, O., Sajwan, K.S., Ornes, W.N., and van Clief, D: Effects of fly ash and sewage sludge amendments on transport of metals in different soils, in *Biogeochemistry of Trace Elements in Coal and Coal Combustion Byproducts,* Sajwan, K.S., Alva, A. K., and Keefer, R. F., Eds., Kluwer Academic/Plenum Publishers, New York, 1999, chap. 12.

34. Bhumbla, G.K., Sekhon, B.S., and Sajwan, K.S., Trace element bioavailability in mine soils treated with sewage sludge and fly ash mixtures, in *Sixth Intern. Conf. Biogeochemistry of Trace Elements, ICOBTE 2001 Conf. Proc.,* Guelph, Ontario, Canada, University of Guelph, 2001, 368.

35. Yuncong L., Zhang M., Stopella P., Bryan, H., and Zhenli, H.E., Influence of fly ash compost application on distribution of metals in soil, water and plant, in *Sixth Intern. Conf. Biogeochemistry of Trace Elements, ICOBTE 2001 Conf. Proc.,* Guelph, Ontario, Canada, University of Guelph, 2001, 374.

36. Clark, R.B., Zeto, S.K., Baligar, V.C., and Ritchey, D., Nickel, lead, cadmium and chromium concentrations in shoots of maize grown in acidic soil amended with coal combustion byproducts, in *Biogeochemistry of Trace Elements in Coal and Coal Combustion Byproducts,* Sajwan, K.S., Alva, A. K., and Keefer, R. F., Eds., Kluwer Academic/Plenum Publishers, New York, 1999, chap.16.

37. Sumner, M.E., and Dudka, S., Fly ash-borne arsenic in the soil-plant system, in *Biogeochemistry of Trace Elements in Coal and Coal Combustion Byproducts,* Sajwan, K.S., Alva, A. K., and Keefer, R. F., Eds., Kluwer Academic/Plenum Publishers, New York, 1999, chap. 17.
38. Rowe, C.L., Hopkins, W.A., and Coffman, V.R., Failed recruitment of southern toads (*Bufo terrestris*) in a trace element-contaminated breeding habitat: direct and indirect effects that may lead to a local population risk, *Arch. Environ. Contam. Toxicol.,* 399, 40, 2001.
39. Kangungo, S.P., Use of fly ash in commercial aromatic plants, in *Indo-Polish Workshop on Fly Ash Management,* Calcutta, R. P. .Das, Ed., RRL Bhubaneswar, CMRI and CFRI Dhanbad, CGCRI Calcutta, 2000, 11.1.

THE EFFECTS OF SWITCHING FROM COAL TO ALTERNATIVE FUELS ON HEAVY METALS EMISSIONS FROM CEMENT MANUFACTURING

Arun B. Mukherjee[1]*, Ursula Kääntee[2], and Ron Zevenhoven[3]

[1]Department of Limnology and Environmental Protection, P.O. Box 62
FIN-00014 University of Helsinki, Finland

[2]Finnsementti Oy, FIN-21600 Parainen, Finland

[3]Energy Engineering and Environmental Protection, Helsinki University
of Technology, P.O. Box 4400, FIN-02015 Espoo, Finland

1. ABSTRACT

The total generation of scrap car tires throughout the world is estimated at 1000 million pieces per year which presents economic, environmental treatment and disposal problems. In the European Union, the estimated generation of scrap car tires is about 180 million per year of which 65% should be recycled by the member states. In Finland, the recovery percentage is about 90. There are many ways scrap car tires can be reused and these may include heat and power production, road construction, landfills, protection of sea shores from waves and so on. A cement production plant at the south-west coast of Finland has replaced traditional fossil fuel (coal and petcoke) by 10% scrap car tires. Car tires contain heavy metals. This study focuses on how toxic elements Hg and Tl can be captured as particulates in presence of Mn and Cr oxides. The shifting of gaseous phases of metals to particulate forms is more beneficial for the ecosystem because metal-containing particulates are more easily seperatable from the gas stream by the emission control equipments.

Chemistry of Trace Elements in Fly Ash, edited by Sajwan *et al.*
Kluwer Academic/Plenum Publishers, 2003

2. INTRODUCTION

2.1 Background

Coal, oil and natural gas are imported to Finland for the generation of heat and power. The energy consumption depends on industrial production and domestic heating, and has increased in Finland from 417270 TJ (tera joule = 10^{12} J) in 1990 to 515807 TJ in 2000.[1] In addition, in recent years special attention has been focused to alternative energy sources, namely renewable wastes and waste tires.

Scarp car tires can be used as a fuel source in cement kilns, pulp and paper industry and also power utilities. Recycling philosophy has developed very quickly and in 1996, recycling of used car tires started in Finland based on the Council of State decision number 1246, October 12, 1995. It has been reported that about 30000 mt (metric tons) of used vehicle tires are produced in a year. It means a number of used vehicle tires are of 2.4 million pieces, whereas in Europe, 250 million used car tires are produced.[2] In Finland, 90% scrap car tires are recycled, whereas the European Union declares scrap car tires as a Apriority waste stream@ and the target of recovery by the member states should be minimal 65%.[3] In the United States, scrap car tires form approximately 1.8 wt % of the total municipal solid wastes and in each year about 150 million mt used car tires are produced with an estimated stock pile of 3000 million tires awaiting disposal or treatment otherwise.[4,5] However, under the European Union legislation, e.g. the Landfill Directive whole used tires will be banned from landfill sites by 2003 and shredded tires or chips by 2006.[6]

In cement production, 30 to 40% of production cost relates to energy, as a result of which alternative fuel such as used car tires has become a key issue for the worldwide cement industry. Cement producers are interested in evaluating to what extent conventional fuels such as coal, oil and petcoke (petroleum coke) can be replaced by alternative fuels, i.e. processed waste materials such as used car tires.[6,7] Nowadays, more than 80 countries produce cement in very energy intensive process that releases NO_x, greenhouse gas, mainly CO_2 and CO, and trace metals. In Finland, two cement facilities (dry process) are producing about 1.3 million mt a^{-1} cement whereas world production was over 70 million mt in 1998-99. The management of cement facilities in Finland realized the benefits of the use of scrap car tires and Finnsementti Oy, situated in Parainen on the south-west coast of Finland near

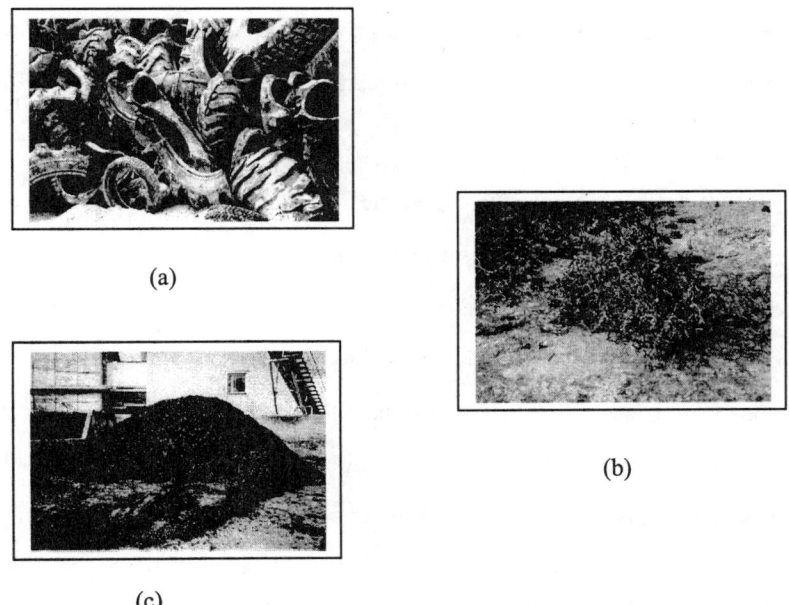

(a)

(b)

(c)

Figure 1. Car tire scrap at the cement plant in Parainen, Finland. (a) received car tire waste (b) iron wires separated from tires and (c) chips of sizes 25 x 25 x 10 mm.[8]

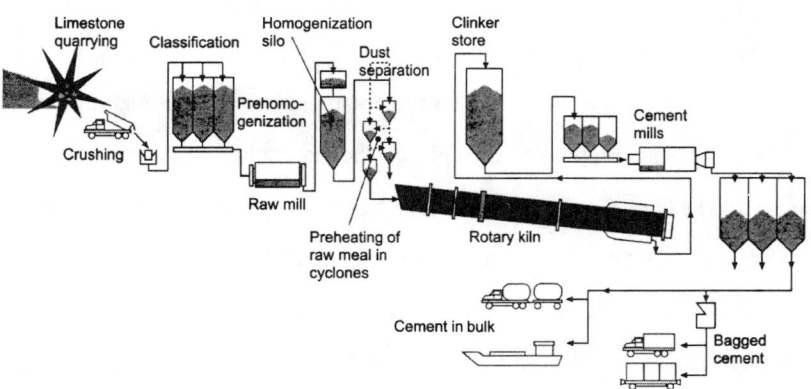

Figure 2. Schematic picture of the cement manufacturing process.

the city of Turku, evaluates the recycling of shredded car tires as a supplement fuel of a 4-stage cyclone preheated rotary kiln (Fig 1).[8] In this study, the trend of trace metals emissions from the stack in particulate and gaseous form during normal operation with coal and petcoke mixture in the process has been compared with the situation when the secondary fuel (10% of the total fuel input) is replaced by shredded car tires.

2.2 Cement manufacturing

Cement is the main component of concrete used in the construction of houses etc. It can be produced by two main processes e.g., dry and wet process. Besides these processes, there are other intermediate process such as semi-dry and semi-wet which are also used. The process depends on raw materials, their form and moisture content. Generally, the required raw materials are limestone, chalk, clay, sand and additives. In Finland, the dry process is used. In this process, less energy is required about 3.3 MJ kg^{-1} clinker than in the wet process, where it is about 5.7 MJ kg^{-1} clinker.[9]

Cement manufacturing consists of raw meal grinding, blending, pre-calcining, clinker burning and cement grinding. In short, limestone and other materials containing calcium, silicon, aluminium and iron oxides are crushed and milled into a raw meal. This raw meal is blended (for instance in blending silos) and is then heated in the pre-heating system to initiate the dissociation of calcium carbonate to calcium oxide and carbon dioxide. A secondary fuel is fed into the preheating system to keep the temperature sufficiently high. The meal then proceeds to the kiln for heating and reaction between calcium oxide and other elements to form calcium silicates and aluminates at a temperature up to 1450 °C. Primary fuel is used to keep the temperature high enough in the burning zone for the chemical reactions to take place. The reaction products leave the kiln as a nodular material called clinker. The clinker will be inter-ground with gypsum, limestone and/or ashes to a fine product called cement.[10] Figure 2 shows a cement manufacturing process from raw material quarrying to the bagging of the cement and the main reactions at different stages inside the rotary kiln are given in Table 1. The waste tire particles are fed into the lower part of the kilns pre-heating system, hereafter referred to as the riser duct.[8]

2.3 Alternative fuels in cement manufacturing

Traditionally, coal is used as a main fuel for manufacturing cement. In the modern society, the increase of waste streams has created negative impact on the environment.

Table1. Main reactions inside the kiln during cement production by dry process [11]

Reaction	Reaction equation	Enthalpy (kJ/kg)
I. Formation of oxides and Decomposing reactions		
Water evaporation	H_2O (l) $\rightarrow$ H_2O (g)	2453
Decomposition of kaolinite	$Al_2O_3 \bullet 2SiO_2 \bullet 2H_2O \rightarrow$ $Al_2O_3 + 2SiO_2 + 2H_2O$	780
Oxidation of carbon	$C + O_2 \rightarrow CO_2$	-33913
Dissociation of $MgCO_3$	$MgCO_3 \rightarrow MgO + CO_2$	1395
Dissociation $CaCO_3$	$CaCO_3 \rightarrow CaO + CO_2$	1780
II. Formation of Intermediates		
Formation of CA	$CaO + Al_2O_3 \rightarrow CaO \bullet Al_2O_3$	-100
Formation of C_2F	$2CaO + Fe_2O_3 \rightarrow 2CaO \bullet Fe_2O_3$	-114
Formation of β-C_2S	$2CaO + SiO_2 \rightarrow 2CaO \bullet SiO_2$	-732
III. Sintering reactions		
Formation of C_4AF	$CA + C_2F + CaO \rightarrow C_4AF$	25
Formation of C_3A	$CA + 2CaO \rightarrow C_3A$	25
Formation of C_3S	β-$C_2S + CaO \rightarrow C_3S$	59

Note: In the cement industry, the following abbreviations are often used to simplify the complex formulas.
$CaO = C$; $SiO_2 = S$; $Al_2O_3 = A$, $Fe_2O_3 = F$

Scrap car tire is one of them. In recent years, scrap tires are found beneficial as a secondary fuel for its chemical as well as its heating values. For these reasons, cement producers recognize car tires as an alternative fuel.

2.4   Objectives of this paper

In this paper, the trend of heavy metals emissions in particulate and gaseous forms during normal use of coal and petcoke (petroleum coke) mixture in the process has been compared with the situation when the secondary fuel (10% of total fuel input) is replaced by shredded car tires. In 1995, a study was conducted for Finnsementti Oy, measuring (amongst other things) the release of heavy metals from stack gases at Parainen when using only traditional fuels such as coal/petcoke and when using car tire scrap as well. This paper is based on these measurements.[12,13] Earlier studies on the use of shredded car tires at the cement facility at Parainen were reported by Karlsson et al.[14] and Kääntee et al.[8]

3. CAR TIRE WASTE AS A FUEL

3.1 Car tire waste

The crucial problems associated with the disposal of vast amount of scrap tires and other types of waste generated annually world wide are well-known.[15] For this reason, special attention has been paid to recycling and preventing the generation of wastes/hazardous wastes in the European Union (EU) as well as in other developed countries. Many industries have focused on clean technology, avoiding or reducing the production of hazardous wastes from the process. In the 1980s, a big problem with used car tires was a lack of encouragement for its applications. Often, used car tires were dumped in landfills. However, strict regulation was implemented by the European Commission in the mid 1990s for handling of scrap car tires. Also, their transportation to the developing countries for disposal will be banned in the near future by the Basel Convention.[16] Still it is not very clear whether scrap car tires are hazardous or non-hazardous wastes. Brazil considers scrap car tires as hazardous waste. However, it may be a waste material, but can be source of clean energy. Its calorific value is higher than the calorific value of coal or petcoke.

Nowadays, the use of waste tires is well established in the sectors of energy, civil engineering, construction and agriculture. Due to increasing fuel prices, special attention has been focused on alternative energy sources including renewable waste materials such as tire-derived fuel (TDF).[17] Beside a high heating value, it is environmentally friendly due to relatively low sulfur content (0.8% wt, dry; coal (bit.):1-5% wt), and high iron concentration (15%-wt, dry).[7, 18] Not but the least, TDF solves a local or regional waste disposal problem. Table 2 indicates the use pattern of TDF in the United States, whereas in Finland 90% is used for road construction and the rest for the maintenance of landfills and energy recovery in the Parainen cement plant.[2] It is also interesting to note that there are many old coal power plants worldwide. In these facilities, operation costs are higher than the newer plants equipped with advanced technology. If these old power plants switch over

Table 2. Use pattern of scrap car tires in the United States and Finland (million tires) [17]

Source	USA 1998	Finland 1997*
	(million car tires)	
Fuel	114	0.71
Cement kilns	38	0.71
Pulp/paper mills	20	0
Tires to energy	25	0
Electric utilities	15	0
Industrial boilers	16	0
Civil Engineering	20	2.71
Products	23	0.45
Ground rubber	15	0
Cut/punched/stamped	8	0
Re-use	not known	0.45
Miscellaneous agriculture	5.5	0
Export	15	0
Total	177	3.87

*Ranta[37]

Note: *For Finland, it is assumed that a car tire weighs 7 kg. Based on it, tons of car tires have been converted into numbers.*

to TDF, the cost of electricity production will be lower, owing to higher heating value than coal. It is expected that these facilities may then compete with modern power plants.

3.2 Application of car tire waste

It took a long time to develop and improve the quality of car tire. Car tire is manufactured from a wide range of compounds including natural and synthetic rubbers plus a large quantity of carbon black as a reinforcing agent and filler.[19] During production, oils are used for mixing the ingredients and to modify physical properties as well as vulcanizing agents such as sulfur, zinc oxide and organic compounds to enable the polymerization reaction at elevated temperature. The general use pattern of waste tire has been mentioned earlier. Besides the energy sector, it also is used to construct artificial reefs to promote fisheries, and a mixture of soil and tire has been considered for geoenvironmental applications such as in-ground barriers and landfill liner systems.[19-21] Attempts have been made to use a clay-tire mixture to absorb petroleum based hydrocarbons, but it was unsuccessful due to swelling pressure on the material causing the permeability of the soil-tire mixture to decrease.[38] Scrap car tires are used also in road paving e.g in the hot asphaltic concrete mixtures. One can use 2 to 6 tires per mt of hot mix asphalt.[22]

A great deal of studies are in progress on the leaching of organic compounds and trace metals from waste tires due to its use in different sectors. In northern Europe, studies have been made on bio-accumulation of organic compounds and trace metals by organisms growing on the tire artificial reef units. Collins et al.[23] could not observe any abnormal presence of benzothiazoles and zinc, a major trace metal in tire manufacturing in species grown inside the tire artificial reef unit. The European Tire Recycling Association (ETRA) has examined a number of civil engineering applications, coastal erosion, tide control, beach reinforcements where whole and shredded tires can be used. But the leaching of organic compounds and trace metals to the aquatic environment or soil compartment is still not quantified.[24,25]

3.3 Properties of the fuels used in cement production

The range of fuels is extremely wide. Traditional kiln fuels are gas, oil or coal. Materials like waste oils, plastics, auto shredder residues, waste tires and sewage sludge are often proposed as alternative fuels for the cement industry. Also all kinds of slaughterhouse residues are offered as fuel nowadays. To be able to use any of these fuels in a cement factory it is necessary to know the composition of the fuel. Table 3 indicates the concentration of heavy metals in different fuels including car tire. But the choice is normally based on price and availability. The energy and ash contents are also important, as are the moisture and volatiles contents. All kinds of varieties from liquid to solids, powdered or as big lumps can be encountered when dealing with alternative fuels, requiring a flexible fuel feeding system. Somehow they should all be fed into the burning chamber of the process. It may be fed directly into the burning zone in the kiln itself or into the pre-heating system for dissociating part of the carbonates from the meal before it enters the kiln for clinker formation. In Table 4 we can see examples of different alternative fuels,

Table 3. Selected trace elements concentration (mg kg^{-1}) in different fuels (modified) [26]

Element	Coal	Oil[#]	Petcoke	Paper-sludge	MSS	Waste wood	Biomass Mix**	Car tire*
As	2.6	0.02	1.1	3.2	8.4	10.0	2.3	0.65
Cd	0.10	0.2	0.2	0.53	3.79	1.4	0.70	<2
Cr	17	3.0	18	18	113	78	26	0.025
Cu	10	2.5	1.8	98	406	135	57	-
Hg	0.11	0.01	0.02	0.24	3.28	0.17	0.06	<0.1
Mn	41	2.5	5.7	6.3	546	92	157	-
Ni	12	120	278.6	10	83	31	16	0.013
Pb	6.7	2.0	2.1	31	260	574	16	0.005
Sb	0.51	0.02	0.6	1.2	4.1	16	1.6	-
Sn	1.4	0.002	0.6	6.2	38	6.4	1.3	-
V	24	180	1560	5	24	10	6	-
Zn	19	4.0	7.0	464	1349	807	133	15300
Caloric value MJ/kg	24.4	32.0	35.0	10.98	13.14	11.86	-	36.0

MSS; Municipal Sewage Sludge

[#]Meij and Pilage[27]

*Åbo Akademi University[28]

**biomass mix consists of green wood, garden waste, straw, road side grass and manure

Table 4. Properties of fuels used in tests [4]

	CAR TYRE RUBBER	COAL-PETCOKE MIX
C (%-wt, dry)	87,0	75,1
H (%-wt, dry)	7,82	4,2
N (%-wt, dry)	0,33	1,7
S (%-wt, dry)	0,80	3,0
O (%-wt, dry)	1,81	4,9
Ash(%-wt, dry)	2,2	11,1
Volatiles (%)	66,6	not analysed
C-fix(%)	31,1	not analysed
H_2O(%, 105°C)	0,73	1,3
LHV(MJ/kg)	35,58	29,71
HHV(MJ/kg)	37,31	28,97

Table 5. Alternative fuel options for the cement industry

Source	Types
Liquid waste fuels	Tar, chemical wastes, distillation residues, waste solvents, used oils, wax suspensions, petrochemical waste, asphalt slurry, paint waste, oil sludge
Solid waste fuels	Petroleum coke (Apetcoke@), paper waste, rubber residues, pulp sludge, used tires, battery cases, plastics residues, wood waste, domestic refuse, rice chaff, refuse derived fuel, nut shells, oil-bearing soils, sewage sludge
Gaseous waste	Landfill gas, pyrolysis gas

separated into three groups. Here, only scrapped car tires are considered. Car tires contain roughly 60% rubber, 25% steel and 15% other materials. It is used after shredding into small pieces (50x50x10 mm). Its heating value is about 37 MJ kg^{-1} (dry) against 30 MJ kg^{-1} (dry) for a typical coal/petcoke mix.[3,4] The properties of the two fuel types used here are given in Table 5.

4.   EXPERIMENTAL PROCEDURE

As described extensively in the measurement reports,[9,10] trace elements emissions were measured separately for the gas phase and the particulate phase. The samples were taken at the stack of the cement plant at flue gas temperatures of approx. 120-130EC. Particulate trace element material was collected by isokinetic sampling on a quartz filter according to the Finnish standard procedure SFS 3866, whilst the gas phase trace elements were sampled and dissolved in a 2 M HNO_3 solution. The material collected on the filter and the HNO_3 solution were analysed at an external laboratory. Sampling duration was of the order of 1-2 hours, the number of samples analysed was 2 and 3 for the tests with and without scrapped car tyres as fuel, respectively.

5. RESULTS AND DISCUSSION

Emissions of heavy metals are subjects to a steady debates. In high temperature processes, heavy metals are released either in non-volatile or volatile form. Their behaviours in a smelting process is basically different from coal combustion or waste incineration or from cement manufacturing process.[29] The fuel and raw materials are the main sources of heavy metals in a combustion system. The release of metals from cement kiln does not depend so much on the concentrations in the raw materials, but chemical mechanisms acting at the elevated temperature. It is an established fact that the physico-chemical behaviour of heavy metals varies strongly from one metal to another. Each different metal behaves in an unique fashion as it is concentrated with raw materials, coal and scrap tire species in the reactor. A metal may react with other metal species and ashes in the reactor to different degrees and it may remain in solid and vaporized form. The vaporized metal species are carried away through the gas stream and recondense as the gas cools. Generally, the vapour condenses homogenously to form new particles, and heterogeneously on the surfaces of the entrained ash and other metal particles.[30] Both processes produce different size of particles. The heterogeneous process tends to favour smaller particles due to with a higher surface area. The concentration of some metals in small particles may be hundred times greater than the original concentration in the feed material. In our study, we have observed this type of behaviour for Hg when fuel is replaced by 10% of scrap tire.

Combustion temperature is an another factor which governs heavy metals released from a process into the atmosphere via the stack. Figure 3 indicates the effect of temperature on selected metals. The vapour pressure generally indicates the quantity of material that vaporizes and subsequently condenses. In the Figure 3 vapour pressure increases with the increase of temperature. But here vapour pressures shown are very small. It means that most of the selected metals are in particulate forms. As noted by Clarke and Sloss[32], metal Mn is in the group 1 and Co, Cr and Tl are in group 2. Manganese does not vaporize whereas elements such as Co, Cr, Tl, U, V, Zn are vaporized at high temperature processes, but they condense on particulates in cooler areas.[33]

The elements Hg and Cd are most critical due to their potential environmental and human health risk. It has been demonstrated that in the industalized regions of the world, mercury loading to the environment has increased by a factor of two to five from the beginning of the industral period (100-150 years ago), to the present day due to atmospheric emission and transboundary deposition of mercury in soils and aquatic environment.[34] Considering car tire scrap as an alternative fuel it is well known that car tires contain more Zn and Cd, but less Hg and As than fossil fuels. Zn volatilizes at 907 $^{\circ}$C (at 1 atm), but due to low temperatures and oxygen potential in the cyclone of the cement plant precalciner system, Zn is released mostly in the particulate form. Due to the temperature in the kiln (1450 $^{\circ}$C), Hg is vaporized and is mostly emitted to the atmosphere as elemental Hg. Figure 4 shows the concentration of heavy metals (particulate and gaseous forms) in the exhaust gases when operating with shredded car tires as secondary fuel as compared to operation with coal/petcoke only. Fe emissions are significantly increased (465 Φg m^{-3}n (dry) @ 11% O$_2$ as compared to 149 Φg m^{-3}n (dry) @ 11% O$_2$) when tire pieces were used as a secondary fuel.

Considering the emission standards and the toxicity of compounds such as Hg and Tl, the fuel switch is very beneficial. It was recently reported by Granite et al.[35,36] that metal oxides such as MnO$_2$, CrO$_2$ and MoS$_2$ have moderate capacities as sorbents for

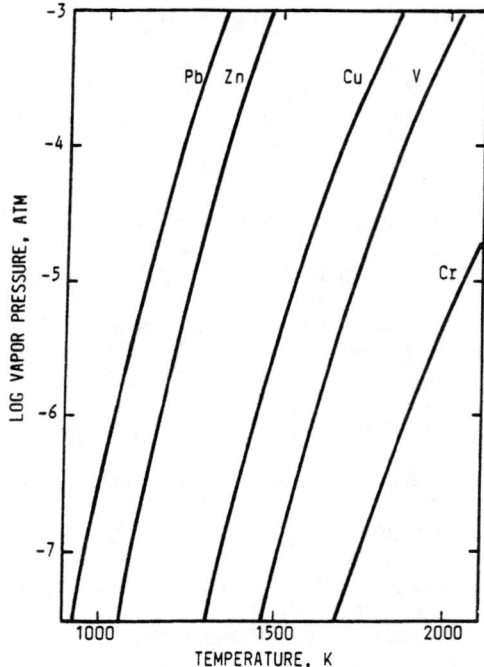

Figure 3. Effects of temperature on the volatility of selected trace elements.[31]

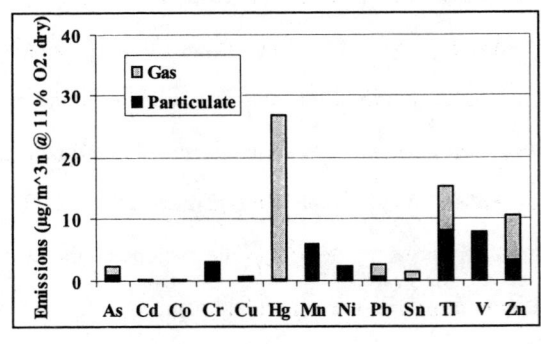

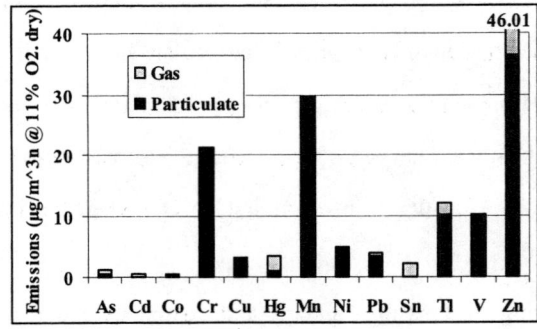

Figure 4. Heavy metals emissions durin operation with coal/petcoke
(TOP) and during operation with car tire scrap as a secondary fuel
(BOTTOM).

Table 6. Trace element emissions from the cement plant and comparison with
emission regulations (this study)

Emission $\mu g/m ;_n$, dry @ 11% O_2	Limit, current and EC proposal	100% coal/petcoke	10% car tire scrap 90% coal/petcoke
Hg	50	27	3
Cd + Tl	50	15	13
As+Cr+Cu+Mn+Ni+Pb+Sb+Sn+V	500	26	76

mercury, which corresponds to our findings. The current standards in Finland as well as the near-future standards as proposed by the European Commission are practically identical and concentrate on three classes, being 1) Hg, 2) Cd+Tl and 3) As+Cr+Cu+Mn+Ni+Pb+Sb+Sn+V. The results given in Figure 4 are combined into these classes in Table 6.

6. CONCLUSIONS

Stack emission measurements at a cement manufacturing plant indicate cleaner processing when car tire scrap is used as an alternative fuel. Although the emissions of Zn and Fe increase, and the emissions of As+Cr+Cu+Mn+Ni+Pb+Sb+Sn+V are tripled (due to Cr and Mn), the emissions of Hg are significantly reduced, as well as those of Cd+Tl in presence of Cr and Mn. There is a strong shift of Hg and Tl from the gaseous phase to the particulates, which are much easier to control and which is also beneficial for the ecosystem.

ACKNOWLEDGEMENTS

Finnsementti OY, Finland, is acknowledged for giving permission for presenting stack emission data from the cement plant, measured in 1995. This paper was presented in the 6th International Conference on Biogeochemistry of Trace Elements in Guelph, Ontario, Canada, July 29 - August 2, 2001.

REFERENCES

1. Statistical year book of Finland. 2000, 193-194.

2. URL: http://www.rengaskierratys.com/e/basic.htm

3. Williams, P. T., Borttrill, R. P., and Cunliffe, A. M. The potential of pyrolyses for recycled used tyres, In: Recycling and Reuse of used Tyres. Dhar, R. K., Limbachiya, M. C. and Paine, K. A. (Eds). MPG Books. Bodmin, UK, 2001, 187-201.

4. National Asphalt Paving Association Scrap tire utilization technologies. Lanham, MD, 1993.

5. Williams, P. T., Waste treatment and disposal, John Wiley & Sons, Chichester, 1998.

6. Davies, R. D., and Worthington, G. S. Use of scrap tyre as a fuel in the cement manufacturing process, See ref. 3, 93-106.

7. Kääntee, U., Zevenhoven, R., Backman, R., and Hupa, M. Process modelling of cement manufacturing using alternative fuels. See ref. 3, 81-92.

8. Kääntee, U., Zevenhoven, T., Backman, R., and Hupa, M. The impact of alternative fuels on the cement manufacturing process. In: Proc of R2000 Recovery, Recycling and Reintegration, June 2000, Toronto (ON), Canada, 2000, 1070-1075 (CD-ROM)

9. IEA Greenhouse Gas R & D Programme (IEA GHG R & D). The reduction of greenhouse gas emissions from the cement industry, Report No. PH3/7, Cheltenham, UK.

10. Alsop, P.A. "Cement Plant Operations Handbook for Dry Process Plants" 2nd Edition, International Cement Review, Dorking, Surrrey (UK) July 1998.

11. Rosemann, H., Theoretische und betriebliche Untersuchungen zum Brennstof-fenergieverbrauch von Zementdrehofenanlagen mit Vorcalcinierung (Theoritical and operational investigation on fuel consumption in rotary kiln cement plants with pre calciner). Schriftenreihe der Zementindustrie 48, Beton Verlag, Dusseldorf, 1987 (in German).

12. Finnsementti Oy, Emission measurements report EVK-19709V-01 (13.9.1995)

13. Finnsementti Oy, Emission measurements report EVK-21067V-01 (16.11.1995)

14. Karlsson, M., Zevenhoven, R., Hupa, M. Laboratory characterisation of waste tyre particles for combustion in a cement klin. In: Proc. of the Finnish-Swedish Flame Days, Naantali, Finland. Sept 3-4, 1996. 18 p.

15. Environment Agency, Tires in the environment, Environment Agency Report, UK, 1998.

16. European Commission, Copy to the Commission, DGXI of the Member States reporting for 1995 to the Secretary of the Basel Convention, 1998

17. Nelson, R. G., Hossain, A. S. M. M., An energetic and economic analysis of using scrap tyres for electricity generation and cement manufacturing, See ref 4,119-127.

18. Zevenhoven, R., Editor, The 1999 Finnish waste-to-energy course, Course material, Helsinki University of Technology, Espoo, Finland,1999.

19. Uniroyal, Rubber chemical selector guide 2000, URL: www.uniyoyalchemical.com/rubchems.htm.

20. Lyons, J. C., Hitchens, D. G., and Monticello, D. A., Recovery of scrap rubber tires from landfill for construction uses, waste disposal by landfill, Green >93, Sarsby, ed, A. A. Balkema, Rotterdam, 1995, 327-334.

21. Park, J. K., Kim, J. Y., and Edil, T. B., Mitigation of organic compound movement in landfills by a layer of shredded tires, Proc 66th Water Envirn Fed Conf, California, 1996.

22. Amirkhanian, S. N., Utilization of crumb rubber in asphaltic concrete mixtures - south Carolina=s experience, See ref 3, 163-174.

23. Collins, K. J., Jensen, A. C., Mallinson, J. J., Smith, I. P., Mudge, S. M. and Russel, A. Scrap tyres for marine construction: Environmental impact. See ref, 3, 149-162.

24. ETRA (European Tire Recycling Association), Civil engineering association in the EU, 2000, 15-17.

25. ETRA (European Tire Recycling Association), Further information, on bales, a new development, 2000, p. 13.

26. Meij, R., Winkel, B. H., and Havinga, H., *Emissies naar van micro-en spoorelementen tijdens bijsstoken van 10% secundaire brandstoffen en biomassa in poederkoolgestookte eenheden in Nederland,* (Emissions of micro- and trace elements from CO-firing 10% secondary fuels and biomass in pulverised coal fired units in the Netherlands) 99530162-KST/MAT 99-6579, Arnhem, The Netherlands, KEMA, 1999, pp. 49, (in Dutch).

27. Meij, R., and Pilage, E., Databank spoorelementen. Deel 6 Emissies bij het stoken van zware stookolie (Data base trace elements Part 6: emissions from firing heavy fuel oil) 63925-KES/WBR 94-3104, Arnhem, The Netherlands, KEMA, 1994, (in Dutch).

28. Abo Akademi University, Combustion Chemistry Research Group, Turku, Finland, Fuel analysis data bank (R. Zevenhoven) (unpublished), 1998.

29. Vogg, H., Braun, H., Metzger, M., and Schneider, J., The specific role of cadmium and mercury in municipal solid waste incineration. Waste Management & Research, 4, 1986, 65-74.

30   Flagan, R. C., Submicron particles from coal combustion, 17th Symposium (International) on Combustion, The Combustion Institute, 1978, p. 97.

31   U.S. EPA, Prediction of the fate of toxic metals in hazardous waste incinerators, U. S. Environmental Protection Agent, Office of Research and Development, Washington, D.C., 1988, 1-3 to 1-14.

32   Clarke, L. B., and Sloss, L. L., Trace elements - emissions from coal combustion and gasification. IEACR/49, London, UK, IEA Coal Research, 1992, 111 pp.

33   Meij, R., Behaviour, control and emissions of trace species by coal-fired power plants in Europe, 58087-KST/MAT 97-6546, Arnhem, Netherlands, KEMA, 1997, 53 pp.

34. Frizgerald, W., Engstrom, D., Mason, R., and Nater, E., The case for atmospheric mercury contamination in remote areas. Environ Sci Technol 32(1), 1998, 1-7.

35. Granite, E.J., Pennline, H.W., Hargis, R.A., ANovel sorbents for mercury removal from flue gas@ Ind. & Eng. Chem. Res. **39**, 2000, 1020-1029

36. Granite, E.J., Pennline, H.W., Haddad, G.J., Hargis, R.A. An investigation of sorbents for mercury removal from flue gas in: Proc. of the 15th Ann. Int. Pittsburgh Coal Conf., Pittsburgh (PA), Sept. 1998 (CD-ROM)

37. Ranta, J., Autonpaloittelujätteen ja rengasromun terminen konversio energiaksi ja raaka-aineeklsi (Thermal conversion of automotive shredder residue into energy and raw materails). VTT Research Notes 1960, Espoo, Finland,1999, 81 pp.

38. Al-Tabbaa, A., and Chifambira, B., and Waterfall, P., Novel and sustainable applications for granulated waste tyre in low permeability subsurface barriers, See ref 3, 175-186.

PRACTICAL AND REGULATORY CHALLENGES IN CONTROLLING TRACE ELEMENT INPUTS TO SOILS FROM LAND APPLICATION OF FLUIDIZED BED COMBUSTION RESIDUES

M. Hope-Simpson[1] and W. Richards[2]

[1]Agrosystems Atlantic, P.O. Box 1674, Truro, Nova Scotia, Canada B2N 5Z5

[2]Nova Scotia Power Inc., Generation Services, Point Aconi Generating Station P.O. Box 1609, Bras d'Or, Nova Scotia, Canada B1Y 3Y6.

1. INTRODUCTION

The 165 MW_e circulating fluidized bed boiler at the Nova Scotia Power Inc. (NSPI) Point Aconi Generating Station, located in eastern Cape Breton Island, Nova Scotia, is Canada's largest fluidized bed unit. Fluidized bed combustion (FBC) allows the burning of high sulphur (S) fuels with *in situ* capture of S (removal of SO_2). The lower operating temperature of the FBC system (870 C versus 2000 C for pulverized coal boilers) permits the combustion of a wide range of 'opportunity' fuels (e.g., petroleum coke <$F, a by-product of the oil refining process. >, low grade coals, tires, and wood waste sludge) in an environmentally acceptable manner. The use of large amounts of a limestone sorbent in FBC results in a high rate of sulphur capture (90%), however, the rate of residue production is approximately two-times higher than that from conventional pulverized coal-fired boilers. While the residue is currently disposed in a fully engineered residue management site, there are environmental, financial, and other incentives for NSPI to find beneficial uses for the residue to minimize placement of the by-product in the landfill.

1.1. Residue Characteristics

The use of limestone for sulphur capture, combined with a lower temperature of combustion in the FBC compared with the pulverized coal fired (PCF) boiler, results in a non-vitrified solids residue with higher CaO and SO_3 levels and lower SiO_2 and Al_2O_3 content than ashes from conventional PCF combustion. The residues are significantly more alkaline and more soluble, but less pozzolanic and lower in fuel-derived trace elements (e.g. As, B, Cd, Cu, Hg, Mo, Pb, Se) than pulverized coal fly ash.[1,2,3] The main constituents of these residues are CaO, $CaSO_4$, and fuel-derived ash (primarily oxides of Fe, Al, and Si). Some unreacted $CaCO_3$ and char derived from the fuel (i.e., devolatilized and pyrolized solid fuel) are also present.[1] The residues are important from an agricultural-use perspective because the CaO has significant liming activity, the $CaSO_4$ represents a highly soluble source of plant available Ca and S, and the fuel-derived ash contains a range of both essential plant micronutrients and other non-essential

Fig. 1. Point Aconi Generating Station (Cape Breton, Nova Scotia) in the background, with the residue landfill site in the foreground.

elements. The use of the term "residue" to describe this by-product importantly distinguishes it from other forms of coal ash, and highlights the fact that only one-third of the material may properly be termed "ash", whereas up to two-thirds is beneficial, sorbent-derived components.

1.2. Research on Agricultural Use

In 1997, a collaborative research initiative between the Nova Scotia Agricultural College (NSAC) and NSPI was launched in order to investigate the potential for using Point Aconi FBC residue as a soil amendment and agricultural liming material. In addition to field testing the residue as a nutrient source and liming material for vegetable (cabbage and rutabaga) and forage crops, the research included a detailed analysis of soils and crops to assess the potential risk to the food chain posed by trace elements present in the residues.

In the United States, agricultural uses for FBC residues have been evaluated for more than 20 years,[4,5,6,7,8,9,10,11,14,15] however, until the recent work in Nova Scotia, little research has been done on agricultural use of this material in Canada. There were a number of conditions in Canada, sufficiently different from those in the U.S., in particular, fuel sources for fluidized bed combustion (influencing residue composition), and regulations and metal criteria for the use of recycled/by-product materials, to justify a Canadian study on agricultural use of this material. In addition, while there is a significant body of literature on bioavailability and fate of metals from pulverized coal-fired (PCF) combustion ashes applied to soils, much less research has been conducted on the potential risk to the food chain posed by trace elements present in the more alkaline FBC residues. The greatest impetus for this work, however, was provided by Cape Breton farmers who, faced with the loss of a trucking subsidy for limestone, contacted NSPI with a view to utilizing Point Aconi residue as a lower cost limestone substitute. NSPI initiated this work to investigate agricultural uses for the residue with the aim of providing a competitive advantage to farmers in the area of the plant, while reducing the amount of by-product destined for disposal.

The Nova Scotia research demonstrated Point Aconi FBC residue to be an effective liming material<$F When the residue is applied to soils at the soil lime requirement, there is a rapid and predictable adjustment of soil pH to target or optimum pH values. FBC residues react more quickly with soil particles than ground agricultural limestone but demonstrate similar liming persistence 2-3 years following application. > with no adverse effects on crop quality or excessive soil loading of metals. In addition to liming activity, the residues were found to have significant fertilizer value. For example, the residues provide significant plant available Ca and S and lesser but important amounts of B, which are advantageous for crops with high requirements for these nutrients (Figures 2 and 3).

While none of the regulated heavy metals present in the residues demonstrated a potential for plant uptake (even at very high loading rates (>50 t/ha)), soil concentrations of elements increase proportionately with application rate. A key requirement for safe and responsible use is therefore ensuring that concentrations of metals in the residue and residue-amended soils do not exceed safe threshold levels.

The research has identified a number of practical and regulatory challenges which need to be addressed in order to derive the greatest benefit from using this material as well as to minimize potential risk. This chapter outlines the nature and importance of these challenges, as well as the progress made to date in developing standards that identify the requirements for safe and effective use of FBC residues as agricultural liming materials.

2. CHALLENGES IN CONTROLLING METAL INPUTS TO SOILS

FBC residues generally contain low concentrations of the heavy metals which are most commonly regulated in waste materials (e.g., Cd, Cr, Cu, Hg, Ni, Pb and Zn). There are, however, a number of potentially toxic elements in the residues, some of which are not captured by existing metal criteria, whose concentrations are variable and highly fuel-dependant. For example, As, Cd, Cu, Se, Pb and Tl (which tend to be present as sulfide minerals associated with the organic fraction in coal[17]) are often higher in residues from combustion of high-S coals. With the combustion of petroleum (pet) coke, residue concentrations of the latter group of elements are often reduced, however, levels of V and Ni, in particular, are generally much higher. Consequently, in 1997 (when the fuel source was domestic (Cape Breton) high-S, eastern bituminous coal), residue element composition is typical of fly ash from high-S coals (elevated As, Se, Pb, and Tl), whereas with the current fuel blend (a mix of pet coke and various offshore bituminous coals), the residue is more characteristic of pet coke-derived ash, with concentrations of Mo, Ni and V increasing in the residues with the proportion of petroleum coke in the fuel mix (Table 1).

The nature of the challenges in controlling metal inputs to soils posed by these elements varies primarily according to whether or not these elements are regulated in soils and soil amendment products in Canada.

2.1. Monitoring for Compliance with Regulatory Criteria

The standards to meet in order to permit the sale of by-products as soil amendments in Canada are the metal standards for products specified in the *Fertilizers Act and Regulations* (FAR) (Table 1). These standards specify the maximum acceptable concentrations in products (mg kg^{-1}) of nine metals (As, Cd, Co, Hg, Mo, Ni, Pb, Se, and Zn), as well as the maximum cumulative additions of these metals in soils (kg ha^{-1}).[18]

Our experience over the past 5 years suggests that meeting FAR product limits on Ni and possibly Mo may present a challenge in marketing pet coke-derived residues, whereas As is the main constraint in marketing residues derived from high-sulphur domestic (Cape Breton) coals

Fig. 2. Yield and quality of cabbage in FBC residue treated plots (split ash treatment) were equal to or better than that of cabbage grown in plots amended with an equivalent rate of dolomitic limestone (treatments applied at the soil lime requirement, split application).

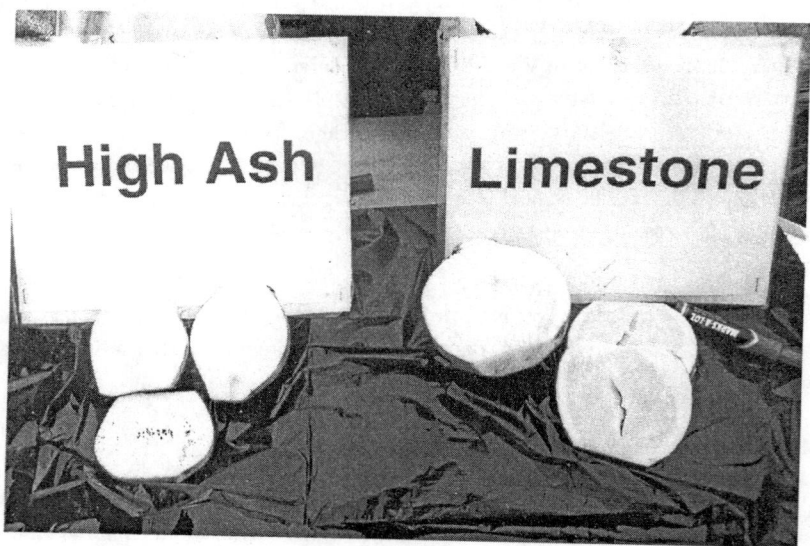

Fig. 3. On the sandy, B-deficient soils in eastern Cape Breton, rutabaga from limestone treated plots showed severe boron deficiency (brownheart), whereas no B-deficiency was found in rutabaga grown in plots treated with an equivalent rate of FBC residue (high ash).

Table 1. Element content of Point Aconi FBC residue in 1997 and 2001, and metal criteria for soil amendment products and soils.

| Element | Point Aconi FBC residue | | Metal criteria | |
	1997[1]	2001[2]	Product[3]	Soil[4]
	---------- mg kg^{-1} ----------		---- mg kg^{-1} ----	----- kg ha^{-1} ------
Aluminum (Al)	25 000	18 000	-	-
Arsenic (As)	110	<2	75	15
Barium (Ba)	290	240	-	-
Beryllium (Be)	<5	<5	-	-
Boron (B)	93	na	-	-
Cadmium (Cd)	1.1	<0.3	20	4
Chromium (Cr)	20	<2	-	-
Cobalt (Co)	8	<1	150	30
Copper (Cu)	70	30	-	-
Iron (Fe)	33 000	11 000	-	-
Lead (Pb)	120	9.9	500	100
Manganese (Mn)	1 100	820	-	-
Molybdenum (Mo)	8	12	20	4
Nickel (Ni)	23	270	180	36
Selenium (Se)	8	<2	14	2.8
Strontium (Sr)	200	160	-	-
Thallium (Tl)	3	<0.1	-	-
Vanadium (V)	35	1400	-	-
Zinc (Zn)	140	69	1 850	370
Mercury (Hg)	na	0.10	5	1

Element concentrations are total sorbed metals (dry weight basis) (EPA Method 3050A for all elements except Hg. Hg was determined using the cold vapour technique (EPA 7471A)).
[1] 1997 - residues derived from combustion of Cape Breton (CBDC-Prince coal).
[2] 2001- residues derived from combustion of 50% petroleum coke, 50% coal.
[3] Product metal criteria - maximum acceptable concentrations of metals (mg kg^{-1}) in fertilizer and supplements (*Fertilizers Act and Regulations*).
[4] Soil metal criteria - maximum acceptable cumulative metal additions to soils (kg ha^{-1}) over a 45 year period (*Fertilizers Act and Regulations*).

(Table 1). As not all of the by-product will meet the low contaminant criteria for soil amendment use, one of the first steps in controlling metal inputs to soils, as well as allowing commercialization of this by-product as a liming material, is to establish a reliable product quality monitoring program for toxic elements in the residues whose concentrations increase with the use of certain fuels in fluidized bed combustion.

Under the FAR, metal limits are based on the assumption of applying a product at an annual rate of 4.4 t ha^{-1} for 45 years. While this application rate is a reasonable average for liming materials, frequency of application would be less, as liming materials are not applied annually but rather, on average, every 5 years <$F The rate and frequency of application of liming materials to relatively low buffer capacity Atlantic region soils is in the range of 2-7 t ha^{-1} pH^{-1} unit every 5 years. > Consequently, metal loadings for this by-product within the specified time frame are likely to be substantially lower than those allowed in the federal regulation. Nonetheless, because soil concentrations of metals increase proportionately with application rate, it is of critical importance to ensure that soil concentrations do not exceed threshold safe levels.

2.2. Elements not Regulated in Soils and Soil Amendment Products

Fossil fuel combustion residues contain varying amounts of at least 15 potentially toxic elements[17,19,20,21], however, only 11 of these are commonly regulated or reported in the literature as posing risks to human or ecosystem health. Tl, V and B are not currently regulated in soils or soil amendment products in Canada. Based on element characteristics such as toxicity, bioavailability/risk pathways, and concentrations in the residues, however, safe thresholds for these elements may be required for the utilization of FBC residues as an agricultural amendment.

2.2.1. Thallium

In the field testing of Point Aconi residue, Tl was flagged because of its toxicity and measurable increase in plants grown on residue-treated soils <$F Tl was the only non-essential element which increased in plant tissues in FBC residue-treated soils. In treatments receiving high rates of residue amendment (33 t ha^{-1}), Tl levels were 0.05 mg kg^{-1} in rutabaga and 0.2 mg kg^{-1} in soils. These levels were significantly higher than Tl levels present in plants (not detectable) and soils (0.1 mg kg^{-1}) in the control treatments.>[16] Soil Tl levels are normally low; the natural range is 0.01 to 2.3 mg kg^{-1} dry weight, however, soils in industrialized areas <$F Areas affected by coal combustion, heavy metal smelting, cement industry and refining processes. > may contain much higher Tl levels (i.e., above 10 mg kg^{-1}).[22] Fractionation experiments on metal contaminated sludges indicate that bioavailability of anthropogenic Tl is high and similar to that of Cd <$F Based on element recovery in the exchangeable fraction (1 hr shaking, extracted with 1 m CH_3COONH_4 at pH 7). >[23] Tl is readily taken up by plants, particularly on sandy, low organic matter soils, and increased plant levels of this element may be highly toxic to both plants and animals. Tl is normally present in FBC residues in very low concentrations but increases with the combustion of certain high-S coals (Table 1). While Tl is not currently regulated in soils or soil amendment products in Canada, the Canadian Council of Ministers of the Environment (CCME) <$F The CCME is the principle vehicle in Canada for interjurisdictional cooperation on environmental issues of national and international concern. Soil quality guidelines developed by the CCME provide equal protection to human health and ecological receptors, and are used to provide a nationally consistent and scientific basis for making decisions regarding the protection of environmental quality in Canada. > have established a soil quality guideline (SQG) limit for Tl of 1 mg kg^{-1} dry weight of soil.[24] While soil levels of Tl found in this study were well below CCME limits, our results demonstrate that detectable increases in plant uptake are possible for this element at soil concentrations below the guideline limit.

2.2.2. Vanadium

V is considered important because of the highly significant increase of this element in the residues with the combustion of pet coke (Table 1). V is one of a number of metals, previously thought to be non-essential and now classified as beneficial to some plants or animals, whose designation as either "essential" or "toxic" depends on concentration.[25] V is relatively non-toxic at normal soil concentrations (58-100 mg kg^{-1}), however, may be toxic to plants at soil solution concentrations above 140 mg kg^{-1}.[22] The CCME have established a SQG limit for V of 130 mg kg^{-1}; unlike the limit for Tl, however, the CCME limit for V does not include a full consideration of all possible exposures and pathways (soil-plant-human).[24] Bioavailability and other characteristics of V are highly dependant on dissolution/precipitation reactions and oxidation state. The most likely types of solubility-controlling solids and oxidation states of V in fossil fuel combustion residues are oxides and V^{+5}, respectively.[26,17] Leachability tests on pet coke-derived residues have shown V to be strongly bound to the solids matrix (predominantly oxides of Ca, Si, and Fe) resulting in leachate metal concentrations which are very low and comparable to concentrations leached from coal-derived residues.[27,26] Inhalation of oxidized forms (V_2O_5) is the main pathway of toxic exposure for humans to V[28], yet occupational exposures to the forms of V present in V-containing petroleum coke-derived residues have not been fully assessed.

2.2.3. Boron

Boron is an essential plant micronutrient and most agronomic plants should contain at least 20 ppm of B for normal growth. Under the *Fertilizers Act*, products represented for use as B fertilizers must have a minimum total B content of 0.02%, or 200 mg kg⁻¹.[29] Boron levels in the soil solution in excess of 20 ppm (hot water soluble B), however, are likely to be phytotoxic.[19] While maximum content of B in soils and soil amendments is not regulated in Canada under the *Fertilizers Act*, the province of Quebec has recommended a B content of less than 100 mg kg⁻¹ (total B) in municipal biosolids applied to soils with 200 mg kg⁻¹ of B as a maximum limit.[30] Soil guideline limits for the amount of B in agricultural soils receiving waste materials are low and in the range of 1 to 1.5 mg kg⁻¹ hot water-soluble B.[31] The total B content of FBC residue varies with coal source, however, the range present in these residues (95-170 mg kg⁻¹) is generally higher than the mean B content of agricultural soils (2-100 mg kg⁻¹).[5] The narrow range between phyto-sufficiency/toxicity that exists for B means that, depending on concentration, B in coal combustion residues may either be beneficial to crops or phytotoxic.[19,32,21] In the Nova Scotia research, boron in FBC residues was considered to be advantageous for the production of cabbage, rutabaga, and legume-based forages on B-deficient soils[16] (Figure 3). B phytotoxicity was likely avoided because of a number of factors, including moderate B content of the residues, the use of B-deficient soils, a high requirement and tolerance for B of the crops, as well as the reduced solubility of B at high soil base saturation with Ca.

For the above elements, reducing hazard of metal contamination from waste/ by-product materials requires not only reliable soil and product quality monitoring, but also identifying thresholds of element concentration that protect the food chain and minimize environmental (phytotoxic, ecotoxic) or occupational exposure risks.

2.3. Variable Calcium Carbonate Equivalence

FBC residues are variable not only in element composition; these materials are also variable with respect to CCE (calcium carbonate equivalent, or neutralizing value). The CCE of FBC residues varies primarily with the CaO content, and may be anywhere from 40-90%; the average is 60%[4]. The CCE of baghouse material (fly ash) from Point Aconi Generating Station is typically between 65-80% <$F Based on lab analyses obtained on Point Aconi residues between 1997-2001. > Because of a lower CCE of FBC residues compared with agricultural limestones (which range in CCE from 90-108%), application rates are increased approximately 20% to provide equivalent acid neutralizing capacity. The agricultural study demonstrated that, provided that loading is adjusted to account for the lower neutralizing value and higher product efficiency <$F Unlike carbonate limestones which contain a range of particle sizes and must be ground in order to react with soil particles, FBC residues predominantly comprise particles of less than 0.15 mm in diameter. This size fraction is considered to be 100% efficient at neutralizing soil acidity. > of the residues compared with agricultural lime, liming response to this material is predictable and similar to agricultural lime. Consequently, variable CCE is not considered a barrier to use of the material as a liming agent. One implication of the variable CCE, however, is that the extent and rate of metal loading to soils will vary according to the amendment application rate. As noted previously, soil metal criteria are typically based on assumptions of "reasonable" amendment loading over a specified time frame, and higher amendment loading rates (associated with a less efficient liming material) would result in limits on maximum permissible metal additions to soils being reached within a shorter time frame. The majority of Point Aconi FBC residues have a CCE >65%, and, provided the residue meets acceptable product criteria, the 20% increased loading required for residues with CCE in the range of 65-80% is not considered to be problematic at typical agricultural liming rates.[16] Nonetheless, variable CCE in by-product liming materials such as FBC residues may present an additional challenge to predicting element loading as well as controlling metal inputs to soils.

3. DEVELOPMENT OF A STANDARD FOR POINT ACONI FBC RESIDUES

In 2001, NSPI, in partnership with Nova Scotia Department of Environment and Labour (NSDEL), commissioned a standard[33] to outline the requirements for Point Aconi FBC residue when used as a soil amendment and agricultural liming material. This interim standard evolved directly from the recommendations from the field testing, and was created to address the above challenges as well as provide quality assurance to agricultural users with respect to performance of the by-product as a liming material. To promote the safest and most beneficial use of this material, the product requirements provided in the standard are supported by recommended residue handling and application procedures, as well as testing and monitoring steps which should be followed before and after land application.

The criteria outlined in the standard for Point Aconi FBC residues differ from those outlined in guidelines for the land application of wastes (e.g., sewage biosolids), which apply in Nova Scotia and most other provinces, in four main respects:

1) the requirements are not regulated by statute, and therefore have no legal authority;
2) the requirements include a wider range of considerations which may be important to users than are found in most regulated standards <$F in addition to safety/environmental quality parameters, there are also performance requirements for attributes such as neutralizing value, product efficiency, and fertilizer value.>;
3) metal standards result in lower metal inputs to soils than metal criteria specified in most provincial waste guidelines and in the *Fertilizers Act and Regulations*; and
4) metal standards include interim criteria for two elements, Tl and V, which are currently not regulated in soil amendment products, as well as recommended limits for B.

3.1. Metal Standards

3.1.1. Elements Regulated in Soil Amendment Products

The metal standards for by-product liming materials established by the Bureau de Normalisation du Québec (BNQ) <$F The BNQ is a Quebec-based standards writing organization (one of five accredited by the Standards Council of Canada) which has developed standards for compost, by-product liming materials and granulated biosolids. >[34] were considered to be the most appropriate standard for this by-product with respect to metals that are currently regulated in soil amendment products. The BNQ metal standards were adopted in the interim standard for Point Aconi FBC Residues without modification.

The BNQ standards specify the maximum acceptable concentrations in products of the nine metals regulated under the FAR (As, Cd, Co, Hg, Mo, Ni, Pb, Se, and Zn) with the addition of limits for Cr and Cu. While the BNQ metal standards are based on the FAR criteria, the approach used to set limits is slightly different. To account for variations in CCE of the various materials and the resulting variable application rates, standards for metals are ratios of liming value to metal concentration, i.e., the ratios are yielded using a CCE set at 50 divided by the maximum values for fertilizers and supplements (Table 2). The total load not to be exceeded is based on limits on metal additions to soil specified in the FAR (Table 1).

Advantages of the BNQ metal standards may be summarized as follows:

- They are consistent with the FAR, so that liming materials meeting the standard can be sold as agricultural liming agents anywhere in Canada. In other words, meeting the

Table 2. Minimum ratio of CCE (%) to metal content, and maximum acceptable metal concentrations in Point Aconi FBC residues with CCE ranging from 50-80%.

Metal	Minimum ratio	Maximum metal concentration in product			
		CCE 50%	CCE 60%	CCE 70%	CCE 80%
		------------------------ mg kg^{-1}, dry weight basis ---------------------			
Arsenic (As)	0.667	(75)[†]	(75)	(75)	(75)
Cadmium (Cd)	2.500	20	24	28	32
Cobalt (Co)	0.333	150	180	210	240
Chromium (Cr)	0.047	1 060	1 271	1 483	1 695
Copper (Cu)	0.066	757	908	1 059	1 210
Mercury (Hg)	10.000	5	6	7	8
Molybdenum (Mo)	2.500	20	24	28	32
Nickel (Ni)	0.278	180	216	252	288
Lead (Pb)	0.100	(500)	(500)	(500)	(500)
Selenium (Se)	3.570	14	17	20	22
Zinc (Zn)	0.027	1 850	2 222	2 593	2 963

[†]values in parentheses are fixed maximum concentrations.
Adapted from: Bureau de Normalization du Québec (BNQ). Liming Materials from Industrial Processes. Standard NQ 0419-090. Bureau de Normalization du Québec, Ste. Foy, QC. 1997.

Table 3. Proposed interim criteria for Tl and B in FBC residues

Element	Maximum metal concentration in product
	------------- mg kg^{-1}, dry weight basis ---------------
Thallium (Tl)[1]	5
Boron (B)[2]	20

[1] USEPA Method 3050A digestion procedure.
[2] Hot-water soluble B.

limits for metals specified in the BNQ standard allows the residue to be sold as a liming agent with potentially no further restrictions on use.

• Because product limits are ratios of liming value to metal content, the standards allow increased metal content only in products with high liming activity (>50% CCE). This approach has the advantage of maintaining flexibility in product standards while providing an effective means of controlling metal inputs to soils. An additional benefit of this approach is that it discourages the use of products with low neutralizing value, which are inevitably less efficient and less cost-effective liming materials.

• The product limits obtained using the ratio approach are more stringent than those adopted in Canadian and foreign standards, even with increasing admissible metal concentrations for by-products with higher neutralizing value.[34] Because of the less frequent application assumed with use of a liming material compared with organic amendments (e.g., compost, sewage-based products), the standards effectively allow a significantly lower metal loading within the specified time frame than the federal regulation.

The FAR soil metal criteria (Table 1) on which the above product criteria are based are recognized in Canada by both federal and provincial jurisdictions, as well as standard-setting organizations such as the CCME and BNQ, as being of fundamental importance to promoting the safe and efficient use of products derived from waste. These soil metal criteria were included in the interim standard for Point Aconi FBC residue to provide benchmark values for monitoring soil quality, and assist in determining the frequency and rate of amendment applications based on metal additions to soil.

3.1.2. Elements not Regulated in Soil Amendment Products

In addition to the product limits indicated in Table 2, interim criteria for thallium and boron in FBC residues are proposed (Table 3), and criteria for V are currently under development. The intent of these criteria is to establish maximum product concentrations of Tl and V and recommended limits for B in FBC residues, to allow a more complete monitoring of the quality of the by-product.

The numerical basis for the criteria for Tl and B was the CCME Interim Assessment Criteria for Soil <$F Interim assessment criteria for soil are largely based on ambient or background concentrations. As these criteria are more stringent than SQG for remediation[35] and SQG for the protection of environmental and human health[24], they represent the most conservative basis for determining acceptable limits at noncontaminated sites. >[35], and product limits were calculated according to methods which have been used by various standard-setting bodies in Canada <$F Agriculture and Agri-Food Canada (AAFC) for fertilizers and supplements, the CCME and BNQ for compost, and the BNQ for by-product liming materials. > to calculate metal limits for soil amendment products[36]. A similar method will likely be used for V.

The metal criteria for elements which are currently not regulated have not yet been endorsed by the wider scientific or regulatory community; hence, they are considered interim criteria. The criteria are, however, scientifically defensible, and based on nationally consistent environmental quality criteria and criteria development methods which are recognized not only in Canada but around the world as being amongst the most protective of the environment. A longer term objective of this initiative may therefore include a more formal, broader based assessment of the suitability of these criteria, and the soil guideline data on which they are based, for soil amendment uses of this and other waste/by-product materials.

4. SUMMARY AND CONCLUSION

Nova Scotia research on the use of Point Aconi FBC residues as agricultural liming materials has identified a number of practical and regulatory challenges in controlling trace element inputs to soils which need to be addressed in order to derive the greatest benefit from using this material as well as minimize potential risk. While FBC residues contain low concentrations of most heavy metals which are commonly regulated in waste materials, there are a number of potentially toxic elements in the residues whose concentrations are variable and highly fuel-dependant. The main challenges in controlling metal inputs to soils presented by the use of this by-product as a soil amendment include:

- Compliance with regulatory criteria for toxic elements (As, Tl, Ni, Mo, V) whose concentrations increase with use of certain fuels in fluidized bed combustion;
- Identifying safe thresholds of concentration for elements, such as Tl, V and B, which are currently not regulated in soils and soil amendment products in Canada; and
- Variable rates of amendment loading, and corresponding metal inputs, with a by-product with variable CCE.

The interim standard commissioned by NSPI in 2001, which outlines the requirements for Point Aconi FBC residues as a liming material, was created to address the above challenges as well as provide quality assurance to agricultural users. Metal standards developed by the Quebec-based BNQ are not only highly protective of the environment but also provide an effective means of controlling metal inputs to soils from by-product liming materials with variable CCE. The BNQ metal standards have been adopted in this standard without modification. In addition, the standard for Point Aconi FBC residues specifies maximum product concentrations of Tl and V (criteria for V are still under development) and recommended limits for B to allow a more complete monitoring of the quality of the by-product. It is expected that this standard for FBC residues will be amended on the basis of discussions currently underway with the scientific and regulatory community, and with improved knowledge and experience with use of the material.

REFERENCES

1.   Anthony, E.J. FBC Waste Disposal: Advanced Ash Disposal Strategies. Proceedings from 49th Purdue University Industrial Waste Conference (May 9-11). Lewis Publishers, Chelsea, Michigan. 1994.
2.   Raask E. Mineral Impurities in Coal Combustion: Behaviour, Problems and Remedial Measures. Hemisphere Pub. Corp. New York. 484 pp. 1985.
3.   Jacques, Whitford and Assoc. Ltd., Trow Consulting Engineers Ltd., and Western Research Institute. Initial Cell Monitoring Program Extension: Nova Scotia Power Inc., Point Aconi Generating Station. Project No. 11139 Final Report. Jacques Whitford and Assoc. Ltd., Dartmouth, NS, 105 pp. 1997.
4.   Stout, W.L., R.C. Sidle, J.L. Hern, and O.L. Bennett. Effects of fluidized bed combustion waste on the Ca, Mg, S, and Zn levels in red clover, tall fescue, oat, and buckwheat. Agron. J. 71:662-666. 1979.
5.   Stout W.L., J.L. Hern, R.F. Korcak, and C.W. Carlson. Manual for Applying Fluidized Bed Combustion Residues to Agricultural Lands. U.S. Department of Agriculture, Agricultural Research Service, ARS-74, 15 pp. 1988.
6.   Korcak, R.F. Fluidized bed material as a calcium source for apples. HortSci. 14:163-164. 1979.
7.   Korcak, R.F. Effects of applied sewage sludge compost and fluidized bed material on apple seedling growth. Commun. Soil Sci. Plant Anal. 11:571-585. 1980.
8.   Korcak, R.F. Fluidized bed material as a lime substitute and calcium source for apple seedlings. J. Environ. Qual. 9:147-151. 1980.
9.   Korcak, R.F. Effectiveness of fluidized bed material as a calcium source for apples. J. Amer. Soc. Hort. Sci. 107:1138-1142. 1982.
10.   Korcak, R.F. Utilization of fluidized bed material as a calcium and sulfur source for apples. Commun. Soil Sci. Plant Anal. 15:879-891. 1984.
11.   Korcak, R.F. Effect of coal combustion waste used as lime substitutes on nutrition of apples on three soils. Plant Soil. 85:437-441. 1985.
12.   Korcak R.F. Fluidized bed material applied at disposal levels: effects on an apple orchard. J. Environ. Qual. 17:469-473. 1988.
13.   Korcak R.F. Inorganic by-product utilization in agricultural situations. In: P.R. Warman, Ed., Alternative Amendments in Agriculture. Proceedings of the 77th Annual Conference of the Agricultural Institute of Canada, NSAC, Truro, Nova Scotia (August 17-20). p. 41-61. 1997.
14.   Bennett, O.L., J.L. Hern, and H.D. Perry. Agricultural uses of atmospheric fluidized bed combustion residue (AFBCR) - a seven year study. Proc. 2nd Ann. Pittsburgh Coal Conf. Sept. 16-20, 1985. pp. 558-577. 1985.
15.   Edwards, J.H., B.D. Horton, A.W. White, Jr. and O.L. Bennett. Fluidized bed combustion residue as an alternative liming material and Ca source. Commun. Soil Sci. Plant Anal. 16:621-637. 1985.
16.   Goodyear, N. and M. Hope-Simpson. Utilization of Point Aconi CFB ash as an Agricultural Soil Amendment. Final Research Report to Nova Scotia Power Inc. Nova Scotia Agricultural College, Truro, NS. October 2000. 201 pp. 2000.
17.   Eary, L.E., R. Dhanpat, S.V. Mattigod, and C.C. Ainsworth. Geochemical factors controlling the mobilization of inorganic constituents from fossil fuel combustion residues: II. Review of the minor elements. J. Environ. Qual. 19:202-214. 1990.
18.   Canadian Food Inspection Agency (CFIA). Standards for Metals in Fertilizers and Supplements. Trade memorandum T-4-93. Plant Products Division, Canadian Food Inspection Agency. 1997.
19.   Baker, D.E., C.S. Baker, and K. Sajwan. Evaluation of coal combustion products as components in disturbed land reclamation by the Baker Soil Test. In Sajwan et al., 1999. Biogeochemistry of Trace Elements in Coal and Coal Combustion Byproducts. Kluwer Academic/Plenum Publishers, New York. 1999.
20.   Carlson, C.L. and D. Adriano. Environmental impacts of coal combustion residues. J. Environ. Qual. 22:227-247. 1993.
21.   Keefer, R.F. Coal ashes-industrial wastes or beneficial by-products. Pp.3-10. In R.F. Keefer and K.S. Sajwan, Eds., Trace elements in coal and combustion residues. CRC Press, Boca Raton, Fla. 1993.
22.   Kabata-Pendias, A. and H. Pendias. Trace Elements in Soils and Plants. 2nd Ed. CRC Press Inc., Boca Raton, Fla. 1991.

23.   Kawasaki, A., R. Kimura and S. Arai. Fractionation of trace elements in wastewater treatment sludges. Commun. Soil Sci. Plant Anal., 31(11-14), 2413-2423. 2000.

24.   Canadian Council of Ministers of the Environment (CCME). Canadian Soil Quality Guidelines for the Protection of Environmental and Human Health: Summary Tables. Canadian Council of Ministers of the Environment, Winnipeg, MB. Available from: http://www.ec.gc.ca/ceqg-rcqe/Soil.pdf

25.   Reilly, C. Metal Contamination of Food. 2nd Ed. Elsevier Science Publishing Co., Inc. New York. 284 pp. 1991.

26.   Jia, L., E.F. Anthony and J.P. Charland. Investigation of vanadium compounds in ashes from a CFBC firing 100% petroleum coke. In: Ash Interactions from Fluidized Bed Conversion of Fossil/Non-Fossil Fuels. Anthony, E.J., and L. Jia, Eds. Proceedings of the 42nd IEA-FBC Meeting, Sydney, NS. May 10-11, 2001.

27.   Fernando, R. The Use of Petroleum Coke in a Coal-Fired Plant (Draft Report). IEA Coal Research, London, UK. 2001.

28.   Lide, D.R. CRC Handbook of Chemistry and Physics, 82nd Ed. CRC Press Inc., Boca Raton, Fla. 2001.

29.   Agriculture and Agri-Food Canada (AAFC). Guidelines to the Fertilizers Act and Regulations. 2nd Ed. Plant Products Division, Food Inspection Directorate, Agriculture and Agri-Food Canada. Nepean, Ont. 1996.

30.   Centre Québecois de Valorisation de la Biomasse (CQVB). Valorisation des boues de stations d'épuration municipales. Le premier colloque Québecois sur la valorisation des boues de stations d'épuration municipales. 18 et 19 Septembre 1990, Hull, QC. 1990.

31.   Ontario Ministry of Environment (MOE) and Ontario Ministry of Agriculture, Food and Rural Affairs (OMAFRA). Guidelines for the Utilization of Biosolids and Other Wastes on Agricultural Land. Ontario Ministry of Environment and Ontario Ministry of Agriculture, Food and Rural Affairs. March 1996.

32.   Ghodrati, M., J.T. Sims, B.L. Vasilas, and S.E. Hendricks. Enhancing the benefits of fly ash as a soil amendment by pre-leaching. Soil Sci. 159:244-252. 1995.

33.   Hope-Simpson, M. Interim Standard for Point Aconi FBC Residue as a Liming Material and Soil Amendment. Prepared for Nova Scotia Power Inc. and Nova Scotia Department of Environment and Labour. Agrosystems Atlantic, Truro, Nova Scotia. 2001.

34.   Bureau de Normalization du Québec (BNQ). Liming Materials from Industrial Processes. Standard NQ 0419-090. Bureau de Normalization du Québec, Ste. Foy, QC. 1997.

35.   Canadian Council of Ministers of the Environment (CCME). Interim Canadian environmental quality criteria for contaminated sites. CCME, Winnipeg. 1991.

36.   Canadian Council of Ministers of the Environment (CCME). Support Document for Compost Quality Criteria – National Standard of Canada (CAN/BNQ 0413-200), The Canadian Council of Ministers of the Environment (CCME) Guidelines and Agriculture and Agri-Food Canada (AAFC) Criteria. Published by the Bureau de Normalization du Québec, Environment Canada and Agriculture and Agri-Food Canada. March, 1996.

THE MINE WATER LEACHING PROCEDURE: EVALUATING THE ENVIRONMENTAL RISK OF BACKFILLING MINES WITH COAL ASH

Paul F. Ziemkiewicz [1], Jennifer S. Simmons [1], and Anna S. Knox [2]

[1] West Virginia Water Research Institute
National Mine Land Reclamation Center
West Virginia University
Morgantown, WV 26506-6064
U.S.A.

[2] Savannah River Ecology Laboratory
University of Georgia
Drawer E
Aiken, SC 29802
U.S.A.

ABSTRACT

Federal and State regulations encourage reduction of industrial waste streams to decrease the acreage consumed by landfills. In particular, applications that resolve environmental problems are recognized by state policy as "beneficial uses." These large-scale projects may involve filling surface and underground coal mines with ash to address hydraulic problems, acid mine drainage, pit backfilling and subsidence. In some states, those mine filling projects classified as beneficial are not subject to industrial waste disposal conditions such as liners, leachate collection and monitoring. Coal Combustion Byproducts (CCBs) are attractive for such applications because they constitute a source of low cost alkalinity and favorable economics resulting from transport back to the mine in otherwise empty coal haulage trucks. The environmental risk of land filling CCBs is generally evaluated by the Toxic Characteristics Leaching Procedure (TCLP) or the Synthetic Precipitation Leaching Procedure (SPLP). However, there is doubt regarding the applicability of these tests to long-term CCBs leaching behavior in groundwater associated with coal mines. The Mine Water Leaching Procedure (MWLP) was developed to provide a site specific risk assessment tool. The MWLP procedure is presented in this chapter as a study case and comparisons with TCLP results were made.

INRODUCTION

Several test procedures have been developed in an attempt to predict the leaching behavior of CCBs. The most widely used procedure is the Toxicity Characteristic Leaching Procedure (TCLP), which was designed by the United States Environmental Protection Agency to "determine the mobility of both organic and inorganic analytes present in liquid, solid and multiphasic wastes."[27] However, this test may only account for fast reactions that take place in short term leaching processes.[28] The results obtained with the TCLP on a limited number of fly and bottom ash samples showed that the range of concentrations of specified constituents was well below the regulatory.[20] These residues are therefore classified as non-hazardous wastes and can be disposed of on land without risk of contaminating groundwaters to the extent of exceeding drinking water standards. However, Bhumbla et al. concluded that, environmental concerns about toxic element release from fly ash amended soils have been lessened by evaluation of data from short-term studies in highly alkaline environments.[7] They point out that all soils in humid regions ultimately become acidic and that the behavior of ash under conditions pertaining over the long term needs to be addressed. Specifically: *"How long will potentially toxic elements in technogenic (i.e. man made) soils made from ashes remain in discrete mineral phases or as components of the glassy fly ash matrix?"* They encouraged a better understanding of the mechanisms and rates of trace element release from CCBs so that behavior over the long term can be predicted. The Mine Water Leaching Procedure (MWLP) was developed to determine the long-term leaching potential of toxic elements present in industrial wastes when placed in specific, saturated environments. Of particular concern is the leaching behavior of these materials when placed in acidic environments associated with reclamation of acidic mine spoils.

A quantitative assessment of leachate composition is crucial to the environmentally sound disposal of solid residues. Currently applied laboratory extraction procedures yield imprecise estimates of field leachate composition, and field studies alone do not provide the causal relationships for the observed behavior. Also, since TCLP and SPLP use standard leaching solutions, they do not predict interactions between the solid waste and components of a specific mine water. Therefore, our research has focused on developing improved laboratory methods for predicting the leaching behavior of CCBs under field conditions. The geochemical reactions of dissolution/precipitation, adsorption/desorption, and oxidation/reduction are recognized as controlling the mobilization of various constituents from solid residues.[2, 24] Therefore, an integrated scheme has been developed that, through an empirical procedure, recognizes the interactions among mine water chemistry and specific CCBs. This sheds light on the fundamental geochemical reactions critical to interpretation and prediction of leachate chemistry and interactions with various geological materials. Such laboratory studies can be conducted for a wide range of conditions, such as pH, complexations, and ionic strength of leachates and porewaters that would be encountered at different field sites. Additionally, laboratory studies are less costly than field studies, and will provide useful and widely applicable data as a variety of CCBs and mine waters are evaluated.

The Mine Water Leaching Procedure (MWLP) was developed to provide a site specific risk assessment tool. The MWLP procedure is presented in this chapter as a study case and comparisons with TCLP results were made.

COAL COMBUSTION BY PRODUCTS

Physical and Chemical Properties

Fly ash is the non-combustible particular matter removed from the flue of coal-fired boilers. It is an amorphous, ferro-aluminosilicate that contains significant quanitities of Fe, Ca, K and Na.[22] It may also be enriched with As, B, Mo, S, Se, Sr and varying concentrations of C. Depending on the elemental composition of the source coal, fly ash may also contain various amounts of other trace elements.

Eastern and Midwestern U.S. coals produce ashes composed of quartz, mullite, hematite and magnetite.[26] These low lime, low S ashes are referred to as class F fly ash.

Table 1. Typical Composition of Class F and C Ashes. *Characterizations based on American Standards and Testing Methods 618.*[3]

Parameter	Class F	Class C
SiO_2	54.9%	39.9%
Al_2O_3	25.8%	16.7%
Fe_2O_3	6.9%	5.8%
CaO	8.7%	24.3%
SO_3	0.6%	3.3%
Moisture Content	0.3%	0.9%
Loss on Ignition (LOI) (@750°C)	2.8%	0.5%
Available alkalis as Na2O	0.5%	0.7%
Specific gravity	2.34	2.67
Fineness, retained on #325 mesh seive	14.0%	8.0%

Class C fly ash is commonly produced from western subbituminous coal and contains the above minerals, as well as, periclase, lime, anhydrite and tricalcium aluminate. Table 1 contains the ASTM definition of the typical Class F and Class C fly ashes.[4]

Bottom ash is the sand-sized material that is collected in the bottom of a dry bottom boiler and boiler slag is the coarse, glassy material collected in the bottom of wet bottom boilers. Physically, both are very different from the powdery fly ash in the flue gas. However, their chemical compositions are very similar and consist primarily of Si, Al, Fe and Ca.[8]

Fluidized bed combustion (FBC) units produce strongly alkaline ash. FBC byproducts: fly ash and spent bed (bottom ash) contain the same type of minerals but in different proportions. Bottom ash is enriched in anhydrite and lime while the fly ash contains more silicon and iron oxides. FBC ashes contain large amounts of gypsum $(CaSO_4)$ and, as much as 25-30% lime.[29] Flue gas desulfurization (FGD) residue consists primarily of gypsum, $Ca(OH)_2$, and unreacted lime. These byproducts are higher in Ca and S and lower in Si, Al, Fe and trace elements than fly ash. In spray drying and lime injection systems, the FGD residue will also contain fly ash. FGD materials resulting from these processes may be enriched in trace elements present in the fly ash. The most important of these trace elements from an environmental standpoint are As, Ba, B, Cd, Cr, Cu, Pb, Hg, Mo, Ni, Se, Sr, V and Zn. However, the trace element concentrations of FGD residues are low. Punshon, et al. reported that B and Cl are the only elements that commonly occur in elevated concentrations in FGD materials.[22]

Although both fly ashes have a small particle size (silt-size) Class F ashes tend to be more permeable than class C ashes due to the tendency of class C ashes to self-cement.[1, 9, 10] Bottom ashes have a sandy to coarse gravelly texture and are highly permeable. At the opposite extreme, fixated FGD solids have very low permeability and the various CCB grouts behave like concrete and are virtually impermeable.

Use of CCBs for Coal Mine Reclamation

According to the American Coal Ash Association, 105 million tons of CCBs were produced by the power generating utilities in 1997. Of that total, 1.68 million tons were used in mining applications.[3] CCBs can be used in mine reclamation for acid mine drainage (AMD) prevention and treatment, subsidence control and surface reclamation.[8] Class C Fly ash and Class F fly ash mixed with lime exhibit self-cementing properties and can be used to cap surfaces, line pavements and isolate acidic materials in the backfill to prevent AMD formation. In addition, highly alkaline CCBs, such as FGD and FBC residues, are used to directly neutralize acidic materials.

A field demonstration of the use of ash for pit floor sealing and surface capping was conducted at the Chaplin Hill Coal Mine near Morgantown, West Virginia. Mine pits in this region had historically produced AMD due to a pyritic pit floor and pyritic materials within the backfill. In 1991 the company began placing a 1 ft layer of FBC ash over the pit floor prior to backfilling. In addition, another 1 ft lift of FBC ash was placed on the graded spoil and compacted prior to topsoil application. Since CCB application no pits on this site have generated AMD.[14]

A 1995 project by the Maryland Department of Natural Resources and the Maryland Department of the Environment demonstrated the use of CCBs for AMD abatement in an underground mine. A grout, consisting of FGD, FBC and Class F fly ash, was injected into the Frazee mine in an attempt to fill the mine void and minimize contact between groundwater and acidic materials in the mine. Unfortunately, the mine size was underestimated and was not completely filled. As a result the mine still produces AMD.[23] Nevertheless, only Ni concentrations increased above background levels.[29]

Fly ash generally has a strong influence on spoil pH.[12] Fly ashes are commonly alkaline but may be neutral or acidic as a function of CaO and various amorphous oxides of Fe, which adhere to the exterior of the fly ash spheres.[19] Alkalinity is released as CaO is liberated from the dissolving alumino-silicate matrix.[12, 16] Alkaline fly ashes normally contain sufficient neutralization capacity to raise the pH of acid soils.[15] While many class F fly ashes have a high past pH (10 to 12), their neutralization potential is low so extremely high additions (up to 625 tonnes of fly ash per hectare) may be needed to sustain a neutral pH in acidic coal mine soils.[17, 18, 21] There have been several case studies on the direct neutralization of AMD using highly alkaline CCBs.[5, 13] In December 1996, the Metikki Coal Corporation began injecting a mixture of FGD, fly ash, AMD metal precipitates and fine coal refuse into its underground mine in Garrett County, Maryland. The injection scheme has significantly reduced the acidity of the mine pool. Otherwise, the principal effect of FGD addition was to increase chloride and sulfate concentrations within flooded parts of the mine.[5]

CCBs are also used during surface reclamation to improve soil physical and chemical properties of acidic minesoils.[9, 25] A study conducted by Dhaliwal and associates compared the properties of a mineland area reclaimed with fly ash to an adjacent area that received no fly ash. Even after 22 years the fly ash treated mine soil had a higher pH and thicker organic horizon than the adjacent untreated area.[11] Bhumbla used a fly ash/rock phosphate mixture as a topsoil substitute and found that fly ash addition improved certain physical conditions of the mine soils and reduced Al, Fe and Mn toxicities.[6]

Implications of CCB Use in Mine Fills

After placement in the mine, bulk amendments, such as CCBs, will weather according to their mineralogy and the chemistry of the mine water. Many mine waters are severely contaminated with Fe, Al, Mn and other trace elements. When CCBs are placed in acidic environments there is a concern over their potential to leach toxic levels of trace elements, particularly As, Ba, B, Cd, Cr, Cu, Pb, Hg, Mo, Ni, Se, Sr, V and Zn, into groundwater. Many of these elements are insoluble under alkaline conditions but may become mobile after the alkalinity has been exhausted. Acidity generation is also finite as controlled by the pytrite concentration of the mine spoil. In some cases, this concern may be relieved by ensuring that adequate alkalinity is added to the acid producing mine spoil to neutralize all potential acid production. The following formula can be applied to estimated required amounts of industrial waste (amendment) for sites where the volume of acid producing materials is known:

Table 2: Water Analysis of Acidic Pit Water Used in the Analysis of Fly Ash. *MWLP was performed on the India ashes using both acid waters. Concentrations are in mg/L.*

Analyte	India Sites Pit 1	India Sites Pit 2	West Virginia Site Seep 1
pH	2.8	2.6	3.6
net acidity	743.7	2128.0	3168.7
Mg	24.1	72.4	743.1
Ca	93.9	192.6	433.1
Fe	39.3	240.3	223.1
Al	86.0	181.6	342.4
Mn	3.0	10.6	278.3
Sb	BDL	BDL	BDL
As	BDL	BDL	BDL
Ba	0.043	0.028	0.243
Be	0.039	0.122	0.013
Cd	0.059	0.013	0.151
Cr	0.023	0.112	0.017
Pb	0.028	0.044	0.022
Se	BDL	0.037	BDL
Ag	BDL	BDL	BDL
Cu	0.047	0.169	BDL
Ni	0.973	3.257	BDL
Tl	BDL	BDL	0.020
Zn	5.730	11.800	5.289
V	0.492	0.013	BDL
B			BDL
Hg	BDL	BDL	10.349

BDL= Below Detection Limit

$$A = \frac{W \times \%S \times 3.125 \times F_s}{\%NNP}$$

Where:

A = required amendment (in tons)
W = amount of waste rock: spoil or tailings to be neutralized (in tons)
%S = per cent sulfur in waste rock, e.g. 2% = 2 tons per 100 tons of rock
%NNP = per cent net neutralization potential of amendment, e.g. %NP-%MPA
F_s = safety factor, e.g. 1.1 = an F_s of 10%

If this formula can be applied and the amendment blended uniformly with the acid forming material, the spoil not become acidic. Three scenarios may, therefore, pertain:

- Amendment is placed in neutral to alkaline groundwater.
- The amendment and spoil remain alkaline until pyrite oxidation ceases.
- The amendment and spoil become acidic.

While toxic elements may be leached under all of these scenarios, acidification of the fly ash is of particular concern. The MWLP was developed to evaluate toxic element leaching potential under each of these scenarios.

CASE STUDY: THE MINE WATER LEACHING PROCEDURE

Materials and methods

Two leaching procedures were used: TCLP and MWLP. The TCLP was conducted according to USEPA's SW 846, Method 1311. The leachant was adjusted to pH 2.88 using acetic acid according to the method.

MWLP was developed to determine the long-term leaching behavior of industrial wastes when placed in contact with the groundwater on a given site. The method sequentially leaches the CCB with a sample of the site's groundwater until the alkalinity is exhausted and the pH of the leachate returns to that of the mine water sample.

In this chapter MWLP results are presented for two fly ashes and three AMD leachants. One fly ash was from a power station in Northern Indian (India fly ash). This was a Class F ash that contained very little alkalinity (Neutralization Potential = 0.52% $CaCO_3$ equivalent). The India fly ash was subjected to MWLP leaching with AMD from two coal mine pits located near the power plant. Table 2 contains the chemical analyses of these AMD samples.

A second fly ash sample was from a power plant in northern West Virginia, USA (West Virginia ash). This was also a class F ash and contained slightly more alkalinity than the India ash (NP = 3.2%). Leaching water used for the MWLP analyses of West Virginia ash was from an acid seep located on a reclaimed surface coal mine (Table 2).

The MWLP is nearly identical for all waste products, differing only in the number of leaching cycles required for alkalinity exhaustion. This point is determined as the pH of the untreated mine water. A general outline of the procedure is included below.

One hundred grams of fly ash were weighed out and transferred into 2-L plastic reaction bottles. To each ash sample was added one of two leachants: mine water or deionized water (control). Three replicates of each ash were prepared. The bottles were then sealed with Parafilm and the lids were secured. Reaction bottles were arranged evenly on a rotating platform, identical to that used in the TCLP test, and rotated end-over-end for 18 hours at 30 rpm.

Following each 18 hour cycle the contents of each bottle were filtered through a 0.7 um glass, borosilicate filter using a stainless steel pressure filtration unit at or below 40 psi. A two liter container was placed under the base of the filtration apparatus to collect the filtrate. The contents of the fly ash + AMD reaction bottle were poured into the top of the pressure cylinder, the lid was secured and N_2 was introduced to pressurize the filtration unit. The pressure was slowly increased to 40 psi until all the liquid was removed from the unit.

Following filtration the unit was dissembled and the filter cake (filter + solids) removed and saved for use in the subsequent cycle.

Five hundred ml of each leachate was collected in two 250 ml bottles. One bottle was sent to an analytical laboratory for pH, acidity and alkalinity determinations using a Brinkman Autotitrator. The other bottle was acidified using 1 ml of 1N nitric acid and sent to the lab for elemental analysis (Sb, As, B, Ba, Cd, Cr, Pb, Hg, Ag, Cu, Ni, Tl, V, Zn, Mo, Fe, Mn, Al, and B) using a high resolution ICP-Mass Spectrometer.

Solids collected during filtration were rinsed back into their corresponding reaction bottles with 2 "fresh" liters of leachant and placed back on the rotating platform for another 18 hour cycle. The leaching-agitation-filtration cycles continued until all alkalinity was removed from the system. The India Fly Ash MWLPs and the West Virginia MWLPs were continued for 3 and 5 cycles, respectively. The number of cycles depended on the alkalinity of the ash.

RESULTS

The results from MWLP analyses performed on India and West Virginia fly ashes are presented in Tables 3-6 and Figures 2-4. Leachate metal concentrations could be attributed to four possible sources; leachant (AMD), released metals from dissolution of the waste matrix; remobilized AMD metal precipitates and remobilized ash metal

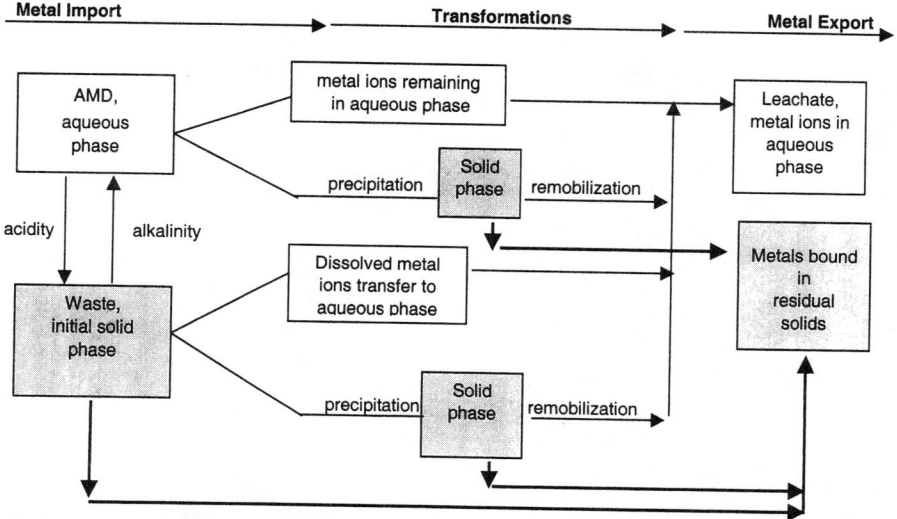

Figure 1. Diagram of the experimental system. *Metals can be imported to the system from either the AMD (aqueous form) or the waste (solid form). Once in the system, metals can either stay in the aqueous phase, precipitate into the solid phase or resolubilize into the aqueous phase.*

precipitates (Figure 1). Since concentrations and volumes of leachant and leachate waters were known it was possible to prepare a series of mass balances. So, by subtracting the elemental masses exported from the system (in the leachate) from the elemental masses imported to the system (in the AMD or leachant), it was possible to determine the net effect of ash addition on the concentration of toxic elements in the leachate. Table 3 shows the cumulative import, export, sequestration and release of Cr during the MWLP. This example indicates that between pH 4.2 and 3.8, the system stops sequestering Cr and begins releasing it.

These cumulative values can be used to make inferences about the leaching behavior of individual waste products as the pH of the leachate changes. For example, table 3 also indicates the percent difference between the mass of Cr imported via leachant (AMD) and the mass of Cr exported via leachate. Negative release values represent sequestration of Cr from the aqueous phase into the solid phase. So, during this part of the leaching process, more Cr is entering the system through the AMD than is exiting the system in the leachate. On the other hand, positive release values indicate Cr concentrations are higher in the leachate than can be attributed to the addition of leachant during each cycle. It is inferred that the additional metal ions were released from the solid phase: CCB and previously precipitated metal oxy-hydroxides.

Table 3. Example of Cr Import/ Export Table for Class F Fly Ash MWLP.
All in, out and release values are cummulative. For example, the value of .180 mg for Cr In at Cycle 5 is the cummulative concentration of Cr in the import water (AMD) for cycles 1-5. The Cr Out is calculated similarly. This permits the calculation of the mg of cr released into the leachate water as a result of the fly ash addition to the AMD.

Cycle	Leachate pH	Cr In (mg)	Cr Out (mg)	Fly ash Release (mg)	% Released
1	4.17	0.044	0.032	-0.012	-28.0%
2	3.82	0.088	0.161	0.073	83.0%
3	3.54	0.132	0.362	0.230	174.3%
4	2.87	0.176	0.547	0.371	210.9%
5	2.92	0.220	0.732	0.512	232.5%
average					134.5%

81

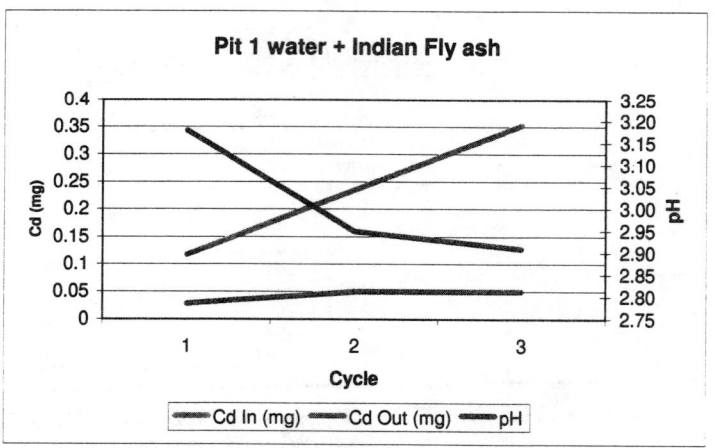

Figure 2. Import/ Export Trend Lines for Cd from Indian Fly Ash Leached With Acidic Pit 1 Water for Three Cycles. *In this example, the import line is above the export line indicating net sequestration of Cd in the solid phase as a result of the fly ash addition. The area between the two lines represents the amount of Cd sequestered in the solid phase.*

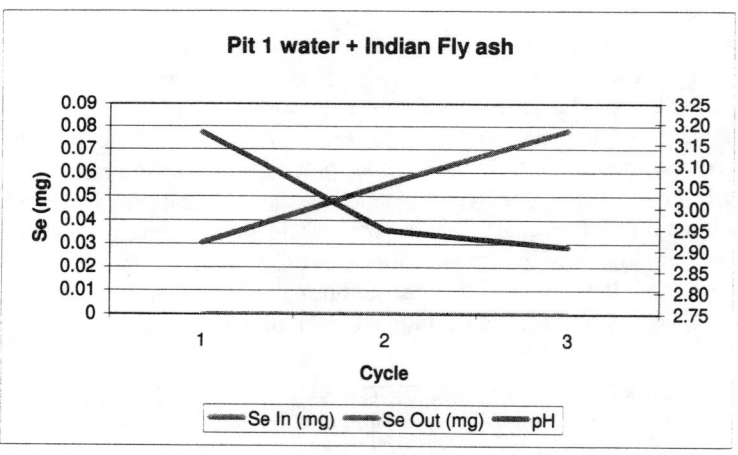

Figure 3. Import/ Export Trend Lines for Se from Indian Fly Ash Leached With Acidic Pit 1 Water for Three Cycles. *In this example, the export line is above the import line indicating a net release of Se from the solid phase as a result of the fly ash addition. The area between the two lines represents the magnitude of Se release. In this example, where the raw water contained no Se, we can assume that it's source is fly ash dissolution.*

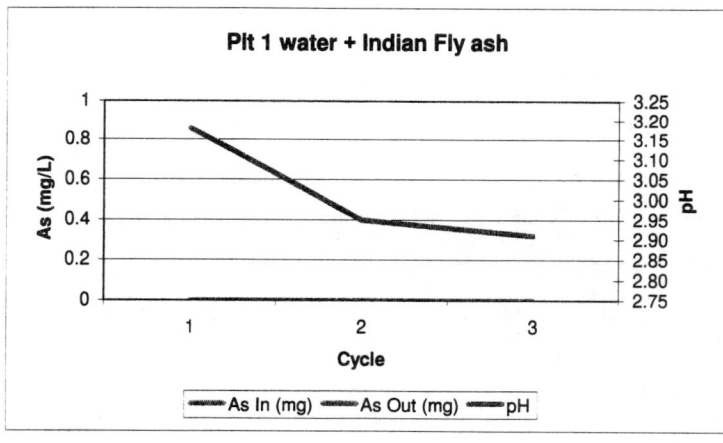

Figure 4. Import/ Export Trend Lines for As from Indian Fly Ash Leached With Acidic Pit 1 Water for Three Cycles. *In this example, there was no arsenic detected in the raw water or any of the three leachates. Therefore, fly ash addition did not result in any As increase in the leachate.*

Sequestration and release trends can also be shown graphically. Figure 2 indicates import/export trends of Cd for the India fly ash. In this graph the import line is above the export line indicating net sequestration of Cd from the acid water into the solid phase. The amount of sequestration is equivalent to the area between the two lines. Figure 3 illustrates the net release of Se from the same fly ash. Net release occurs when the export line is above the import line. In some instances, waste product additions had no effect on leachate metal concentrations. Figure 4 indicates that As was present in neither the raw AMD nor the leachate.

The MWLP also enables estimation of the overall affect of industrial waste additions to mine water while tracking leaching behavior over time and through various pH ranges. There are five possible trends for metal leaching behavior: 1) net sequestration, 2) net release, 3) sequestration then release, 4) release then sequestration, or 5) no effect.

Table 4 summarizes leaching behavior of all analytes when the two fly ashes were subjected to the MWLP. The West Virginia fly ash had no effect on the leachate concentrations of Sb, V, Ag and Hg. Iron was the only element that underwent net sequestration over all leaching cycles. On the other hand, As, B, Ba, Be, Cu, Ni, Mn, Pb, Se, Tl and Zn were released throughout the leaching period. Aluminum and Cr were initially sequestered but switched to net release after the pH reached 3.8 in cycle 2. Cd showed the opposite trend: net release during the first four cycles and then net sequestration in the final cycle. Table 4 also indicates the cycle and the pH where the change occurred.

The leaching behavior of the India fly ash varied with the two AMD leachants. Pit 1 AMD had a slightly higher pH (2.8 vs. 2.6) but significantly less acidity (676 mg/L vs. 2,128 mg/L). Nonetheless, eleven of the elements behaved similarly in both leachants. Antimony, As, Ag and Hg concentrations in leachate were unaffected by fly ash addition in both leachants while Pb was sequestered in both. Aluminum, Ba, Cu, Cr, Mn and Ni all underwent net release in both leachants. In both leachants Tl was initially released then sequestered. It is perhaps significant that none of the elements were sequestered then released from the India fly ash. This may be explained by the lack of alkalinity in this ash and the near absence of metal oxy-hydroxide precipitates relative to the West Virginia fly ash. Another important difference between the West Virginia and

Table 4. Summary of sequestration/release trends for the two fly ashes and three AMD leachants.
The leach cycles and pH are noted where a change in trend was noted.

	West Virginia, class F fly ash, seep 1 AMD					Cycle	pH
			trend			of change	of change
	no effect	seqestration	release	seq/release	release/ seq		
Sb	X						
V	X						
Ag	X						
Hg	X						
Fe		X					
Zn			X				
Ba			X				
Mn			X				
B			X				
As			X				
Cu			X				
Pb			X				
Ni			X				
Se			X				
Be			X				
Al				X		2	3.8
Cr				X		2	3.8
Cd					X	5	2.9
Tl					X	2	3.8
	India, class F fly ash, Pit 1 AMD					Cycle	pH
			trend			of change	of change
	no effect	seqestration	release	seq/release	release/ seq		
Sb	X						
As	X						
Ag	X						
Hg	X						
V		X					
Pb		X					
Cd		X					
Ba			X				
Mn			X				
Cu			X				
Ni			X				
Se			X				
Be			X				
Al			X				
Cr			X				
Fe				X		3	2.9
Tl					X	2	3.0
Zn					X	3	2.9
	India, class F fly ash, Pit 2 AMD					Cycle	pH
			trend			of change	of change
	no effect	seqestration	release	seq/release	release/ seq		
Sb	X						
As	X						
Ag	X						
Hg	X						
Be	X						
Zn		X					
Pb		X					
Fe			X				
Ba			X				
Mn			X				
V			X				
Cu			X				
Ni			X				
Al			X				
Cr			X				
Cd					X	2	2.6
Se					X	2	2.6
Tl					X	3	2.5

Table 5. Comparison of leachate concentrations from the West Virginia Class F Fly ash using TCLP and MWLP results. *MWLP values shown are the elemental concentrations of the filtrates minus the concentrations in the leachant. Negative values indicate concentrations less than the leachant (AMD) and positive values indicate values higher than the leachant. For calculation purposes, leachant values below detection limits were assigned the detection limit value. All values are in mg/L.*

West Virginia Fly ash	method detect. limit	RCRA TCLP Limit	TCLP	Leachant (AMD)	MWLP Cycle 1	Cycle 2	Cycle 3	Cycle 4	Cycle 5
					(leachate concentration - AMD concentration)				
pH			4.1	3.6	4.2	3.8	3.5	3.0	2.9
est. acidity			221.3	3023.4	-521.2	459.9	191.6	316.0	221.9
alkalinity			0.0	0.0	0.0	0.0	0.0	0.0	0.0
acid-alk			221.3	3023.4	-521.2	459.9	191.6	316.0	221.9
Mg	0.1		15.3	743.1	172.5	252.9	171.5	144.9	137.8
Ca	0.1		443.7	433.1	309.5	262.2	261.3	171.1	169.1
Fe	0.1		0.7	223.1	-216.3	-215.5	-210.8	-176.7	-163.9
Al	0.1		38.5	342.4	-8.2	147.6	109.0	112.2	87.5
Mn	0.1		0.8	278.3	63.8	123.4	83.3	70.1	71.1
Sb	0.005	1	0.015	0.005	BDL	BDL	BDL	BDL	BDL
As	0.005	5	0.135	0.005	0.017	0.016	0.020	0.015	0.011
B	0.01		6.460	0.243	4.467	0.607	0.417	0.287	0.267
Ba	0.005	100	0.128	0.013	0.063	0.049	0.067	0.048	0.046
Be	0.005	0.007	0.012	0.151	0.010	0.015	0.022	0.027	0.023
Cd	0.005	1	BDL	0.017	0.006	-0.001	-0.001	-0.002	-0.002
Cr	0.005	5	0.315	0.022	-0.006	0.043	0.079	0.071	0.071
Pb	0.005	5	BDL	0.005	BDL	0.012	0.027	0.021	0.026
Hg	0.005	0.2	BDL	0.005	BDL	BDL	BDL	BDL	BDL
Se	0.005	1	0.072	0.005	0.091	0.047	0.078	0.055	0.011
Ag	0.005	5	BDL	0.005	BDL	BDL	BDL	BDL	BDL
Cu	0.005		0.103	0.020	0.099	0.085	0.306	0.041	0.028
Ni	0.005	70	0.132	5.289	0.279	0.100	0.197	0.385	0.310
Tl	0.005	7	0.013	0.005	0.012	BDL	BDL	BDL	BDL
V	0.005		0.114	0.005	BDL	BDL	BDL	BDL	BDL
Zn	0.005		0.337	10.349	0.215	0.546	0.496	0.906	0.648

RCRA = Resource Conservation and Recovery Act
BDL= Below Detection Limit

85

Table 6. Comparison of leachate concentrations from the India Class F Fly ash using TCLP and MWLP results. *MWLP values shown are the elemental concentrations of the filtrates minus the concentrations of the raw water. Negative values indicate concentrations less than the leachant (AMD) and positive values indicate values higher than the leachant. For calculation purposes, leachant values below detection limits were assigned the detection limit value. All values are in mg/L.*

India Fly ash	method detect. limit	RCRA TCLP Limit	TCLP	Pit 1 water Leachant (AMD)	MWLP Cycle 1	MWLP Cycle 2 [leachate]-[AMD]	MWLP Cycle 3	RCRA TCLP Limit	Pit 2 water Leachant (AMD)	MWLP Cycle 1 [leachate]-[AMD]	MWLP Cycle 2	MWLP Cycle 3
pH			3.5	2.8	3.2	3.0	2.9	3.5	2.6	2.9	2.6	2.5
est. acidity			115.2	675.8	82.5	91.2	94.1	115.2	1812.1	327.0	303.0	376.8
alkalinity			0.0	0.0	0.0	16.3	0.0	0.0	0.0	0.0	0.0	0.0
acid-alk			115.2	675.8	82.5	75.0	94.1	115.2	1812.1	327.0	303.0	376.8
Mg	0.1		2.9	24.1	5.1	4.1	5.7	2.9	72.4	11.6	11.2	16.2
Ca	0.1		23.4	93.9	32.1	18.3	19.7	23.4	192.6	65.4	35.4	44.7
Fe	0.1		0.8	39.3	-20.7	-5.1	5.4	0.8	240.3	19.0	28.7	43.9
Al	0.1		17.0	86.0	34.3	24.3	18.5	17.0	181.6	61.0	41.3	44.6
Mn	0.1		0.4	3.0	0.6	0.3	1.1	0.4	10.6	2.2	1.4	2.4
Sb	0.005	1	BDL	0.005	BDL	BDL	BDL	0.005	0.005	BDL	BDL	BDL
As	0.005	5	0.013	0.005	BDL	BDL	BDL	0.013	0.005	0.071	0.030	0.060
Ba	0.005	100	1.005	0.043	0.090	0.068	0.064	1.005	0.028	0.166	0.045	0.050
Be	0.005	0.007	BDL	0.039	0.035	0.026	0.026	0.005	0.122	0.028	-0.056	-0.004
Cd	0.005	1	BDL	0.059	-0.045	-0.048	BDL	0.005	0.013	0.011	0.000	0.000
Cr	0.005	5	0.019	0.023	0.048	0.025	0.029	0.019	0.112	0.062	0.014	0.038
Pb	0.005	5	BDL	0.028	BDL	-0.020	BDL	0.005	0.044	BDL	BDL	BDL
Hg	0.005	0.2	BDL	0.005	BDL	BDL	BDL	0.005	0.005	BDL	BDL	BDL
Se	0.005	1	0.021	0.005	0.010	0.007	0.007	0.021	0.037	0.013	-0.017	-0.009
Ag	0.005	5	BDL	0.005	BDL	BDL	BDL	0.005	0.005	BDL	BDL	BDL
Cu	0.005		0.129	0.047	0.362	-0.034	0.075	0.129	0.169	0.310	0.019	0.056
Ni	0.005	70	0.112	0.973	1.013	0.906	0.719	0.112	3.257	0.489	-0.326	0.028
Tl	0.005	7	BDL	0.005	0.002	BDL	BDL	0.005	0.005	0.008	0.002	BDL
V	0.005		0.137	0.492	-0.462	-0.456	-0.433	0.137	0.013	0.272	0.142	0.164
Zn	0.005		0.349	5.730	0.406	0.041	-0.446	0.349	11.800	-1.093	-4.387	-2.558

RCRA = Resource Conservation and Recovery Act
BDL= Below Detection Limit

86

India ashes was the tendency to release As from the former and its absence from the ash and leachant in the later.

Tables 5 and 6 compare elemental concentrations of TCLP and MWLP leachates for the two fly ashes. The MWLP results represent the differences between leachant (AMD) concentrations and leachate concentrations. Thus a negative value represents net sequestration and positive values, net release. The West Virginia fly ash contained more alkalinity than the India fly ash and required five rather than three cycles to return to the pH of the AMD leachant. No elemental concentration in the TCLP tests exceeded USEPA limits per the Resource Conservation and Recovery Act of 1976 (40 CFR section 261). While the results of TCLP and MWLP are not directly comparable since MWLP uses AMD as the leachant, the later, nonetheless, indicated a trend toward increasing elemental release with successive leach cycles and with decreasing pH.

DISCUSSION

The MWLP differs from TCLP in two ways. First, TCLP, uses standard synthetic extraction fluids, titrated to various pH ranges with acetic acid, MWLP uses water from the intended application site. It is expected to provide a more accurate simulation of field conditions than TCLP and accounts for chemical interactions between ions released from the CCB and those in the mine water. Second, TCLP uses a single 18 hr leach cycle while MWLP continues leaching until all alkalinity is exhausted. In the case of many CCBs the TCLP stops while the pH is still strongly alkaline. While the intent of the CCB application may be to neutralize mine water acidity, the TCLP sheds no light on situations where re-acidification of the CCB mass is a possibility.

While MWLP is meant to simulate the likely chemical products resulting from exposure of a given CCB to a particular mine water, in its current configuration, it does not simulate reducing conditions. Additionally, MWLP simulates many years of weathering in a short period (roughly 32 pore water exchanges per leach cycle). It is important to remember that field concentrations of contaminants will be strongly influenced by the method of CCB placement, its volume, groundwater gradients and spoil quality. Therefore, the concentrations yielded by the test are not expected to estimate concentrations under field conditions.

It is clear that MWLP and TCLP yield very different results. For example, as pH and acidity change through the MWLP cycles, various elements appear in the leachate. In some instances, an element will appear for one or two cycles then drop below detection limits. This could indicate that an element is being sequestered or it could mean that its soluble fraction has been leached out of the fly ash. In either case, MWLP will highlight elements that may become mobile and the pH range in which it is likely to occur.

It is understood that the short 18 hour cycle time may not allow many intermediate mineral phases to come into equilibrium. Accounting for these phenomena and their significance over the long term are yet to be determined.

The MWLP is helpful in predicting the long term leaching behavior of wastes placed in acid environments. Many metals, such as Al, Cu, and Pb may not be leached from the waste until the leaching fluid becomes very acidic. In the case of highly alkaline wastes this may not happen until dozens of pore water exchanges have occurred,Comparisons of TCLP and MWLP leachates showed that in many cases different toxic element mobilities were indicated by MWLP and TCLP. The relationship of these results to field observations has yet to be made. While this project used strongly acidic mine waters and the most acidic TCLP leachant, it is important to note that many mine waters will be circumneutral and/or alkaline. They would, doubtless, yield very different results than those reported in this study.

The benefits of MWLP are most apparent when dealing with alkaline waste products, where a single leaching cycle will not exhaust the alkalinity in the system. In these cases several leaching cycles are necessary in order to understand the leaching behavior of these wastes as its alkalinity is exhausted and the leachate becomes acidic. This is not meant to imply that all CCB minefills will become acidic. Rather, it allows the user to evaluate leaching behavior through a range of pH conditions controlled by the pH of the leachant and the inherent alkalinity of the CCB.

CONCLUSIONS

The Mine Water Leaching Procedure (MWLP) was developed to determine the long-term leaching behavior of industrial wastes in groundwater found on the potential application site. The method sequentially leaches the fly ash with a sample of the site's groundwater until the alkalinity is exhausted and the pH of the leaching solution is reestablished at its pre treatment level.

Results of comparisons between TCLP and MWLP indicates that MWLP provides additional information for planners regarding the likelihood of toxic element release from CCB minefills and the pH ranges which must be maintained in order to minimize the risk of further groundwater contamination. Given these precautions, MWLP can be used to enhance the environmental performance of many industrial waste applications in mine reclamation and water treatment.

Import/export calculations showed the extent to which metals entering the leaching system via AMD were sequestered as solid phase precipitates. It appeared that precipitation and remobilization was a significant pathway for most metals.

A single MWLP cycle represents about 32 pore water exchanges. If one could estimate groundwater flux through the CCB mass, it would be possible to estimate the years required under field conditions to effect a single pore water replacement. This would allow placement of a temporal axis on the leachate data and further enhance the ability to predict risks associated with CCB placement in mine fills.

Elemental concentrations obtained via MWLP are unlikely to reflect actual field concentrations. They will also be influenced by the method of CCB placement, its hydraulic conductivity, the ability of the surrounding mine spoil to sequester toxic elements, adjacent ground water quality and gradients. Nevertheless, MWLP is expected to provide an important component of the overall risk assessment picture.

REFERENCES

1. Adriano, D.C., A.L. Page, A.A. Elseewi, A.C. Chang, and I Straughan. 1980. "Utilization and Disposal Of Fly Ash and Other Coal Residues in Terrestrial Ecosystems: A Review." *Journal of Environmental Quality.* 9:333-344.
2. Ainsworth, C.C., and D. Rai. 1987. *Chemical Characterization of Fossil Fuel Wastes*, Report EA-5321, Electric Power Research Institute, Palo, California, USA.
3. American Coal Ash Association. 1998. *Innovative Applications Of Coal Combustion Products (CCPs).* Alexandria, Virginia: American Coal Ash Association, Inc.
4. American Society for Testing and Materials (ASTM). 1988. *Annual Book of ASTM Standards.* Section 4: Construction, Vol. 04.01: Cement, Lime, Gypsum.
5. Ashby, James C. 2001. "Injecting Alkaline Lime Sludge and FGD Material Into Underground Mines For Acid Abatement." In *Proceedings of the 22nd West Virginia Surface Mine Drainage Task Force Symposium,* Morgantown, West Virginia, USA. April 3-4.

6. Bhumbla, D.K. 1991. "Ameliorative Effect of Fly Ashes." Ph.D. dissertation. West Virginia University, Morgantown, West Virginia, USA.

7. Bhumbla, D.K., R.N. Singh, and R.F. Keefer. 2000. "Coal Combustion By-Product Utilization For Land Reclamation." In *Reclamation Of Drastically Disturbed Lands*, edited by R.I. Barnhisel, R.G. Darmody, and W.I. Daniels. Madison, Wisconsin, USA: American Society of Agronomy, Inc., Crop Science Society of America, Inc., and Soil Science Society of America, Inc.

8. Butalia, Tarunjit S., and William E. Wolfe. 2000. "Market Opportunities For Utilization Of Ohio Flue Gas Desulfurization (FGD) and Other Coal Combustion Products (CCPs), Volume 1—Executive Summary." Available at http://ccohio.eng.ohio-state.edu/ccpohio/Marketing/Volume1.PDF

9. Capp, J.P. 1978. "Power Plant Fly Ash Utilization For Land Reclamation in Eastern United States." In *Reclamation Of Drastically Disturbed Lands*, edited by F.W. Schaller and P. Sutton. Madison, Wisconsin, USA: American Society of Agronomy, Inc.

10. Chang, A.L., L.J. Lund, A.L. Page, and J.E. Warneke. 1977. "Physical Properties Of Fly Ash Amended Soils." *Journal of Environmental Quality.* 6:267-270.

11. Dhaliwal, S.S., R.N. Singh, D.K. Bhumbla, P. Saini, and R.F. Keefer. 1995. "Effects of Weathering On Trace Metal Distribution In Fly Ash Amended Mine Soils." In *Proceedings of the 11th International Symposium On Use and Management Of Coal Combustion By-Prodcts (CCBs), Vol. 2*, Orlando, Florida. American Coal Ash Association, Washington, DC, and Electric Power Research Institute, Palo Alto, California, USA. 57-1 to 57-11.

12. Elseewi, A.A., I.R. Straughan, and A.L. Page. 1980. "Sequential Cropping Of Fly Ash Amended Soils: Effects On Soil Chemical Properties and Yield and Elemental Composition Of Plants." *The Science of the Total Environment.* 15:247-259.

13. EPRI, Allegheny Energy Supply and US Department of Energy. 2001. *Omega Mine Injection Program: Monongalia County, West Virginia.* #1004032.

14. Hamric, R. 1993. "Utilization of CFB Ash In Reclamation To Prevent Post-Mining AMD." In *Proceedings of 14th Annual West Virginia Surface Mine Drainage Task Force Symposium.* Morgantown, West Virginia, USA. April 7-8.

15. Hodgson, D.K., and R. Holliday. 1966. "The Agronomic Properties Of Pulverized Fuel Ash." *Chemistry and Industry.* 20:785-790.

16. Hodgson, D.K., and D.A. Brown. 1982. "Nuetralization and Dissolution Of High-Calcium Fly Ash." *Journal of Environmental Quality.* 11:93-98.

17. Jastrow, J.D., C.A. Zimmerman, A.J. Dvorak, and R.R. Hinchman. 1981. "Plant Growth and Trace Element Uptake On Acidic Coal Refuse Amended With Lime Or Fly Ash." *Journal of Environmental Quality.* 10:154-160.

18. Keefer, R.F., R.N. Singh, O.L. Bennett, and D.J. Horvath. 1983. "Chemical Composition Of Plants and Soils From Vegetated Mine Soils.*" In Proceeding of the 1983 Symposium for Surface Mine Hydrology, Sedimentology, and Reclamation.* University of Kentucky, Lexington, Kentcky, USA.

19. Martens, D.C. 1971. "Availability Of Plants Nutrients in Fly Ash." *Compost Science.* 12(6):15-18.

20. Muraka, I.P., S.V. Mattigod, and R.F. Keefer. 1993. "An Overview of Electric Power Institute (EPRI) Research Related To Effective Management of Coal Combustion Residues." In *Trace Elements in Coal and Combustion Residues*, edited by R.F. Keefer and K.S. Sajwan, Lewis Publishers, Boca Raton, Florida, USA.

21. Phung, H.T., L.J. Lund, and A.L. Page. 1978. "Potential Use Of Fly Ash As a Liming Material." In *Environmental Chemistry and Cycling Processes*, edited by D.C. Adrinao, and L. Brisbin. U.S. Department of Commerce, Springfield, Virginia, USA: Tech. Inform. Cent. Publ. CONF-760429.

22. Punshon, T., A.S. Knox, D.C. Adriano, J.C. Seaman, and J.T. Weber. 1999. "Flue Gas Desulfurization (FGD) Residue: Potential Applications and Environmental Issues." In *Biogeochemistry Of Trace Elements In Coal Combustion Byproducts,* edited by Sajwan et al. New York: Kluwer Academic/Plenum Publishers.

23. Rafalko, L., and P. Petzrick. 1999. "The Western Maryland Coal Combustion By-product/Acid Mine Drainage Initiative, The Winding Ridge Project." In *Proceeding: The 13th International Symposium on Use and Management of Coal Combustion Products Volume 3 (TR-111829-V3).* January 1999, Orlando, Florida paper 70: pp 70-1 to 70-16.

24. Roy, W.R., and R.A. Griffin. 1982. "A Proposed Classification System For Coal Fly Ash In Mulitdisciplinary Research." *Journal of Environmental Quality.* 11, 563.

25. Singh, R.N., D.K. Bhumbla, R.F. Keefer, and D.J. Horvath. 1992. "Improving Crop Production By Altering Chemical Properties Of Mineland With Industrial Wastes." In *Proceedings of the International Symposium on Nutrient Management*, edited by M.S. Bajwa and P.S. Sidhu. Punjab Agricultural University, Ludhiana, India. 366-380.

26. Tishmack, Jody K. 1996. "Bulk Chemical and Mineral Characteristics of Coal Combustion By-Products (CCB)." In *Proceedings of: Coal Combustion By-Products Associated With Coal Mining—Interactive Forum: Southern Illinois University at Carbondale*, October 29-31.

27. U.S. Environmental Protection Agency. 1992. Toxicity Characteristic Leaching Procedure. SW 846, Method 1311.
http://www.epa.gov/epaoswer/hazwaste/test/main.html

28. Yan, Jinying, Luis Moreno, and Ivars Neretnieks. 2000. "The Long-Term Acid Neutralizing Capacity Of Steel Slag." Department of Chemical Engineering and Technology. Royal Institute of Technology. SE-100 44 Stockholm, Sweden: Elsevier Science Ltd.

29. Ziemkiewicz, P.F. and Black, D.C. 2000. "Disposal and use of coal combustion byproducts in mined environments." Paper presented at the *2000 ICARD Conference*, May 21-24 2000, Denver CO, USA.

MOISTURE RETENTION AND HYDRAULIC CONDUCTIVITY OF COARSE-TEXTURED SOILS AMENDED WITH COAL COMBUSTION FLY ASH

J.C. Seaman*, S.A. Aburime**, B.P. Jackson*, and T. Punshon*

*Savannah River Ecology Laboratory, The University of Georgia, Drawer E, Aiken, SC 29802, USA
**Dept. of Engineering, Clark Atlanta University, 223 James P. Brawley Dr. Atlanta, GA 30314, USA

1. ABSTRACT

Previous studies have suggested that an increase in water holding capacity or matric potential (ψ) may result from the addition of coal combustion fly ash (FA) to coarse-textured soils, but common laboratory techniques for evaluating such characteristics can be time-consuming and difficult to replicate. Therefore, centrifuge-based methods were used to assess the matric potential (ψ) and hydraulic conductivity (K) as a function of the degree of saturation for a coarse-textured surface soil from the Southeastern US that was amended with acidic FA at application rates ranging from 0-15% (wt/wt). In repacked columns, a low ionic strength rainwater surrogate was used as the leaching solution. For comparison, similar amounts of standard clays and sand (i.e., kaolinite (KA); montmorillonite (MONT); and ottawa sand (OS)) were added to the test soil to demonstrate the sensitivity of the centrifuge-based methods. The water dispersible clay (WDC) content, an indicator of the susceptibility of the soil clay to dispersion, was also evaluated for the amended soils. A minor increase in matric potential was observed only at the highest FA application rates, while saturated K (K_{sat}) actually increased and then leveled off with increasing FA addition. In contrast, the matric potential and K for the other tested amendments was altered in the expected manner. KA and MONT decreased HC and increased matric potential at a given moisture content, while OS addition increased the soil HC and decreased the water holding capacity. Consistent respective trends were also evident in the particle size analyses of the amended soil. The seemingly inconsistent behavior observed for the FA amended columns may reflect changes in pore-water composition resulting from soluble FA components that increased the background ionic strength of the soil solution for the readily dispersive surface soil, as column effluents were generally less turbid with increasing FA addition. Changes observed in WDC for the various amendments support such a mechanism as the dispersible clay decreased and the ionic strength increased significantly for the FA amended soils.

2. INTRODUCTION

Previous studies have suggested that land disposal of coal combustion FA can improve the water holding capacity of coarse-textured soils and the drainage properties of fine textured soils because of the dominance of silt-sized particles.[1-5] In a recent example, Adriano and Webber[6] found that high application rates of fly ash (0, 280, 560, 1120 Mg h[-1]) improved the water holding capacity and increased plant available water for turfgrass plots without deleteriously impacting infiltration. However, Ishak et al.[7] observed a significant reduction in the K_{sat} of a coarse-textured soil with even moderate FA amendment levels (50 Mg h[-1]).

High FA application rates necessary to enhance soil water holding capacity, however, are likely to increase the potential for plant toxicity problems due to boron and environmental hazards associated with other trace elements (As, Ba, Mo, Pb, Se, Sr) that may pollute surface water or leach into groundwater.[8, 9] In spite of the potential hazards associated with its use on cropland, alkaline FA can be an effective liming agent and reduce the toxic effects of Al and Mn for acidic soils.[1, 9] Utilization of such materials as soil amendments is further complicated by their tremendous variability in chemical composition, in addition to the variability in the physical and chemical properties of soils to which they are applied, making it difficult to develop standardized management practices that ensure safe utilization.

Studies evaluating the environmental impact associated with the land application of FA often focus on the solubility, plant toxicity, and migration of ash-derived trace elements.[9-11] However, a better understanding of the influence of FA addition on the physical properties of amended soils is also critical to the development of standardized management practices. The lack of practical laboratory protocols for studying moisture retention and solute transport under variably saturated conditions has often hindered our understanding of such processes. Centrifuge-based techniques have gained recent acceptance as a means of rapidly testing the matric potential and hydraulic conductivity of porous media, both strong non-linear functions of soil water content.[12-15] One commercially available instrument, known as the Unsaturated Flow Apparatus (UFA; UFA Ventures Inc.) consists of a modified centrifuge and rotor system (Figure 1). The UFA enables the investigator to centrifuge repacked or intact soil columns while continuously adding a leaching solution via an infusion pump to the top of the columns in a controlled manner, and collecting the column effluent. When determining the matric potential of a sample, the infusion pumps are not used.

In determining the equivalent pressure head resulting from centrifugation:

$$P=\frac{\rho\omega^2}{2g}(r_1^2-r_0^2)$$

where P is the average pressure of the sample in cm of H_2O, g is the acceleration due to gravity (981 cm s[-2]), ρ is the density of the sample fluid (g cm[-3]), ω is the rotational rate (rad s[-1]), r_1 and r_2 are the radial distances from the point of rotation to the top and bottom of the sample (cm), respectively.[16] The matric potential for a material is then determined by first saturating the sample and then spinning it in the UFA at progressively faster rotation speeds, periodically stopping to measure the water content of the sample after it has reached steady state for a given centrifuge speed. Using the UFA, the ψ curve from 0.04 to 6 Bar can be generated for either two or four columns in approximately 10 days.

In a similar manner to ψ, the steady-state unsaturated hydraulic conductivity (K_{unsat}) can be determined using the following Darcy relationship:

$$q=-K(\psi)\left[d\psi/dr-\rho\omega^2 r\right]$$

where q is the fluid flux density (cm), K is the hydraulic conductivity, r is the distance from the rotation axis, ρ is the fluid density (gm cm[-3]), ω is the rotation speed

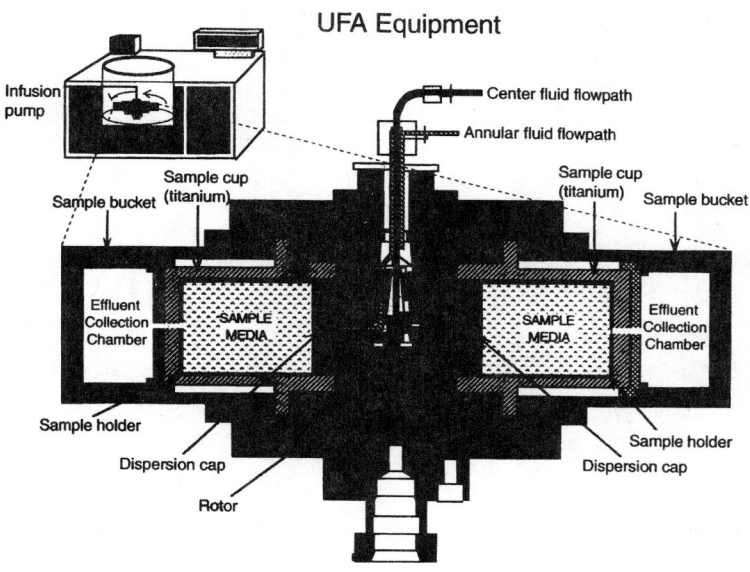

UFA Equipment

Figure 1. Schematic diagram of the Unsaturated Flow Apparatus (UFA Ventures, Inc.).

(radian/sec), $d\psi/dr$ is the matric potential gradient, and $\rho\omega\ r$ is the centrifugal force. At sufficient rotational speed and mass flux, the impact of matric potential is much less significant than the centrifugal force, i.e., $\ll \rho\omega\ r$.[17, 18] When rearranged the hydraulic conductivity (K) can be expressed as a function of matric potential (ψ) or in the present case as function of volumetric water content (θ):

$$K(\theta) = q / \rho\omega^2 r$$

In contrast to matric potential measurements, an infusion pump is used to continually introduce a leaching solution while the column is spinning. The centrifuge rotation speed is progressively increased, while the pumping rate is decreased in a controlled manner to control inlet flux over a range of decreasing water contents. As before, the column is periodically removed from the centrifuge rotor to empty the effluent collection cup and weigh the column to determine the water content at a given K or inlet mass flux.[16, 17] The K_{unsat} for two soil columns at 12 different moisture levels can be determined in approximately four days using the UFA. In addition to measuring the hydraulic properties of a given porous media, the UFA can also be used to conduct unsaturated solute transport experiments by introducing the compound of interest with the leaching solution once the desired degree of saturation has been achieved and then retaining the effluent for subsequent analysis.[16, 19]

The objective of the current study was to evaluate the impact of FA addition on the hydraulic properties (i.e., matric potential and hydraulic conductivity) of a coarse-textured soil using a modified centrifuge system, namely the UFA. Experimental conditions were chosen to mimic the field application of such material, including incorporation within the surface soil horizon and subsequent exposure to low ionic strength wetting solutions typical of precipitation.

3. MATERIALS AND METHODS

A coarse-textured surface soil (> 90% sand) collected from a mixed coniferous/deciduous forest (0-15 cm depth) on the Department of Energy's Savannah

Table 1. Select physical and chemical properties of the surface soil and coal combustion fly ash (FA) used in the current study.

Horizon	A_p	FA
Sand (%)*	91.0	
Silt (%)	3.49	
Clay (%)	5.51	
Organic Carbon (g 100 g^{-1})**	0.59	
pH-water***	5.48 ± 0.11	4.31 ± 0.07
pH-1.0 M CKCl***	3.98 ± 0.02	4.35 ± 0.05
EC (µS/cm)***	10.3 ± 1.5	3520 ± 170
Clay Mineralogy****	k, HIV, goe, gibb	
BET Surface Area*****		$m^2 g^{-1}$
Surface Soil		2.40
Fly Ash		1.76
Kaolinite		11.18
Montmorillonite		86.3

*Hydrometer Method[31]

**Dry Combustion Method[32]

***2/1 water:soil ratio for pH

****Based on XRD: k = kaolinite, HIV = hydroxy-interlayered vermiculite, goe = goethite, gibb = gibbsite

*****BET Surface Area[33]

River Site, located near Aiken, SC, and stored in a field moist state at 4 °C was used in the current study (Table 1). The clay mineralogy of the soil material consisted mainly of kaolinite, hydroxy-interlayered vermiculite (HIV), and gibbsite. The coal combustion FA used in the present study is likely a Class F, acidic ash, with limited nutrient value and liming capacity.[20] For comparison, clays such as kaolinite (kaolin, KGA-1A, 11.8 m^2 g^{-1}) and montmorillonite (montmorillonite, SAZ-1, 86.8 m^2 g^{-1}), and ottawa sand were also tested as amendments to demonstrate the utility of the UFA method. These treatments will be referred to as KA, MONT and OS, respectively. The appropriate amount (0, 5, 10, and 15% wt/wt) of FA or an alternate soil amendment was thoroughly mixed with the soil in a plastic bag before packing the UFA columns to a uniform bulk densities of 1.78 ± 0.02 gm cm^{-3} for the control and FA treatments, with bulk densities decreasing with increasing amendment levels for the clay treatments. The treatment levels correspond to 0, 50, and 100 Mg ha^{-1} assuming a 15 cm field incorporation depth. Repacked columns were deemed appropriate for the current study because of the necessity to mimic amendment incorporation after application in the field. Select properties of the treated soils are presented in Table 2.

Since the composition of the leaching solution can alter water retention and hydraulic conductivity[21, 22], a surrogate artificial rainwater solution (ARW) typical of precipitation in the southeastern US was used in the current study (Table 3).[23]   For

Table 2. Select physical and chemical properties of the soil/amendment mixtures evaluated in the current study.

Treatment	Clay	Silt	Sand	pH_{DIW}	pH_{KCl}	EC (µS/cm)
Control Soil	5.51	3.49	91.00	5.48±0.11	3.98±0.02	10.03±1.51
5% FA	5.19	6.40	88.41	5.11±0.06	4.39±0.08	279±25
10% FA	5.74	7.50	86.77	5.41±0.08	4.69±0.04	506±47
15% FA	6.13	9.80	84.07	5.56±0.04	4.95±0.01	650±95
5% Kaolinite	7.99	2.94	89.07	5.43±0.08	4.02±0.08	12.9±1.5
10% Kaolinite	11.39	3.68	84.93	5.22±0.01	4.00±0.06	15.1±3.4
5% Montmorillonite	7.34	4.02	88.64	5.90±0.11	4.10±0.07	13.8±0.1
5% Ottawa Sand	4.52	2.29	93.19	5.47±0.15	4.08±0.01	10.3±1.5

Table 3. Composition of the Artificial Rainwater (ARW) based on precipitation samples collected on the Savannah River Site located near Aiken, SC.[23]

Analyte	Concentration (mg L^{-1})
Ca	0.082
K	0.042
Mg	0.033
Na	0.199
NH$_4$	0.139
Cl	0.351
NO$_3$	0.711
SO$_4$	1.29

comparison, 0.001 M CaSO$_4$ was also used as a leaching solution for measuring K$_{unsat}$ in an effort to reduce possible clay dispersion and swelling that may occur under low ionic strength conditions.

3.1. Matric Potential

Field moist soil was packed into a UFA column and then saturated with the ARW. Once saturated, the columns were placed in the centrifuge and spun at progressively higher rotation rates and durations to extract water under different equivalent tensions. At predetermined time intervals for a give rotation speed, the rotor was removed from the centrifuge and the columns were weighed to determine the amount of soil moisture that had been extracted. At the end of the final centrifuge episode, the soil was removed from the UFA and the final water content was determined gravimetrically after heating overnight at 60°C. The calculated water tension in bars was then plotted against the water content for the soil to produce the characteristic matric potential curve (see Equation 1).

3.2. Saturated Hydraulic Conductivity

The K$_{sat}$ of the amended soils was determined using ARW as the leaching solutions for three replicates of each soil amendment treatment using a constant-head permeameter in an upflow configuration.[16, 21] An analysis of variance (ANOVA) test was performed to confirm treatment differences, and then a Tukey's multiple comparison test was conducted to define the least significant difference between sample means at the $\alpha =$ 0.05 significance level.

3.3. Unsaturated Hydraulic Conductivity

The K$_{unsat}$ of the amended soil was determined in the UFA using either ARW or 0.001 M CaSO$_4$ as the leaching solution.[16, 17] After a given column experiment was completed, the soil materials were removed from the UFA column, and the final water content was determined gravimetrically after heating a portion of the soil to 60 °C.

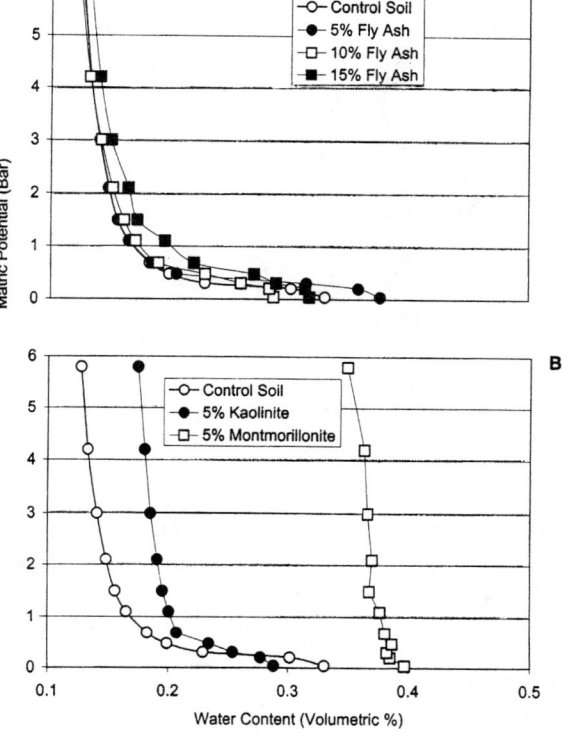

Figure 2. Matric potential as determined by the UFA method for coarse-textured soil amended for FA (A) and other soil materials (B).

3.4. Water Dispersible Clay

A batch study was conducted to determine the impact of FA and the mineral standard treatments on the relative dispersibility of the clay fraction from the study soil. Treatments were created by mixing the field-moist soil with FA or the standard clays at application rates of 0, 50, 100 and 150 g kg^{-1} (0, 5, 10, and 15% wt/wt). Triplicate five-gram samples of soil treatments were then weighed into centrifuge tubes and shaken 4 hrs with 35 mL of deionized water (DIW). After shaking, the water-dispersible clay (WDC) was measured using a modified micro-pipette method, and the pH and electrical conductivity of the remaining suspensions were also determined.[24, 25] As above, an ANOVA test was performed to confirm treatment differences, and then a Tukey's multiple comparison test was conducted to define the least significant difference between sample means at the $\alpha = 0.05$ significance level.

4. RESULTS AND DISCUSSION

4.1. Matric Potential

The soil matric potential increased only moderately for the FA-amended soils, even at the 15% application rate (Figure 2A). The marginal increase in water holding capacity led us to test additional materials that would alter the soil texture in a predicted manner to demonstrate the sensitivity of the UFA method. The addition of either KA or

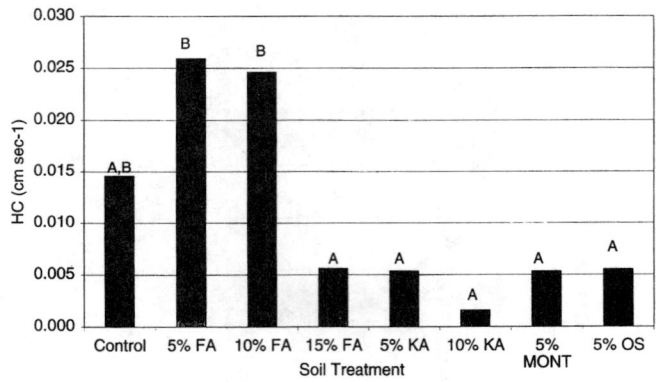

Figure 3. K_{sat} for the coarse textured soil amended with FA and other soil materials. Treatment means with the same letter are not significantly different at $\alpha = 0.05$.

MONT enhanced the water holding capacity of the soil, especially for the MONT, even though the clays were not size fractionated prior to addition (Figure 2B). Changes in soil texture were also evident in the particle size data for the amendment treatments, with only a slight increase in the clay and silt content observed with increasing FA addition (Table 2). Adding 5% OS to the column resulted in only a slight decrease in water holding capacity when compared to the control (data not shown).

4.2. Saturated Hydraulic Conductivity

The K_{sat} increased slightly for the two lowest FA amendment levels when compared to the other amendment treatments, KA, MONT, OS (Figure 3). In some respects, this seems inconsistent with even a minor increase in water holding capacity. However, the column effluents for the control soil were turbid, indicative of the inherent dispersibility of the clay fraction. The clay fraction in surface soil horizons from the Southeastern US tends to be dispersive due to the low ionic strength of the pore solution, the moderate intrinsic sodium adsorption ratios (SAR), and organic coatings on clay size minerals which increase net-negative surface charge.[26-29] The electric conductivity (EC) increased and the turbidity decreased for the column effluents with increasing FA addition (data not shown). This suggests that solubilized salts from the FA may help flocculate the soil clays and reduce column plugging with ARW leaching. With increasing FA addition, however, it appears that the increase in pore solution ionic strength may be offset by the alteration in soil texture, resulting in a lower hydraulic conductivity, even with less clay dispersion.

4.3. Unsaturated Hydraulic Conductivity

The K_{unsat} results were quite variable and highly dependent on the saturating solution. In general, the position of the $K(\theta)$ curve is controlled by the mean pore size, while the shape is more a function of the pore-size distribution.[30] A lower column water content was observed for the $CaSO_4$ leaching solution when compared to the ARW at a given hydraulic conductivity, i.e. inlet flux and rotation speed, indicative of an increase in the mean pore size of the material (Figure 4A). This was presumably due to the greater soil flocculation when exposed to the higher ionic strength solution as the ARW effluents were quite turbid for all amendment treatments, even the FA. When $CaSO_4$ was used as the leaching solution the effluents were clear (NTU < 1) and there was little difference in the K_{unsat} curves for the control soil and the FA amendments, regardless of application rate (Figure 4B). An increase in back pressure that disabled the infusion pump was often

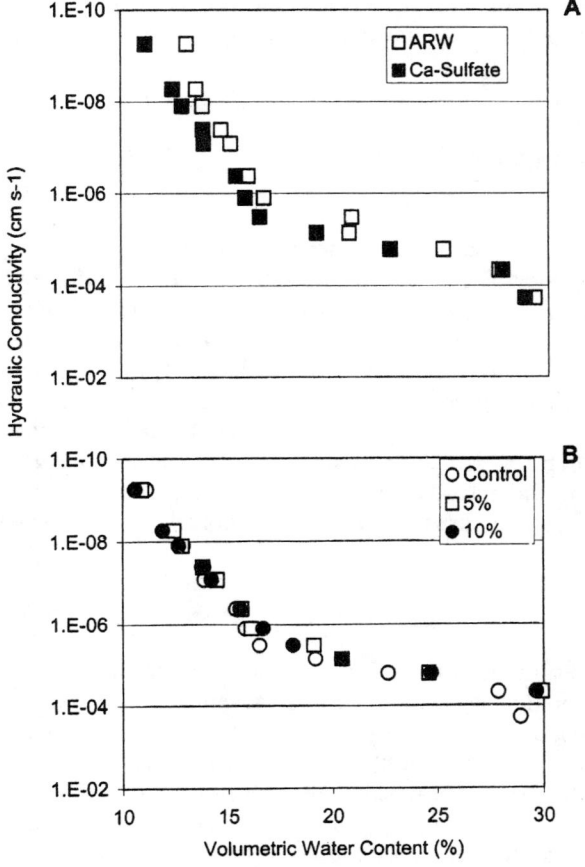

Figure 4. K$_{unsat}$ for the coarse-textured soil leached with either ARW or 0.001 M CaSO$_4$ (A), and the soil amended with FA and leached with 0.001 M CaSO$_4$ (B).

Table 4. Water dispersible clay and other suspension characteristics for soil amended with FA and other test amendments. Treatment means with the same letter are not significantly different at $\alpha = 0.05$.

Treatment	WDC	EC	pH
	(g 100 gm^{-1} soil)	(μS cm^{-1})	
Control Soil	1.45 (A,B)	21.6 (A)	5.46 (A)
5% FA	0.38 (C)	175 (B)	5.31 (B)
10% FA	0.40 (C)	257 (C)	5.64 (C)
5% Kaolinite	3.97 (D)	12.6 (A)	5.55 (A,C)
10% Kaolinite	3.32 (A)	14.3 (A)	5.55 (A,C)
5% Montmorillonite	1.39 (A,B)	12.1 (A)	6.04 (D)
5% Ottawa Sand	1.09 (A)	17.5 (A)	5.59 (C)

observed for the clay amendments when leaching with ARW, especially the MONT treatments, indicative of severe column plugging that precluded completion of the experiment.

4.4 Water Dispersible Clay

Batch clay dispersion experiments were conducted to explain the impact of the tested amendments on clay flocculation/dispersion and subsequent changes in the soil matric potential and hydraulic conductivity. FA addition significantly reduced clay dispersion as indicted by WDC when compared to other treatments (Table 4), presumably resulting from the release of soluble salts that reduced colloidal stability as indicated by the higher electrical conductivity of the FA amended suspension. Apparently the higher ionic strength was sufficient to overcome any alterations in texture, except for the highest FA amendment level, 15% FA (see Figure 2 and Table 2). Such an increase in soluble salts was not apparent in the particle size data (Table 2) because of the greater dilution inherent in the hydrometer method compared to the WDC procedure (i.e., 20/1 vs. 7/1) and the utilization of a strong dispersing agent, sodium-hexametaphosphate.

5. CONCLUSIONS

The current results suggest that FA addition was marginally effective at increasing the water holding capacity, i.e., matric potential, of the coarse textured test soil. In addition, the lowest FA amendment rates (5 and 10%) actually increased the K_{sat} of the soil, through the addition of soluble salts that reduced clay dispersion when compared to the control or clay treatments. Apparently the increase in soluble salts was offset at the highest treatment level, 15% FA. Such a mechanism is further supported by the fact that little difference in K_{unsat} was observed for the FA treatments in comparison with the control soil when a high ionic strength 0.001 M $CaSO_4$ was used as the leaching solution. The initial pore water composition may have a lesser impact on matric potential determination when compared to hydraulic conductivity experiments because the limited ability to alter soil chemistry with such a small volume of pore solution. These results have obvious implications to the common practice of mixing flue-gas desulfurization gypsum (i.e., $CaSO_4$) with FA before land disposal.

Additional treatments such as KA, MONT, and OS were used to demonstrate the utility of the UFA methods in a predictable manner. In contrast to FA, the clay amendments clearly increased the matric potential, decreased saturated conductivity, and often resulted in severe column plugging that precluded measuring the unsaturated hydraulic conductivity using ARW.

The current study clearly demonstrates the utility of centrifuge-based techniques, and more specifically the UFA, in evaluating the hydraulic properties of porous media; however, the results observed for the FA amendments must be viewed with caution because of the tremendous variability in coal combustion fly ash and the soils to which they are applied. Companion studies are underway to test the influence of FA aging/equilibration before and after soil application on the resulting soil physical properties (ψ and $K(\theta)$), as well as the impact of the initial pore solution. In addition, similar studies are underway to test the influence of a range of coal combustion materials that are typical of the inherent variability encountered for FA materials. Also, field experiments are underway to correlate the UFA-based matric potential results to that of the undisturbed soil at the sampling site using soil tensiometers.

ACKNOWLEDGEMENTS

The authors would like to acknowledge the laboratory assistance of J. Logan and J. McIntosh. This research was supported by Financial Assistance Award Number DE-FC09-96SR18546 from the DOE to the University of Georgia Research Foundation.

REFERENCES

1. Adriano, D. C., Page, A. L., Elseewi, A. A., Chang, A. C. and Straughan, I., Utilization and disposal of fly ash and other coal residues in terrestrial ecosystems: A review, *J. Environ. Qual.*, 11, 563, 1980.

2. Chang, A. C., Lund, L. J., Page, A. L. and Warneke, J. E., Physical properties of fly ash-amended soils, *J. Environ. Qual.*, 6, 267, 1977.

3. Jacobs, L. W., Erickson, A. E., Berti, W. R. and MacKellar, B. M., Improving crop yield potentials of coarse textured soils with coarse fly ash amendments, *Proc. Ninth International Ash Use Symposium*, 3, 59, 1991.

4. Ghodrati, M., Sims, J. T. and Vasilas, B. L., Evaluation of fly ash as a soil amendment for the atlantic coastal plain: I. Soil Hydraulic properties and elemental leaching, *Water, Air, and Soil Pollution*, 81, 349, 1995.

5. Punshon, T., Adriano, D. and Weber, J., Restoration of eroded land using coal fly ash and biosolids, TR-113940, EPRI, 1999.

6. Adriano, D. C. and Weber, J. T., Influence of Fly Ash on Soil Physical Properties and Turfgrass Establishment, *J. Environ. Qual.*, 30, 596, 2001.

7. Ishak, C. F., Seaman, J. C., Miller, W. P. and Sumner, M., Contaminant mobility in soil amended with fly ash and flue-gas desulfurization gypsum: Intact soil cores and repacked columns, *Water, Air, and Soil Pollution*, In Press, 2002.

8. Carlson, C. C. and Adriano, D. C., Environmental impacts of coal combustion residues, *J. Envorn. Qual.*, 22, 227, 1993.

9. Keefer, R. F., Chapter 1: Coal ashes- Industrial wastes or beneficial byproducts?, in *Trace elements in coal and coal combustion residues*, R. F. Keefer and K. S. Sajwan, Eds., Lewis Publishers, Ann Arbor, 1993, 3.

10. Kukier, U., M.E. Sumner, and W.P. Miller, Boron release from fly ash and its uptake by corn, *J. Environ. Qual.*, 23, 596, 1994.

11. Sims, J. T., B.L. Vasilas, and M. Ghodrati, Evaluation of Fly Ash as a soil amendment for the Atlantic Coastal Plain: II. Soil chemical properties and plant growth, *Water, Air, and Soil Pollution*, 81, 363, 1995.

12. Nimmo, J. R., Stonestrom, D. A. and Akstin, K. C., The feasibility of recharge rate determinations using the steady-state centrifuge method, *Soil Sci. Soc. Am. J.*, 58, 49, 1994.

13. Nimmo, J. R., Akstin, K. C. and Mello, K. A., Improved apparatus for measuring hydraulic conductivity at low water content, *Soil Sci. Soc. Am. J.,* 56, 1758, 1992.

14. Wright, J., Conca, J. L. and Chen, X., Hydrostatigraphy and recharge distributions from direct measurements of hydraulic conductivity using the UFA method, PNL-9424, Pacific Northwest Laboratory, 1994.

15. Khaleel, R., Relyea, J. F. and Conca, J. L., Evaluation of van Genuchten-Maulem relationships to estimate unsaturated hydraulic conductivity at low water contents, *Water Resour. Res.,* 31, 2659, 1995.

16. UFA Method Technical Procedures, UFA Ventures, Inc. Richland, WA 99352.

17. ASTM, Test method for determining unsaturated and saturated hydraulic conductivity in porous media by steady-state centrifugation, D18.21, 2000.

18. Nimmo, J. R., Rubin, J. and Hammermeister, D. P., Unsaturated flow in a centrifugal field: Measurement of hydraulic conductivity and testing of Darcy's Law, *Water Resour. Res.,* 23, 124, 1987.

19. Gamerdinger, A. P. and Kaplan, D. I., Application of a continuous-flow centrifugation method for solute transport in disturbed, unsaturated sediments and illustration of mobile-immobile water, *Water Resour. Res.,* 36, 1747, 2000.

20. Mattigod, S. V., Rai, D., Eary, L. E. and Ainsworth, C. C., Geochemical factors controlling the mobilization of inorganic constituents from fossil fuel combustion residues: I. Review of major elements, *J. Environ. Qual.,* 19, 188, 1990.

21. Klute, A. and Dirksen, C., Chapter 28: Hydraulic conductivity and diffusivity: Laboratory methods, in *Methods of Soil Analysis, Part 1. Physical and Mineralogical Methods,* A. Klute, Ed., American Society of Agronomy, Madison, WI, 1986, 687.

22. Klute, A., Water retention: Laboratory methods, in *Methods of Soil Analysis, Part 1. Physical and Mineralogical Methods-Agronomy Monograph no. 9,* A. Klute, Ed., American Society of Agronomy-Soil Science Socity of America, Madison, WI, 1986, 635.

23. Strom, R. N. and Kaback, D. S., *SRP Baseline Hydrogeologic Investigation: Aquifer Characterization Groundwater Geochemistry of the Savannah River Site and Vicinity (U),* Westinghouse Savannah River Company, Environmental Sciences Section, 98, 1992.

24. Burt, R., Reinsch, T. G. and Miller, W. P., A micro-pipette method for water dispersible clay, *Commun. Soil Plant Anal.,* 24, 2531, 1993.

25. Miller, W. P. and Miller, D. M., A micro-pipette method for soil mechanical analysis, *Commun. Soil Plant Anal.,* 18, 1, 1987.

26.   Kaplan, D. I., Bertsch, P. M., Adriano, D. C. and Miller, W. P., Soil-borne colloids as influenced by water flow and organic carbon, *Environ. Sci. Technol.,* 27, 1193, 1993.

27.   Kretzschmar, R., Robarge, W. P. and Weeds, S. B., Flocculation of kaolinitic soil clays: Effect of humic substances and iron oxides, *Soil Sci. Soc. Am. J.,* 57, 1277, 1993.

28.   Miller, W. P. and Radcliffe, D. E., Soil crusting in the southeastern US, *Soil crusting: Chemical and physical processes,* 233, 1992.

29.   Kaplan, D. I., Sumner, M. E., Bertsch, P. M. and Adriano, D. C., Chemical conditions conducive to the release of mobile colloids from ultisol profiles, *Soil Sci. Soc. Am. J.,* 60, 269, 1996.

30.   Conca, J. L. and Wright, J., The UFA method for characterization of vadose zone behavior, in *Vadose Zone: Science and Technology Solutions,* on CD, B. B. Looney, B. B. and Falta, R. W., Eds., Battelle Press, Columbus, 2000,

31.   Gee, G. W. and Bauder, J. W., Chapter 15: Particle Size Analysis, in *Methods of Soil Analysis, Part 1: Physical and Mineralogical Methods, 2nd Edition,* 9, Ed., A. Klute, American Society of Agronomy, Madison, WI, 1986, 383.

32.   Nelson, D. W. and Sommers, L. E., Total carbon, organic carbon, organic matter, in *Methods of Soil Analysis,* 2,Eds., Page, A. L., Miller, R. H. and Keeney, D. R., Eds., American Society of Agronomy, Madison, WI, 1982, .

33.   Carter, D. L., Mortland, M. M. and Kemper, W. D., Specific Surface, in *Methods of Soil Analysis, Part 1. Physical and Mineralogical Methods,* A. Klute, Ed., American Society of Agronomy, Madison, WI, 1986, 413.

GEOCHEMISTRY OF AN ABANDONED LANDFILL CONTAINING COAL COMBUSTION WASTE: IMPLICATIONS FOR REMEDIATION

Christopher Barton[1], Linda Paddock[2], Christopher Romanek[2,3], and John Seaman[2]

[1]USDA Forest Service, Center for Forested Wetlands Research, c/o Savannah River Ecology Laboratory, Drawer E, Aiken, SC 29802

[2]University of Georgia, Savannah River Ecology Laboratory, Drawer E, Aiken SC 29802

[3]Department of Geology, University of Georgia, Athens, GA 30602

1. ABSTRACT

The 488-D Ash Basin (488-DAB) is an unlined, earthen landfill containing approximately one million tons of dry ash and coal reject material at the U.S. Department of Energy's Savannah River Site, SC. The pyritic nature of the coal rejects has resulted in the formation of acidic drainage (AD), which has contributed to groundwater deterioration and threatened biota in adjacent wetlands. Establishment of a dry cover is being examined as a remedial alternative for reducing AD generation within this system by minimizing the contact of oxygen and water to the waste material. To determine the potential benefit of a cover on pore water chemistry, a series of flow-through column experiments were performed under varying environmental conditions using materials from the site. The experiment was designed to demonstrate the influence of temperature, gaseous composition (dissolved nitrogen vs. oxygen), and flow regime (continuous flow vs. episodic wetting/drying) on effluent chemistry. Results indicated that the fluid composition (e.g., pH, redox, elemental composition) was closely associated to dissolved and/or gaseous oxygen content and wetting regime. Given these conditions, the use of a dry cover could reduce the production of acid lechate over time, pending that it retards or eliminates fluid and oxygen transport to the subsurface.

Chemistry of Trace Elements in Fly Ash, edited by Sajwan *et al.*
Kluwer Academic/Plenum Publishers, 2003

2. INTRODUCTION

The degradation of water resources from coal mining and coal combustion activities is a problem of global significance. Acidic drainage (AD), a low pH water enriched in iron, aluminum, sulfate and trace elements (e.g., lead, selenium, arsenic, mercury, zinc), is formed upon exposure of pyrite to the oxidizing forces of air and water. Pyrite (FeS_2), the most common sulfide mineral on Earth, is often found within coal seams and their associated geologic strata. Once disturbed or extracted during the mining process, pyrite oxidation and AD generation may begin and can continue for thousands of years. Moreover, since pyrite is found within coal seams, AD is not solely a "mining" problem and may occur anywhere that pyrite enriched coal or its byproducts are stored. In the United States alone, approximately 20,000 km of streams and over 72,000 ha of lakes and reservoirs are impacted by AD[1].

The oxidation of sulfide minerals and formation of AD is a complex process involving hydrolysis, redox, and microbial reactions[2]. The general stoichiometry can be described by the following reactions:

$$2FeS_{2(s)} + 7O_2 + 2H_2O \rightarrow 2Fe^{2+} + 4SO_4^{2-} + 4H^+ \qquad [1]$$

$$Fe^{2+} + 0.25O_2 + H^+ \rightarrow Fe^{3+} + 0.5H_2O \qquad [2]$$

$$Fe^{3+} + 3H_2O \rightarrow Fe(OH)_{3(s)} + 3H^+ \qquad [3]$$

$$FeS_{2(s)} + 14Fe^{3+} + 8H_2O \rightarrow 15Fe^{2+} + 2SO_4^{2-} + 16H^+ \qquad [4]$$

where iron sulfide and other mixed-metal sulfides decompose upon exposure to the atmosphere, producing ferrous iron, sulfate and proton acidity [1]. The partial oxidation of ferrous to ferric iron consumes some protons [2]. However, ferric iron may act as an electron acceptor and contribute to additional acid production through hydrolysis [3] and/or further pyrite oxidation [4]. Subsequently, acids produced from the oxidation may dissolve minerals and mobilize metals from materials found in the surrounding environment (coal, ash, soil etc.).

The oxidation of ferrous to ferric iron was determined by Singer and Stumm[3] to be the rate-limiting step in the generation of AD. Further, these researchers showed that the oxidation of Fe^{2+} is pH dependent and extremely slow at pH ≈ 3.0. However, the bacteria *Thiobacillus thiooxidans*, a sulfur-oxidizing bacteria, and *Thiobacillus ferrooxidans*, a chemosynthetic iron-oxidizing bacteria, were found to act as catalysts and accelerate the oxidation of ferrous sulfide to ferric sulfate at pH levels below 4.0^4. Singer and Stumm[3] demonstrated that the rate of Fe^{2+} oxidation in an untreated mine sample was 10^6 times greater than that observed in a sterilized sample. As such, the rate of pyrite oxidation may be influenced not only by redox conditions, but also by environmental conditions that may affect the microbial communities (i.e., temperature and pH).

Several innovative techniques have been proposed to reduce AD generation by limiting the exposure of pyrite to air and water. Inundation is a basic application of this concept. The diffusion coefficient of dissolved oxygen in water is approximately 10,000 times lower than that in the gas phase. As such, oxygen-consuming reactions are greatly reduced upon flooding of pyrite-bearing materials and bacterially enhanced pyrite oxidation is diminished. However, equation [4] (above) indicates that oxidation of pyrite may continue as long as ferric iron is present. Hence, long-term stagnation or bacterially mediated hypoxia must also occur to maintain a reducing environment and hold iron in the ferrous state.

A dry cover is an innovative technology that utilizes a layer of soil, compost or other organic material (peat, hay, straw, sawdust) above mining waste to deplete oxygen through bacterial consumption[5]. The organic waste may also inhibit oxidation by removal of Fe^{3+} from solution through complexation, and the formation of pyrite-Fe^{2+}-humate complexes[6]. Changes in the particle size distribution of the cover over that of the mining waste may also be engineered to enhance water storativity. As such, water percolation through the waste and exposure to the pyrite could be decreased. Under such a scenario, the use of vegetation may also be included to aid in the removal of water through

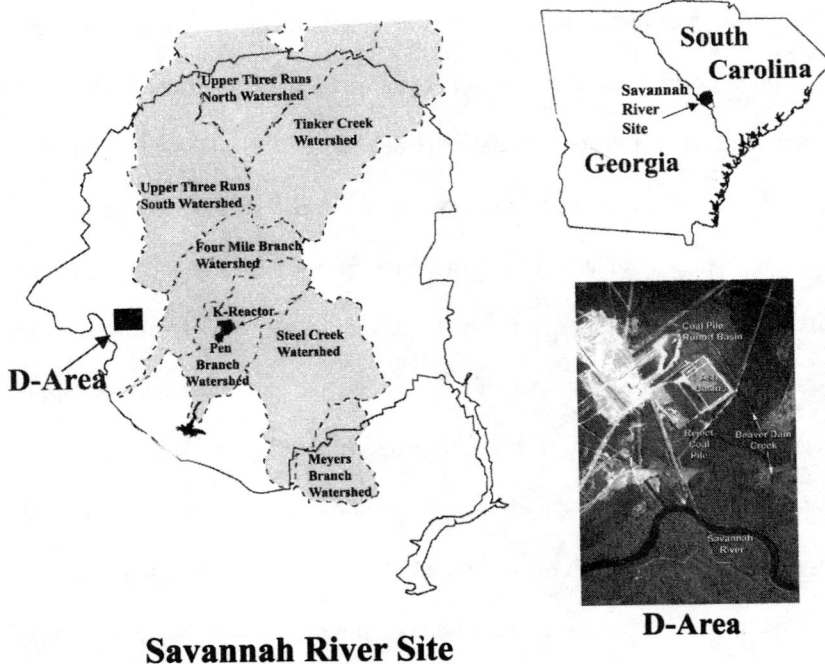

Savannah River Site

D-Area

Figure 1. D-Area power plant and vicinity on the Savannah River Site, SC.

evapotranspiration[7]. On a large scale, the net effect of these techniques is to cut off the

oxygen source. On a microscopic scale, these techniques may also aid in preventing

oxidation by altering the surface chemistry of pyrite. Several researchers have shown that

soluble organic acids from the breakdown of litter, and colloidal silicate and phosphate

salts from soils may act as passivating agents through the formation of a surface coating,

which may render pyrite impenetrable to oxidative attack[6,8,9].

Establishment of a dry cover is being examined as a remedial alternative for

reducing AD generation within the 488-D Ash Basin (488-DAB) at the U.S. Department

of Energy's Savannah River Site, SC. In order to determine the potential benefits of a

cover, a series of column experiments was conducted to determine the effect of

temperature and wetting regime on surficial materials collected from the 488-DAB. Data

from these experiments are used to predict the chemistry of lechate anticipated under

various field conditions.

3. MATERIALS AND METHODS

3.1. Site Description and Background

The 488-DAB is an unlined, earthen containment basin located on the Savannah River Site, SC that received sluiced fly ash, dry fly ash and coal reject material from the early 1950's to the mid 1990's (Figure 1). Non hazardous wastes deposited in the basin contain metals typical of fly ash and coal (As, Co, Cr, Cu, Fe, Mn, Ni and Zn). The 488-DAB is ~1,800' x 600' x 18' in size and contains ~19 x 10^6 ft^3 of waste material. The basin was constructed on the existing land surface at ~35m (msl) and is ~9m above the Savannah River. The present surface of the basin is at ~40 to 37' (msl), sloping gently to the west and it is filled with waste, except in the far western reach where surface waters collect during rainfall events.

The extent of water-saturated material in 488-DAB is not known, nor if any communication exists between basin waste and the local water table. Based on limited soil borings, the waste seems to be variably saturated, having both wet and dry zones that vary with depth. The source of this water may be meteoric infiltration or lateral groundwater flow from adjacent basins (488-1D, -2D and -4D). Hydraulic anisotropies may have been created when various materials were introduced to the basin. Alternatively, post-depositional features such as diagenetic "hard pans" may have developed as materials weathered over time, creating local anisotropies

3.2. Column Construction

A 2 x 1 x 1 m deep trench was excavated in the center of 488-DAB to retrieve material for the column study. Six horizons were noted in a cross-section of the trench; of these, a coal rubble zone (A) and fly ash residue zone (D) best represented the character of the waste and were chosen for additional study. In the field, zone A consisted of poorly-sorted, pebble-sized pieces of coal and pyrite, while zone D contained a relatively uniform fine-grained material. Approximately 1 m^3 of material was shoveled from zone A and D into individual plastic bags and subsequently placed inside large plastic containers to minimize atmospheric interaction. The containers were transported to the laboratory where subsamples were collected for thin sectioning. The remaining material was disaggregated by hand and stored for later use.

Flow-through columns were made using clear PVC pipe of approximately 15 cm in diameter and 25 cm in length (~4500 cm^3). Attached to the ends of each cylinder were PVC plates that contained a circular groove, to ensure a snug fit, and ports for influent or effluent flow. The base plate was permanently fixed with epoxy, while the top plate was removable and contained a compression o-ring for an airtight seal. Each column was connected to fluid and gas reservoirs with tubing. Columns were flushed with fluid from the bottom up to minimize air-filled void space, and from top to bottom with gas to facilitate the expulsion of liquid and accelerate the drying process. Preliminary experiments revealed that the waste material could be dried to ambient water saturation (~12% by wt.) within a day at a gas flushing rate of 1 L per minute.

3.3. Flow-through Experiment Procedures

Waste material from the selected horizons was packed in columns. Nylon screens (600-mesh) were placed on each end of a column to prevent mass loss. After loading, the lid was securely fastened to the column base plate by connection with metal rods and bolts. Gas-equilibrated water was pumped independently into the columns from a common reservoir at 2.5 mL min^{-1} with Barnant metering pumps. The flow rate was set to be similar to measured infiltration rates for waste at 488-DAB (Smail, pers. comm.). Depending upon the treatment, gas-equilibration of the influent was achieved by continuous sparging of the inlet reservoir with either breathing quality oxygen or zero grade nitrogen gas. Ultra purification of the nitrogen was achieved by passing the gas over hot copper in a reduction furnace prior its introduction in an experiment. Non-potable ground water pumped from a local artesian aquifer was stored in fifty-liter carboys and pre-equilibrated to the appropriate temperature and gaseous conditions before being pumped into the columns. Prior to an experimental run, each column setup was examined for leakage, settling, and mass loss by flushing gas-equilibrated fluid through the system.

Four experimental runs were conducted sequentially, each run consisted of four independent columns that were leached for approximately 35 days. In each run, two columns contained zone A material (columns 1 and 2) and two columns contained zone D material (columns 3 and 4). In each run, columns 1 and 3 were flushed continually with fluid while columns 2 and 4 experienced a periodic wetting (5 days) and drying (2 days) cycle flow regime. Based on preliminary tests, gas flow rate was set at 1 L per minute. Flow rates were monitored and adjusted periodically during each experiment. Column experiments were conducted at two different temperatures, 12.7° and 29.4°C, to simulate winter and summer soil temperatures, respectively. Maintenance of the experimental conditions was achieved by housing the entire column set-up, including gas and liquid reservoirs, in a large environmental chamber.

The treatment sequence for the experimental runs was: 1) O_2-equilibrated fluid at 12.7° C (Run 1), 2) N_2-equilibrated fluid at 12.7°C (Run 2), 3) N_2-equilibrated fluid at 29.4°C (Run 3), and 4) O_2-equilibrated fluid at 29.4°C (Run 4). When a column experiment was initiated, effluent was collected every 4 hours for the first 48 hours, and daily thereafter. During each sampling event, effluent pH, Eh, temperature, and EC were recorded, measurements of total Fe and Fe^{2+} were performed (HACH test kit), and three samples of fluid were collected for laboratory analysis. Twenty milliliters of effluent was filtered (0.45 micron acetate syringe filter) into a glass scintillation vial preloaded with 100 microliters of trace metal grade nitric acid for dissolved metal analysis, 60 mL of unfiltered effluent was stored in an amber glass vial for measurement of acidity, and approximately 20 mL of unfiltered effluent was collected for sulfate (SO_4) analysis.

3.4. Analytical Methods

Measurements of pH, temperature, and EC were made in the environmental chamber using a HI 991301 Hanna pH meter and probe. Eh was measured separately using an ORP probe and meter. Calibration checks were performed on all probes once per week for quality control.

Acidity titration was performed using an ABU901 autoburette and TimTalk 9 computer software. The ICP-OES analyses were performed on a Perkin-Elmer Optima 4500DV Optical Emission Spectrometer. Total Fe, Fe^{2+} and sulfate analysis were conducted using HACH test kits and a DR/890 Colorimeter. Samples for iron and sulfate determination were filtered with 0.1 micron and 0.22 micron syringe filters, respectively, prior to analysis. Geochemical modeling of aqueous-phase chemical equilibria was performed with the MINTEQA4 computer program[10]. Measured pH and Eh values were used as model inputs in the computer simulation.

Thin sections were made by Mineral Optics Laboratory (Wilder, Vermont) to determine the texture, fabric and mineralogy of zone A and D materials prior to the experiments. Percent silt and clay were determined on samples of zone A and D starting material using the hydrometer method of Gee and Bauder[11]. The XRD analyses were conducted on powder mounts of zone A and D materials prior to and at the conclusion of each experiment using a XRD diffractometer (X2 Advanced Diffraction System, Scintag Inc.) with Co $K\alpha$ radiation. A TGA 2950 thermogravimetric analyzer (TA Instruments, New Castle, DE) was used for TG and DTG characterizations.

4. RESULTS AND DISCUSSION

4.1. X-Ray Diffraction and Thermal Analysis

Surficial waste material at the 488-DAB experienced fluctuating saturated/unsaturated conditions that promoted the incongruent precipitation/dissolution of minerals. During dry periods of the summer, yellow efflorescences erupting from below the surface were observed and were a common superficial feature on the basin. In addition, white globules consisting of a hard inner shell and soft outer coating with the consistency of powder were abundant in the upper 1 cm of the surface horizon. Diffraction patterns of the yellow efflorescences showed sharp peaks at 8.25, 5.45 and

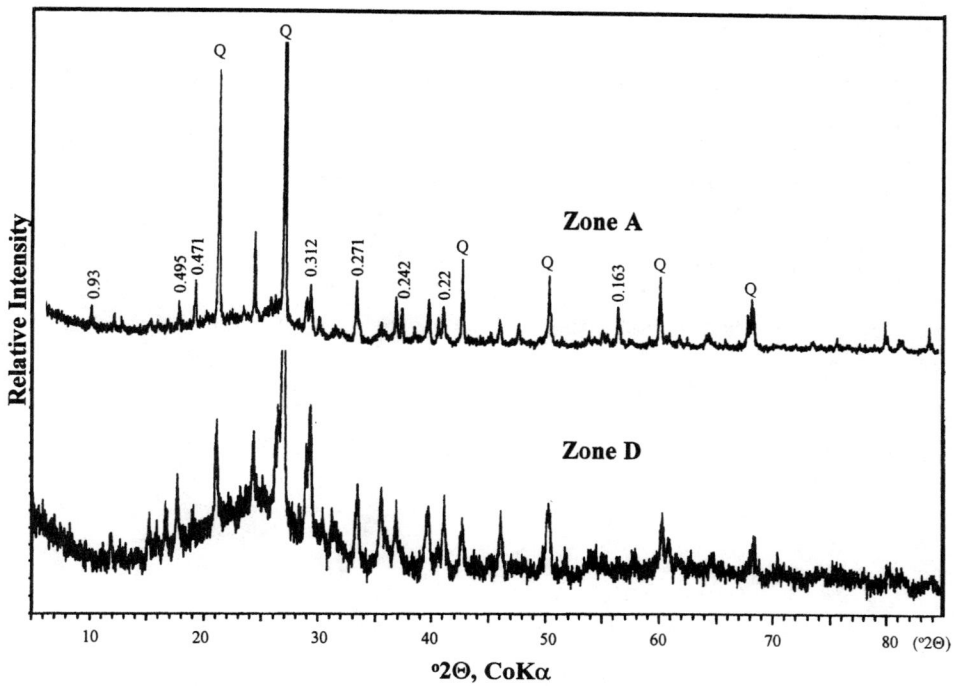

Figure 2. X-ray powder diffraction patterns of waste material from two depths in the 488-DAB.

2.76-Å, which are consistent with the iron sulfate mineral coquimbite ($Fe_2^{III}(SO_4)_3 \cdot 9H_2O$). The precipitation of iron sulfate minerals is a commonly associated product of an evaporative system that has accumulated dissolved species from pyrite oxidation[12]. As noted by Jambor et al[13], the subsequent dissolution of these salts during rain events represents an additional source of acidity and a major contributor to the contaminant pool of these systems and adjacent environments.

Two minerals associated with the white globular material were identified by XRD. Sharp diffraction peaks at 7.63, 4.28, 3.06 and 2.87- Å characteristic of gypsum ($CaSO_4 \cdot 2H_2O$) were exhibited from the external powder coating. The inner hard material showed reflections at 3.03, 2.49 and 2.28- Å, which are consistent with that of calcite ($CaCO_3$). Apparently, the acidic conditions of the basin have contributed to the dissolution of limestone particles that were co-mingled within the waste material. As with the iron salts, recrystalization of limestone as a sulfate salt likely occurred on the surface of the calcite minerals during evaporative periods.

113

The XRD patterns of basin material collected from zones A and D exhibited a very similar array of peaks (Figure 2). Both horizons displayed sharp peaks at 3.34, 4.25, 2.45, 2.28 and 1.81- Å, which are consistent with quartz; and at 3.12, 2.71, 2.42, 2.2 and 1.63- Å suggesting the presence of pyrite. Although it is difficult to assess due to potential peak overlap, the reflection at 4.95- Å may also suggest the presence of goethite (α-FeOOH) in samples from both horizons. Material from zone A also displayed weak reflections at 9.3, 4.71 and 3.69- Å, which is indicative of basaluminite ($Al(SO_4)(OH)_{10}$). Insoluble aluminum hydroxysulfates may precipitate from acid sulfate solutions when buffered to a relatively high pH by carbonate materials, however, a pH near 5.0 must be achieved for this to occur (pK$_1$ for Al-hydrolysis is 5.0[14]). Although the presence of calcite was verified in this zone, samples analyzed for pH rarely exceeded a value greater than 2.5. It is possible such minerals formed in the local environment when pH was buffer by the release of alkalinity.

Thermogravimetric (TG) analysis of zone A material displayed a weight loss of $\approx$ 5% between 25 and 300°C, followed by weight losses of $\approx$ 9 and 3% in the 400 to 500 and 750 to 900°C regions, respectively (Figure 3). Derivative thermogravimetry (DTG) indicated an inflection at 150°C, which is typical for the dehydration of gypsum[15]. Weight loss (TG) and strong DTG inflections at 450 and 500°C correspond to the oxidation of pyrite to hematite[16,17]. A broad and weak inflection at 800°C corresponding to the weight loss in the 750 to 900°C region is likely indicative of the decomposition of carbonates (calcite) and evolution of CO_2[15]. Another weak DTG inflection is exhibited at 250°C may be the result of dehydroxylation of poorly crystalline goethite, lepidocrocite, or akaganéite[18] and/or an Al-hydroxide mineral[19].

Material from zone D showed only $\approx$1% weight loss in the 25 to 300°C range and no significant DTG inflections. A weight loss of $\approx$5% occurred at the 300 to 600°C range with sharp DTG inflections at 350 and 500°C. The inflection at 350°C is likely due to the dehydroxylation of goethite[20], while the 500°C inflection is attributable to the oxidation

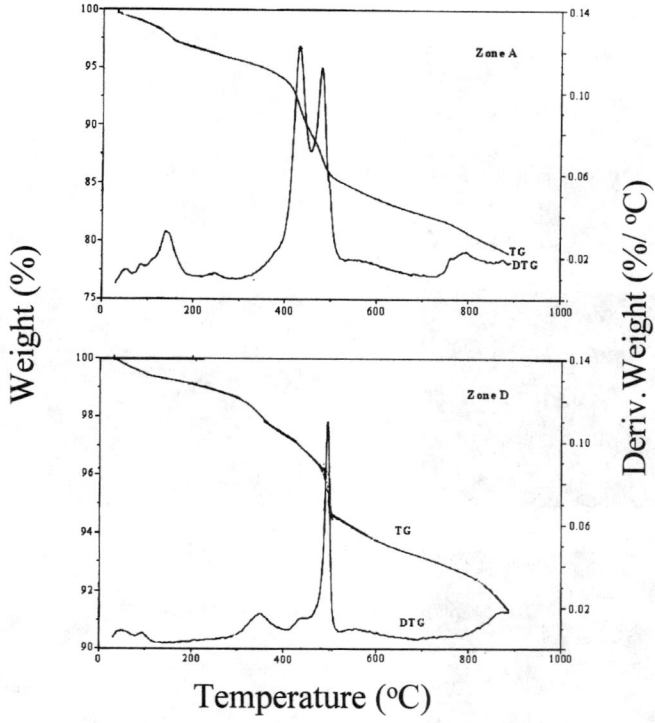

Figure 3. TG and DTG of waste material from two depths in the 488-DAB.

of pyrite to hematite as noted for zone A. Weight loss of ≈3% was observed in the 600 to 900°C region, but no significant DTG inflections were detected.

Based upon these results, it is likely that materials in the upper part of the basin are enriched to some extent with carbonate materials, while subsurface zones are primarily dominated with pyrite and secondary precipitates that resulted from its' oxidation. Apparently, gypsum precipitation is a surface phenomenon on the basin due to the relative abundance of carbonates in that zone. In addition, further evidence to support the presence of Al-hydroxides in zone A was revealed by the thermal analysis. Given that calcite was detected in zone A by both XRD and TG, the possibility for localized areas containing enough buffering potential to support a mineral such as basaluminite exists. However, gypsum and the Al-mineral phase will likely be short-lived on the 488-DAB as carbonate alkalinity is consumed over time.

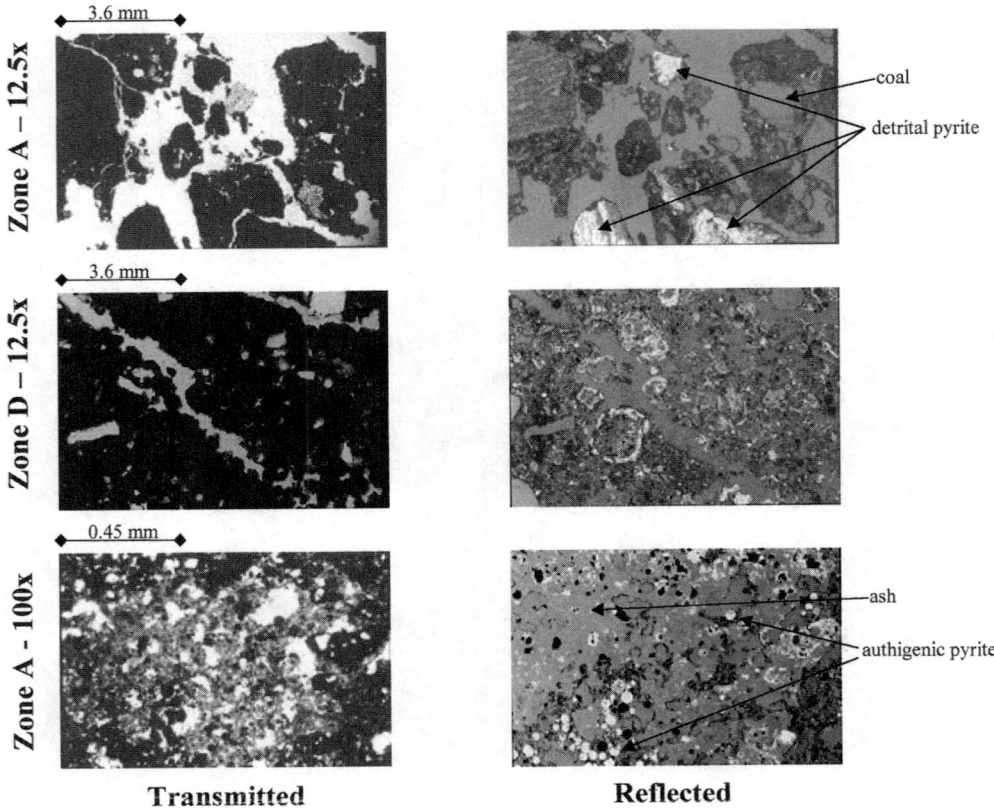

Transmitted   **Reflected**

Figure 4. Thin section from zones A and D of the 488-DAB in transmitted and reflected light under 1.25x and 10.0x magnification.

Thermal analysis and XRD on samples collected from each column after the leaching procedure indicated that no major changes in mineralogy occurred during the course of the experiment.

4.2. Particle Size and Thin Section Observations

Bulk samples collected from the basin indicated that zone A and D contained 41.4 and 6.3% course fragments (>2mm diameter), respectively. Texture analysis indicated that zone A material was composed of 7.1% clay, 14.2% silt and 78.1% sand, while zone D contained 18% clay, 29.9% silt and 52.1% sand.

Thin sections of starting material for zones A and D are shown in Figure 4. Zone A contains large clasts of coal and detrital pyrite, shown as brown and brassy material in

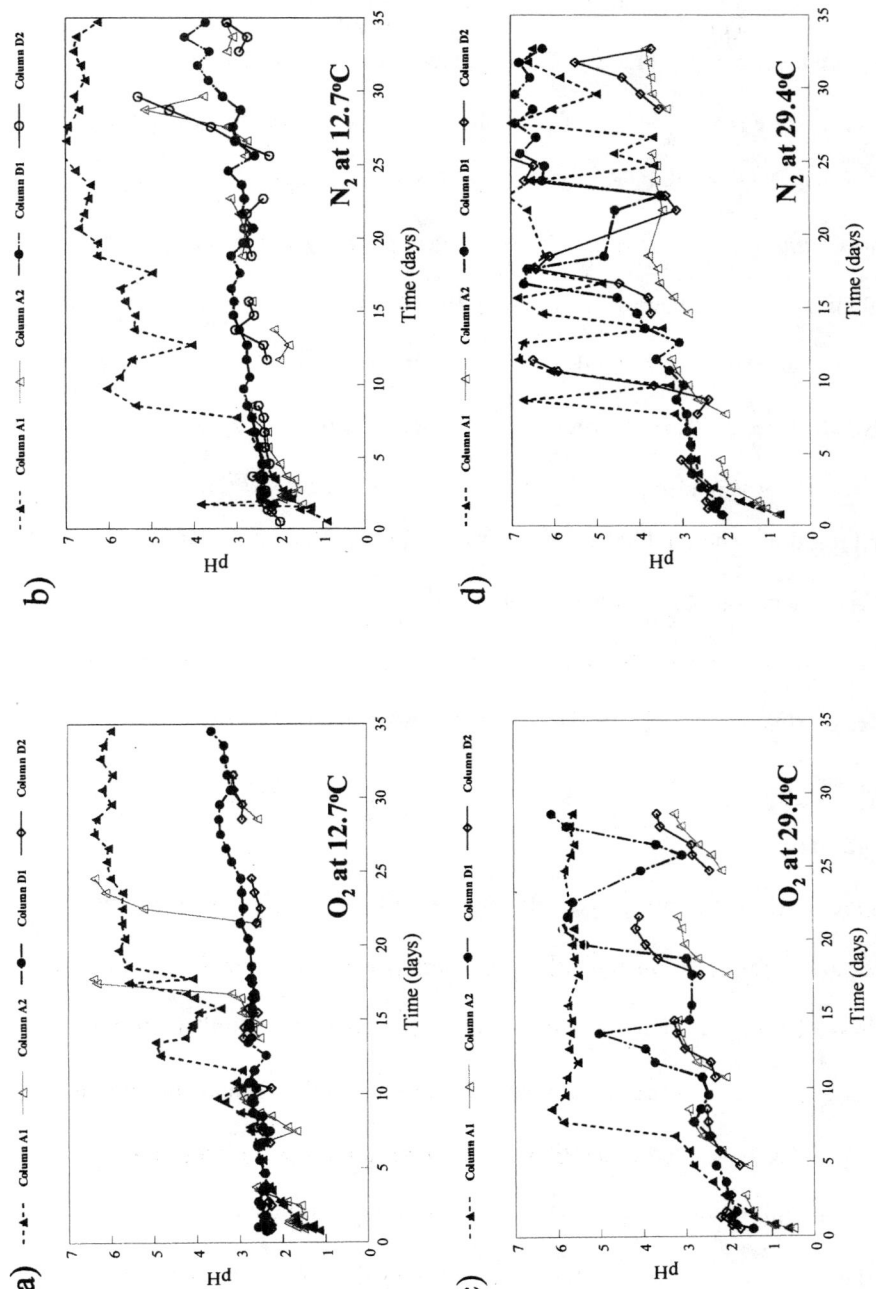

Figure 5. Effluent pH levels from zone A and D columns at (a) 12.7 °C in equilibrium with oxygen, (b) 12.7 °C in equilibrium with nitrogen, (c) 29.4 °C in equilibrium with oxygen, and (d) 29.4 °C in equilibrium with nitrogen.

reflected light, respectively, and as black grains in transmitted light. White areas in transmitted light are pore space, occasional mineral grains, or country rock.

Zone D predominantly contains fine-grained material that cannot be distinguished at relatively low magnification (12.5x). Under higher magnification (100x), a fine-grained brassy texture is noted under reflected light that is pervasive throughout the section. This material is either detrital or authigenic pyrite. At high magnification, spherical brassy grains are observed and these are most probably authigenic pyrite grains that formed through the weathering of overlying detrital pyrite grains. The circular cross-section is indicative of pyrite framboids that commonly occur in marine sediments where dissolved SO_4 diffuses into microenvironments that are slightly to strongly reducing. It is probable that pyrite was oxidized at the surface, either inorganically or through biomediated reactions, and the liberated SO_4 was transported with infiltrating waters to the subsurface where microbial Fe- and sulfate-reduction were active. Analyses are presently underway to measure $^{34}S/^{32}S$ ratios in these grains, as relatively low ratios are a strong indicator of biological sulfate reduction in terrestrial environments.

4.3 Column Eluent Chemistry

4.3.1 Eh and pH

A plot of pH over time is presented in Figure 5 for each experimental run. For column A1 (coal rubble, continuous flush), pH generally increased from ~1.0 to 6.0 over the first two weeks and then remained relatively constant thereafter. Similar low pH values (pH ~ 1 or lower) have been observed in sulfide-rich tailings impoundments and mine waste sites that have been exposed to atmospheric conditions for long periods of time[21,22]. For column A2 (coal rubble, periodic flush), pH generally increased from ~1.0 to ~2.5 within the first 3 days and increased to ~ > 3.0 by the end of the experiment. Occasional transient events were noted in the lower temperature columns where pH increased to ~6.0 over a relatively short time interval. For the fly ash residue columns (D1 and D2), effluent pH was initially higher at ~2.5 and less variable over the course of

118

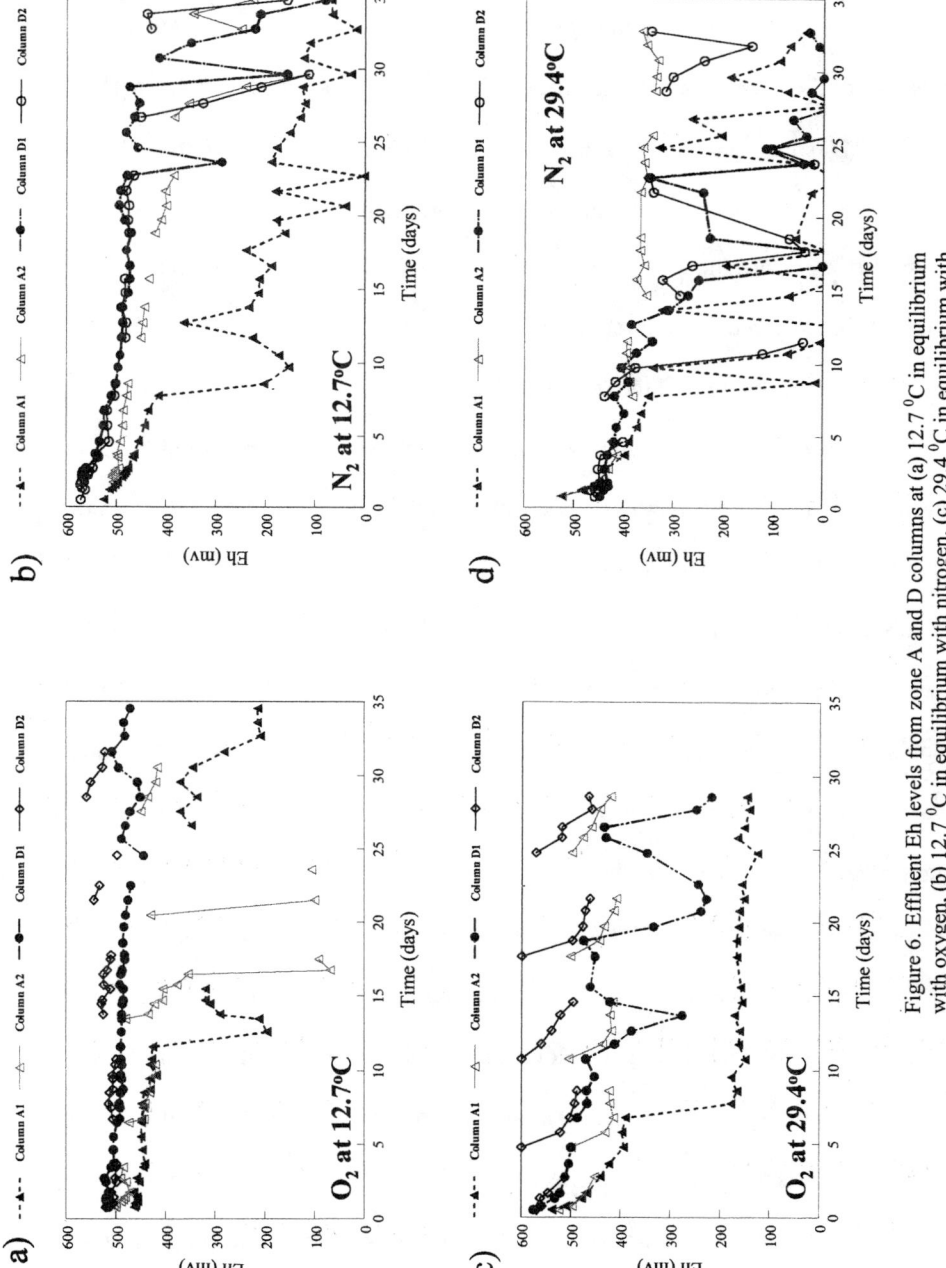

Figure 6. Effluent Eh levels from zone A and D columns at (a) 12.7 °C in equilibrium with oxygen, (b) 12.7 °C in equilibrium with nitrogen, (c) 29.4 °C in equilibrium with oxygen, and (d) 29.4 °C in equilibrium with nitrogen.

the experiments. At the low temperature, zone D effluent pH rarely exceeded ~3.0-3.5, regardless of flow regime. However, the higher temperature produced several events where the pH exceeded ~6.0 in both the oxygen and nitrogen equilibrated columns.

Effluent Eh for the column experiments are presented in Figure 6. For column A1, effluent Eh was relatively constant at ~450 mv for the first week then decreased abruptly (< 200 mv) thereafter. For column A2, effluent Eh remained relatively high at ~400-500 mv, except for several transient low Eh spikes (~100 mv) that were coincident with spikes in pH. For the fly ash residue columns (D1 and D2), effluent Eh remained relatively higher than that observed from the coal rubble material and exhibited less fluctuation for the duration of the experiments.

A general inverse correlation was observed between Eh and pH for all four experimental runs and this relationship is similar to that noted elsewhere for acid mine drainage[23,24]. The relationship between Eh and pH was consistent ($r^2 = 0.78$) both within columns of an experimental run (e.g., among flow regimes for both the coal rubble and fly ash residues) and among experimental runs (e.g., for different fluid compositions and temperatures), although slight differences in the Y-intercept were noted which could be explained by a temperature dependent reaction mechanism (Figure 7). This observation suggests that the basic chemical mechanism responsible for the production of column effluent was similar among treatments. In addition, the closeness of the curve fit between runs may suggest that the driver for pyrite oxidation in these sediments is one of a physicochemical nature rather than biological. As indicated in the introduction (equations [2] and [3]), ferric iron may act as an electron acceptor and contribute to acid production through hydrolysis and/or as a direct mechanism for pyrite oxidation. Further examination of Figure 7 reveals two dominating clusters of points; one in the pH 2 to 3 range with Eh values between 400 and 500 mv, and another in the pH 5 to 7 range with Eh values less than 200 mv. The clusters may represent the transition of iron from the Fe^{3+} valence state (low pH, high Eh) to the Fe^{2+} state (circumneutral pH, low Eh). With

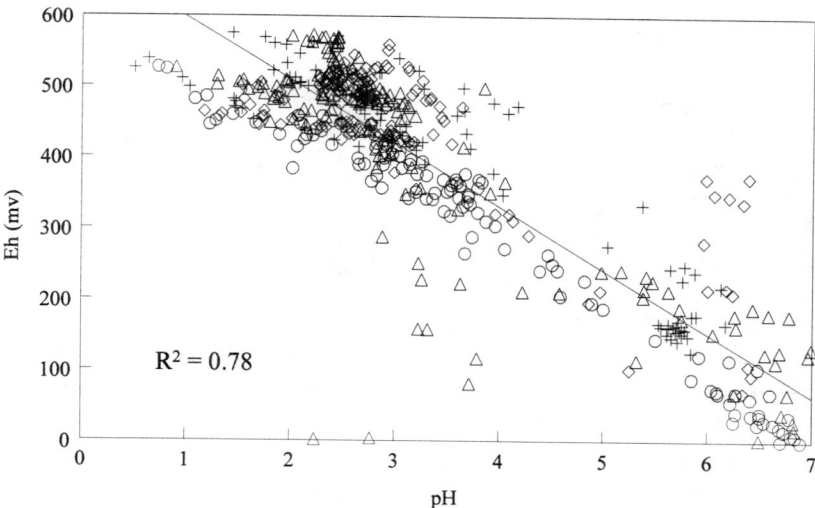

Figure 7. Relationship between pH and Eh in column effluents.

the removal of ferric iron from the system (via reduction or precipitation) acid generation

and potential pyrite oxidation is lowered and waters make the transition from acidic to

circumneutral.

4.3.2 Major Element Composition

The discharge of sulfur from column A1 exhibited similar patterns throughout the

experiment, regardless of the temperature or gas phase (Figures 8 – 11). In general, the S

content gradually fell from an initial concentration near 10,000 mg L^{-1} to 100 mg L^{-1} after

one week of flushing. Subsequent effluent S concentrations remained relatively flat

between 100 mg L^{-1} and 50 mg L^{-1}. Systematic decreases were observed in individual

flushing cycles of runs that experienced periodic wetting and drying conditions and this

was consistent with the pH-Eh data. Peaks and valleys observed in continuously flushed

columns may be indicative of the highly reactive material of the basin, but could also be

attributable to changes in the solution flow path within the column. Sulfur concentrations

from D1 eluents (all runs) followed the same pattern as those described for column A1

except that the initial concentrations were significantly lower in the deeper horizon

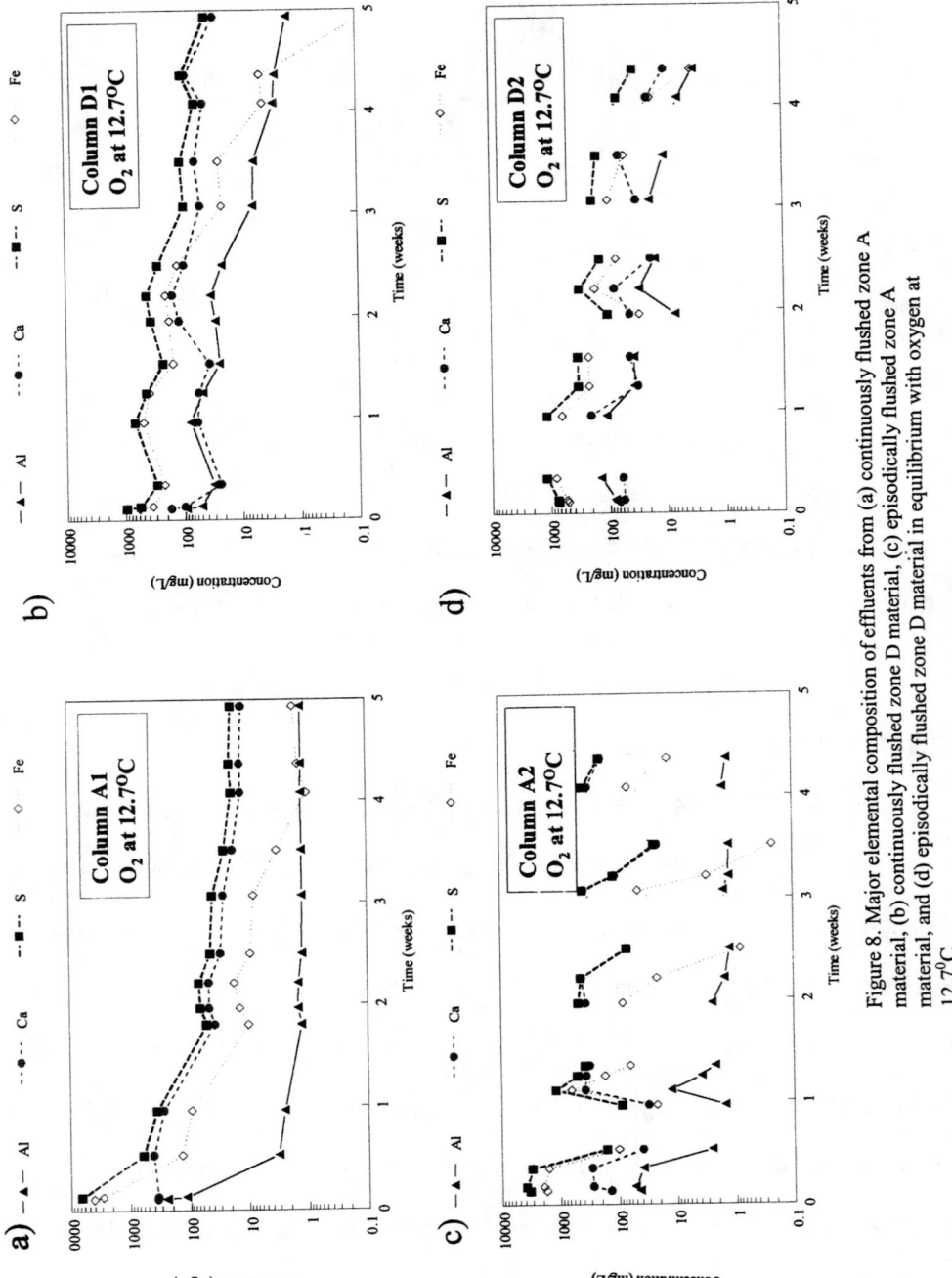

Figure 8. Major elemental composition of effluents from (a) continuously flushed zone A material, (b) continuously flushed zone D material, (c) episodically flushed zone A material, and (d) episodically flushed zone D material in equilibrium with oxygen at 12.7°C.

122

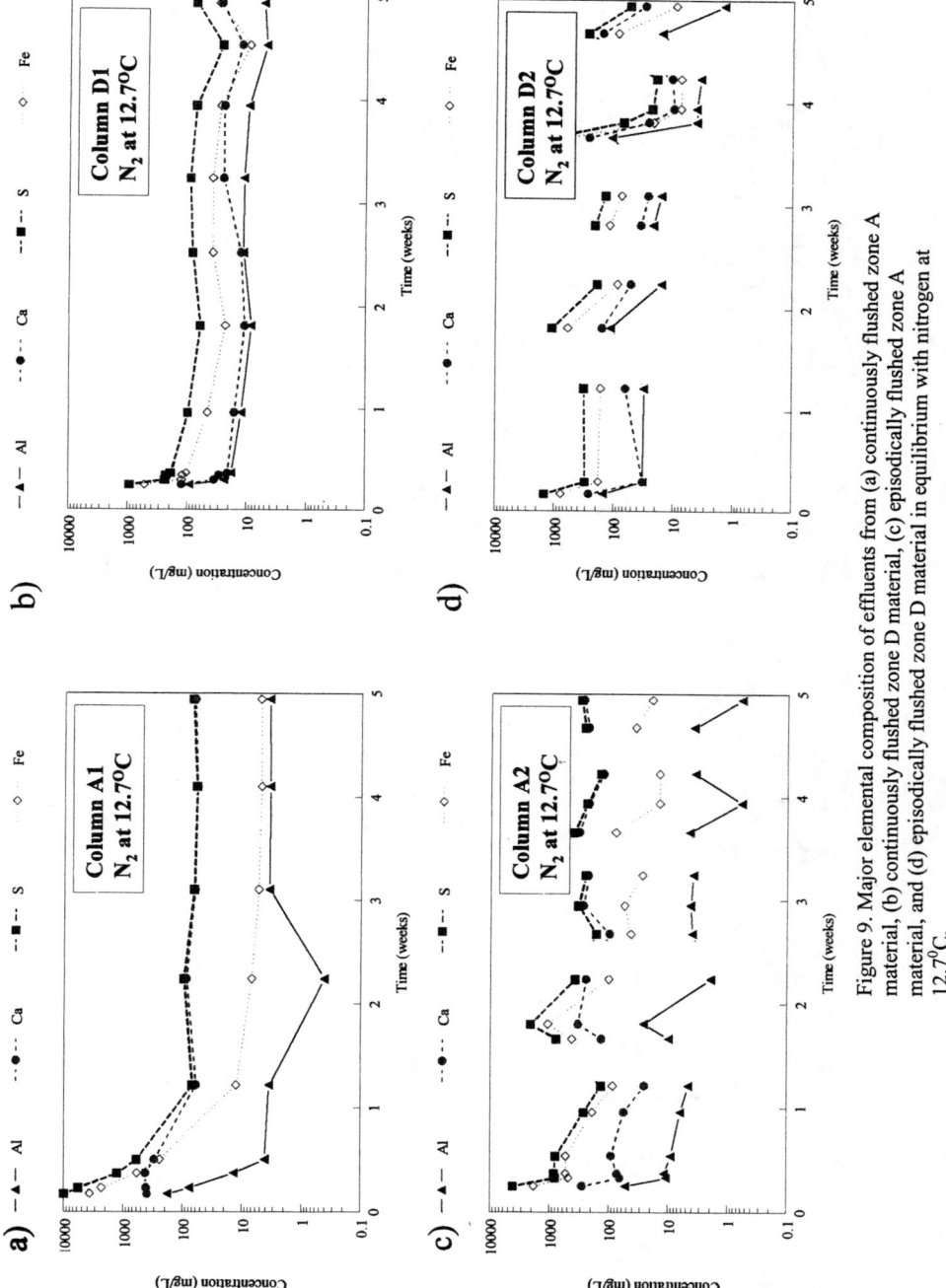

Figure 9. Major elemental composition of effluents from (a) continuously flushed zone A material, (b) continuously flushed zone D material, (c) episodically flushed zone A material, and (d) episodically flushed zone D material in equilibrium with nitrogen at 12.7°C.

123

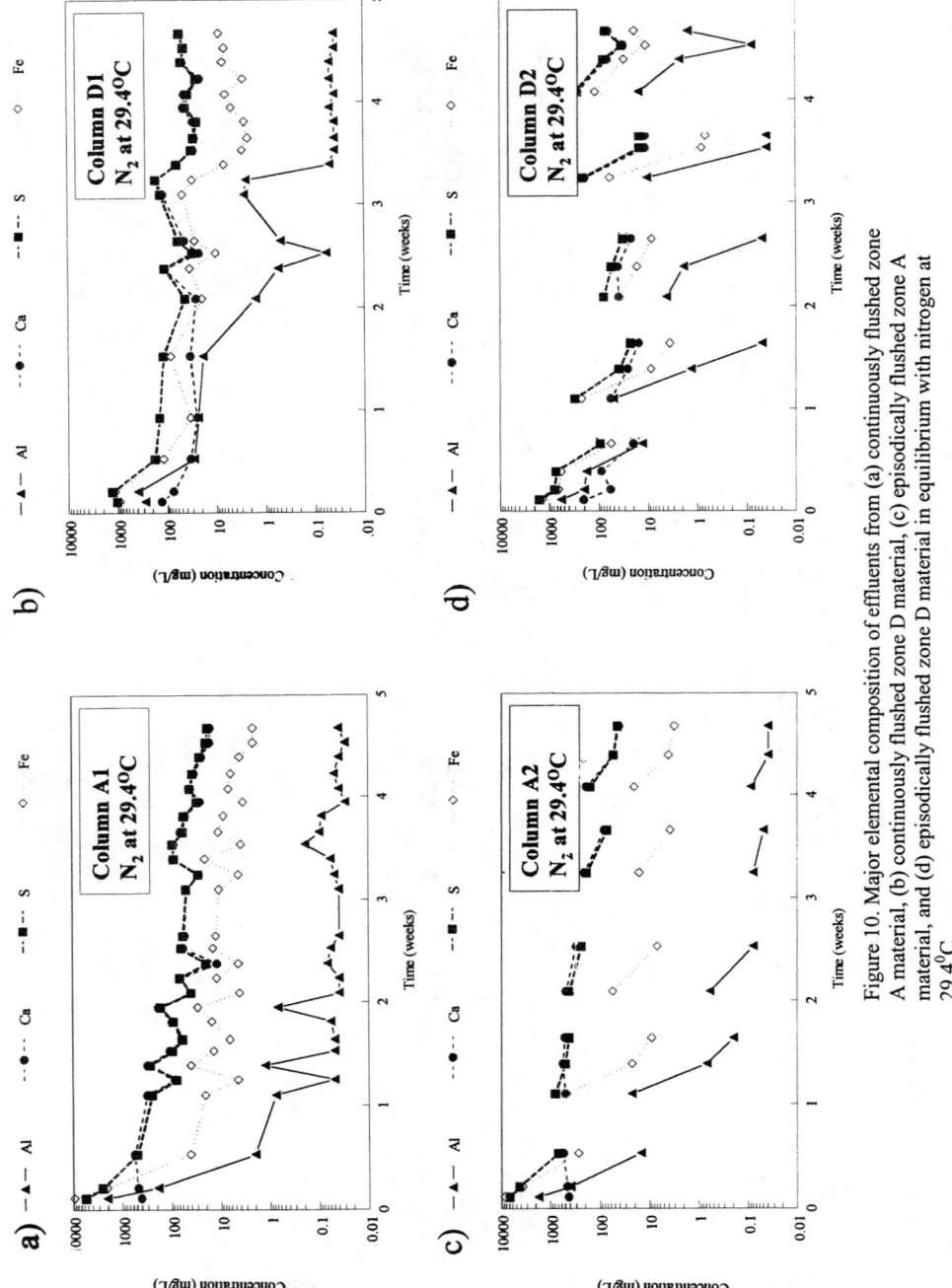

Figure 10. Major elemental composition of effluents from (a) continuously flushed zone A material, (b) continuously flushed zone D material, (c) episodically flushed zone A material, and (d) episodically flushed zone D material in equilibrium with nitrogen at 29.4°C.

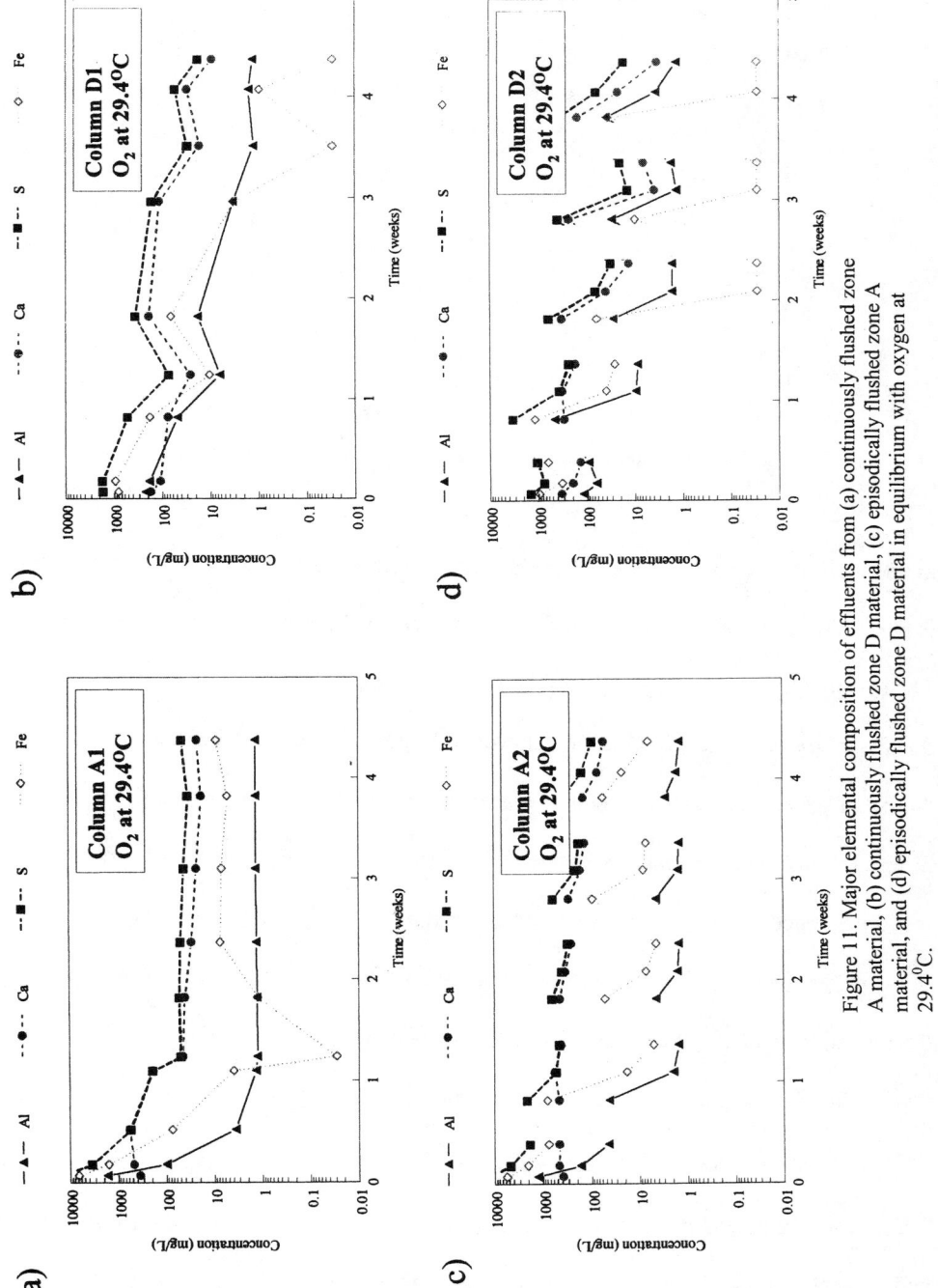

Figure 11. Major elemental composition of effluents from (a) continuously flushed zone A material, (b) continuously flushed zone D material, (c) episodically flushed zone A material, and (d) episodically flushed zone D material in equilibrium with oxygen at 29.4°C.

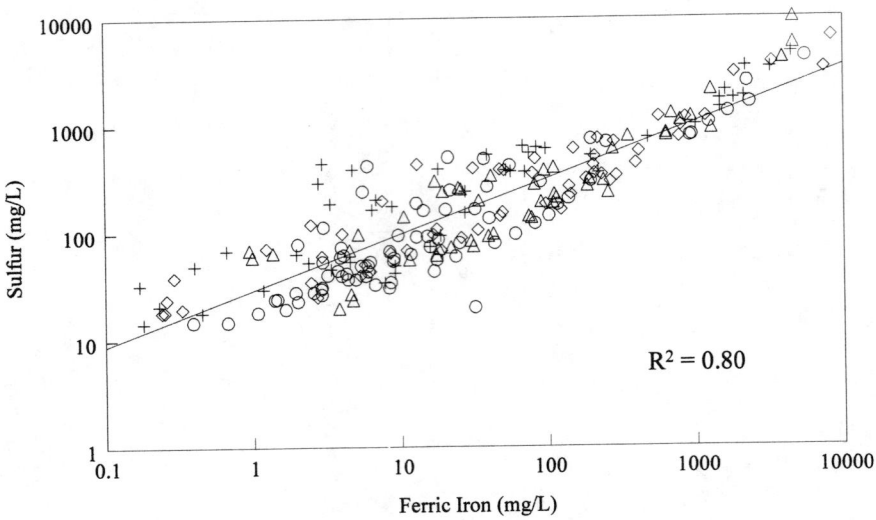

Figure 12. Relationship between sulfur and ferric iron in column effluents.

material (1,000 vs. 10,000 mg L^{-1}). The higher initial S content in the column A1 material

is likely a result of the dissolution of evaporative salts that formed on the basin's surface.

The episodic wetting and drying in columns A2 and D2 showed the same general

downward trend in S levels with time, however, concentrations tended to exhibit a

"rebound" effect after each successive dry-down period. The magnitude of this "rebound"

differed slightly among columns and between runs, but the general result was similar in

that the S concentrations at the beginning and end of a wetting cycle were lower than that

of the preceding cycle. Even though the pH of these episodic events reached levels above

6.0 toward the end of a cycle, the pH value at the beginning of a cycle rarely exceeded a

level of 3.0 for both columns (A2 and D2) during all runs (Figure 5). As such, mobility of

the elements was highest at the beginning of a cycle.

Initial calcium concentrations in effluents were ≤ 400 mg L^{-1} for all runs. By the

end of week 1, Ca concentrations from zone A columns dropped to levels similar to that

of S and tracked S concentration thereafter. In zone D columns, S approached Ca by the

end of the experiment. The approach was kinetically favored at the higher temperature.

The strong association between Ca and S suggests that these elemental concentrations may be controlled by the solubility of a solid or mineral of similar stoichiometry, most probably gypsum[25]. The episodic flushing of column D2 yielded a Ca "rebound" effect similar to that of column A2.

Iron concentrations for column A1 eluents were near 10,000 mg L^{-1} at the beginning of each experimental run, and rapidly fell to levels approaching 10 mg L^{-1} by the end of the first week. Subsequent effluent Fe concentrations remained in the 1 to 10 mg L^{-1} range. Although lower in concentration, column A1 elution curves for Fe maintained a similar pattern to that observed for S and Ca for all runs. Column D1 eluents were similar to that of A1 for all runs. Once again, the eluent pattern for the columns experiencing episodic wetting exhibited a "rebound" effect, and mimicked that of sulfur. A plot of ferric iron ($Fe^{3+)}$ vs. sulfur resulted in a positive correlation ($r^2 = 0.80$) for all four experimental runs (Figure 12). This observation reemphasizes the possibility that a basic chemical mechanism responsible for the production of column eluent was similar among all treatments. In addition, the closeness of the curve fit between runs may suggest that Fe^{3+} is the principal oxidizing agent leading to the formation of AD from the 488-DAB material. As such, maintenance of the iron redox couple at a level favoring the ferrous state ($\sim < 150$ mv) may be an essential factor for the control of pyrite oxidation in these sediments.

Initial eluent aluminum concentrations for column A1 were less than 300 mg L^{-1} for runs 1 and 2 (Figures 8a and 9a), and greater than 1,000 mg L^{-1} for runs 3 and 4 (Figures 10a and 11a). The difference between the concentration levels is most likely attributable to initial pH conditions of the runs. The warmer runs (3 and 4) produced a more acidic initial solution (~ 0.3 pH units lower), which likely increased the solubility of aluminum in those columns. By the end of the first week, A1 columns generally exhibited an eluent pH > 4.0 (Figure 5) and aluminum concentrations were ≤ 1.0 mg L^{-1}. Runs 1, 2 and 4 for column A1 maintained an eluent concentration around 1.0 mg L^{-1} for the duration of the experiment, however, run 3 (N_2 at 29.4°C) exhibited a further decrease to

~ 0.05 mg L^{-1} (Figure 10a). We suspect that the lower aluminum concentration in run 3 is

attributable to enhanced microbial metabolism under anaerobic conditions, which

resulted in a higher pH within the column and enhanced Al precipitation.

4.3.3 Trace Element Composition

Examination of the trace element data from ICP analysis revealed four elements

(As, Cu, Se, and Zn) with appreciable quantities in the eluent samples. Graphs depicting

the change in concentration of these elements over the course of the experiment are

presented in Figures 13 –16. The concentration of zinc was relatively stable during the

experiment regardless of the temperature, gas phase, or column material. Zinc

concentrations fluctuated from 1 mg L^{-1} at the beginning of a run to ~ 0.1 mg L^{-1} at the

end. The rebound effect observed with the major elements did occur for Zn in columns

A2 and D2.

Copper concentration in columns A1 and D1 follows a similar pattern to that of

zinc during the first week, but concentrations tend to decline further to the 0.01 mg L^{-1}

level through the remainder of the experiment. The "rebound" of copper is exhibited in

episodically flushed column A2 when oxygen is present (Figures 13c and 16c), however,

the response does not occur in the presence of nitrogen (Figures 14c and 15c). In

addition, the final effluent concentration in the A2 column is generally lower when

leached in the more reducing N_2 environment. Column D2, on the other hand, exhibits

the rebound for all runs but is more expressed at the higher temperature with a general

order of magnitude change in concentration during each solution pulse phase. This

suggest that a slight reduction in the water oxidation state may deter Cu leaching from the

zone A coal rubble material, whereas continual Cu releases from the zone D ash zone are

likely to continue with each successive wet/dry period.

Selenium concentrations in runs 2 through 4 were below detection for columns

A1 and D1 after one week of continual leaching (Figures 14a,b-16a,b), while in run 1 (O_2

128

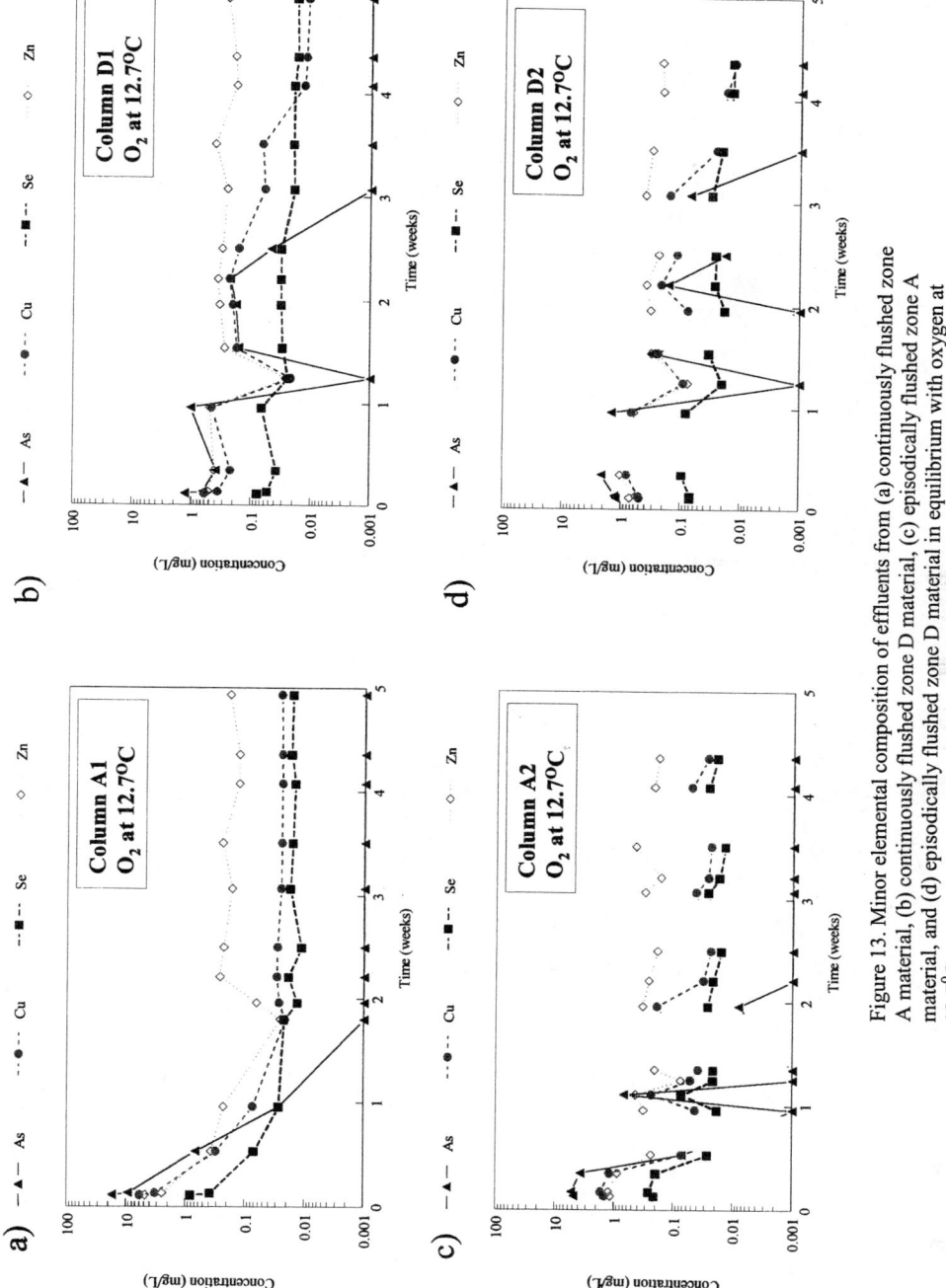

Figure 13. Minor elemental composition of effluents from (a) continuously flushed zone A material, (b) continuously flushed zone D material, (c) episodically flushed zone A material, and (d) episodically flushed zone D material in equilibrium with oxygen at 12.7°C.

129

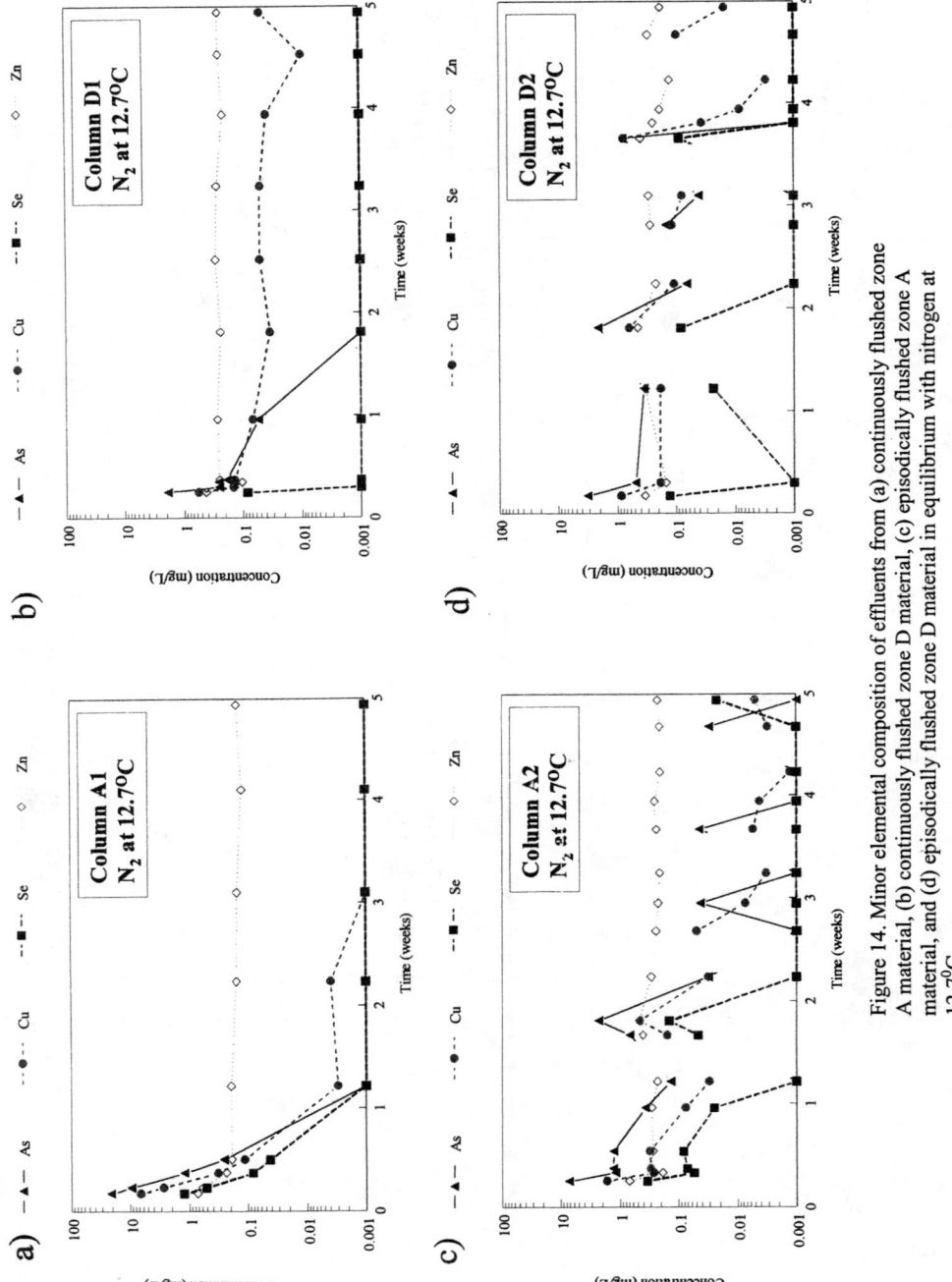

Figure 14. Minor elemental composition of effluents from (a) continuously flushed zone A material, (b) continuously flushed zone D material, (c) episodically flushed zone A material, and (d) episodically flushed zone D material in equilibrium with nitrogen at 12.7°C.

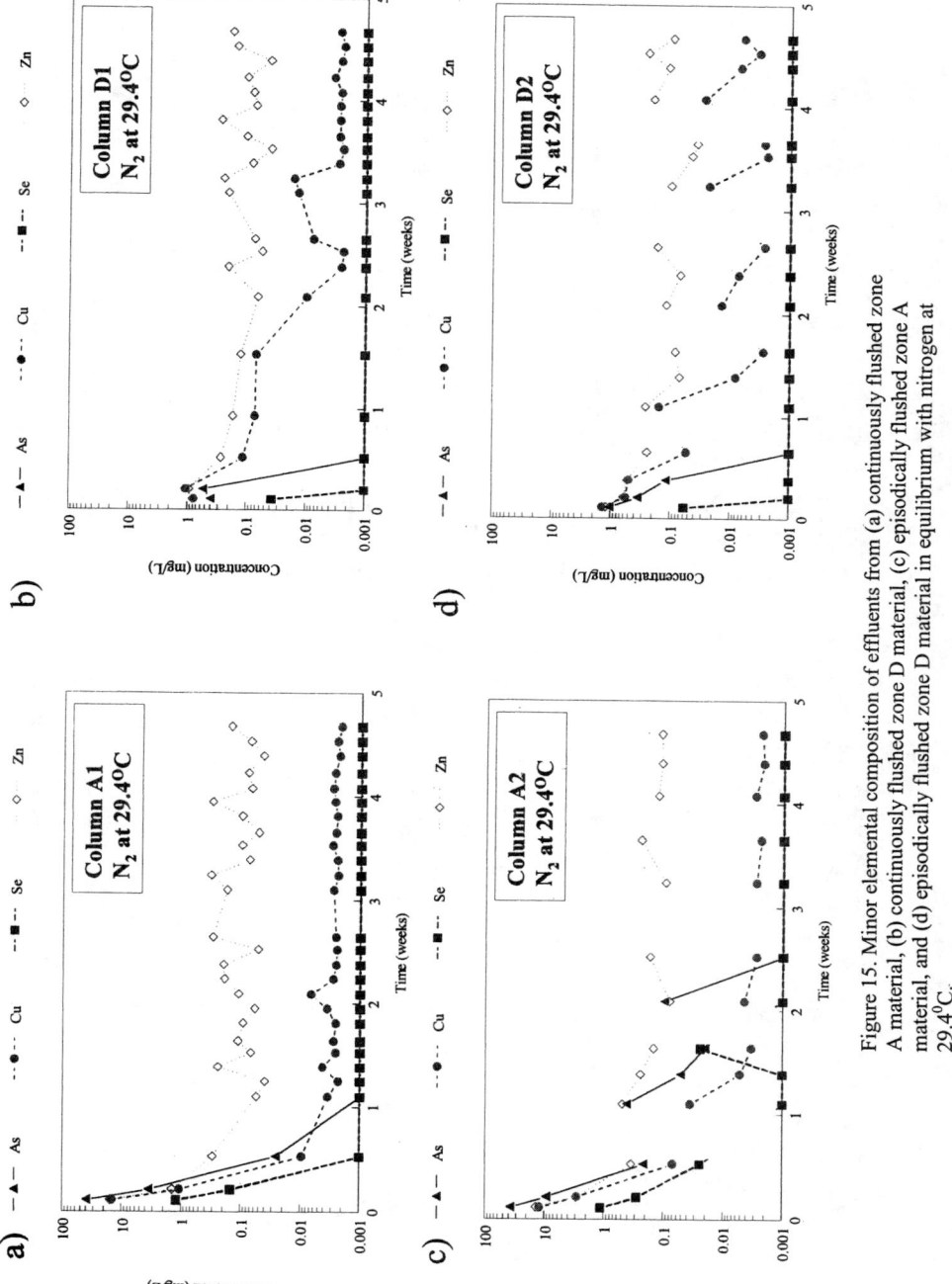

Figure 15. Minor elemental composition of effluents from (a) continuously flushed zone A material, (b) continuously flushed zone D material, (c) episodically flushed zone A material, and (d) episodically flushed zone D material in equilibrium with nitrogen at 29.4°C.

131

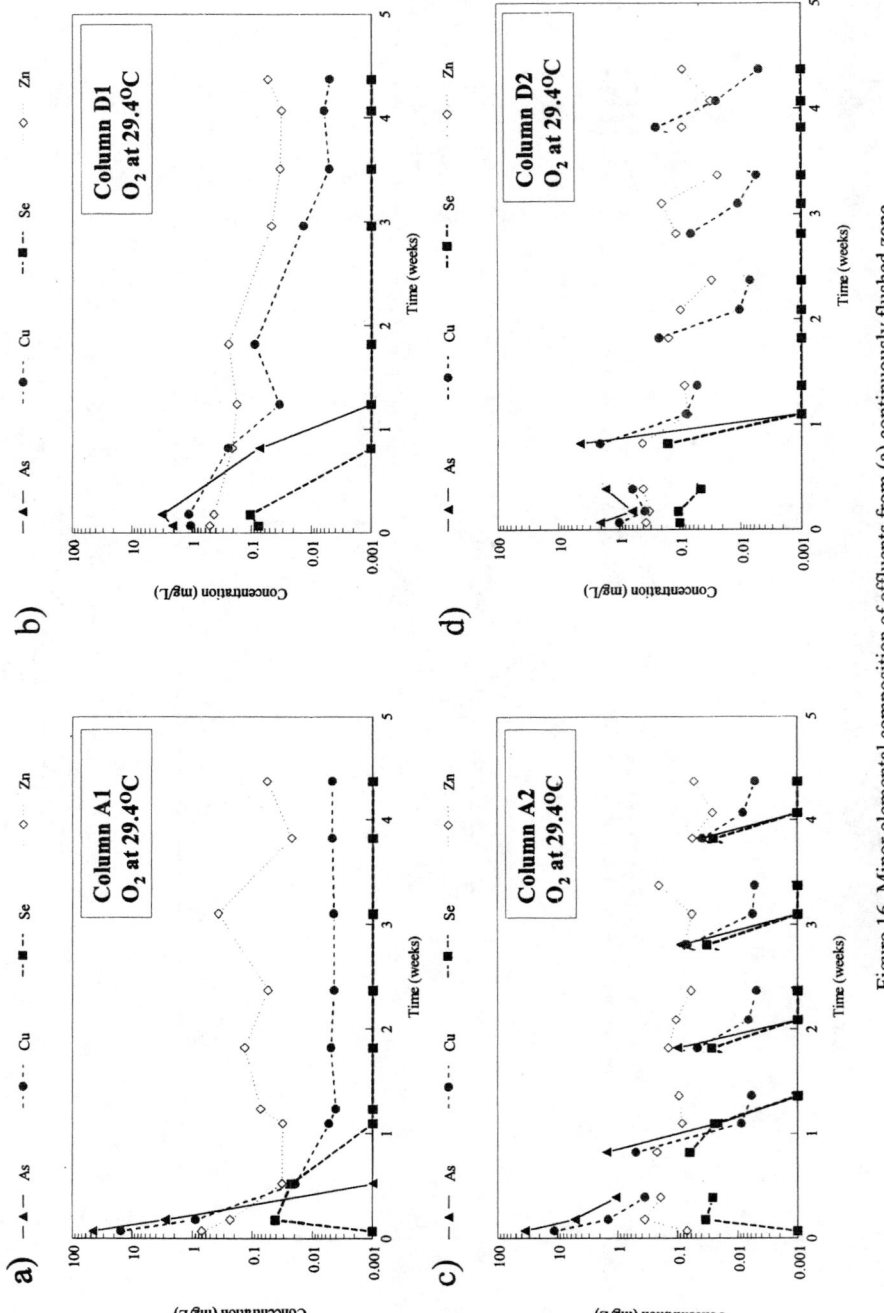

Figure 16. Minor elemental composition of effluents from (a) continuously flushed zone A material, (b) continuously flushed zone D material, (c) episodically flushed zone A material, and (d) episodically flushed zone D material in equilibrium with oxygen at 29.4°C.

132

at 12.7^{0}C) Se concentrations were slightly above detection ~ 0.03 mg L^{-1} in both columns (Figure 13a,b). Selenium mobility in soils is controlled by pH and redox conditions[26], but the difference between these factors over the four runs was not significant enough to elicit the observed response. Relationships to other chemical constituents that may explain the Se behavior in run 1 were not observed. As such, the enhanced Se mobility was probably due to a mobile Se-colloidal fraction that formed under the specific environmental conditions of this run. Selenium concentration in the eluents from episodic flushing (columns A2 and D2) in the presence of N$_2$ (Figures 14c,d and 15c,d) did not differ significantly from that in the saturated columns. Some fluctuation in Se concentration was observed in the lower temperature run during episodic flushing (Figure 14), however, a "rebound" was not observed for all pulses. The episodically flushed columns saturated with oxygen at a low temperature exhibited a nearly identical pattern to those described for the saturated columns in run 1 (Figure 13). Once again, concentration levels did not fall below ~ 0.03 mg L^{-1}, and the rebound between pulses were minimal during this run. At the higher temperature, the rebound was clearly evident in the zone A coal rubble column (A2) (Figure 16c) but missing in the deeper zone D ash column (D2) (Figure 16d).

Arsenic showed a very similar pattern to that described for selenium. Initial As concentrations were > 10 mg L^{-1} and ~ 1 mg L^{-1} in zones A and D, respectively. In the saturated columns, As concentrations fell below detection after one week of leaching for all runs. Arsenic in the episodically flushed columns exhibited a wide array of responses with changes in environmental conditions. At low temperature, eluted As was below detection in zone A materials within two weeks of leaching in the presence of O$_2$ (Figure 13c), but exhibited a rebound to levels of ~ 0.03 mg L^{-1} with each successive pulse in the presence of N$_2$ (Figure 14c). Eluent As concentrations from zone D1 materials at this temperature were below detection by week four regardless of the gas phase (Figures 13d and 14d). Conversely, the higher temperature elicited quite a different response. Eluent

As levels from the zone D2 columns (N_2 and O_2) were below detection by the second pulse phase of the run (Figures 15d and 16d). In the zone A1 columns under higher temperature, As concentrations eventually fell below detection during week 2 in the presence of N_2 (Figure 15c), but were maintained well above detection with each successive pulse in the presence of O_2 (Figure 16c).

The response of As and Se in zone A under the presence of the differing gases at the higher temperature (Figures 15 and 16) is very informative from the standpoint of evaluating the usefulness of a cover for controlling AD release. For demonstration purposes, we may assume that these two runs represent the ambient state of the 488-DAB basin during the summer. Under saturated conditions there would be essentially no release of these toxic substances from the surface materials once an initial pulse of fluid has migrated through the subsurface. This initial contaminant pulse is probably attributable to the dissolution of evaporative salts. Under the changing conditions associated with intermittent rains at 488-DAB, the pulse is reestablished through each successive wet/dry period. As such, the waste is a long-term source of potential contamination via evaporation and precipitation. However, if there were a slight reduction in the oxidation state of water that comes in contact with the coal rubble material (e.g. N_2 column experiments) Se and As mobility would be inhibited. The use of a cover, particularly one enriched with organic matter, could generate this response. Not only would the cover provide a buffer against direct exposure to the oxidizing forces of water and air, but microbial oxygen demand within the cover could also result in lowering the oxidizing potential of rainwater prior to contact with the waste material.

4.4 Mineral Solubility

The solution saturation indices (SI) of the column eluents with respect to various minerals were evaluated using the equilibrium geochemical speciation model MINTEQA2[10]. The SI (logQ/K), where Q = ion activity product and K = solubility product constant, was calculated for each column after one and four weeks of leaching

(Tables 1 and 2). A SI of zero indicates equilibrium, a negative value undersaturation, and a positive value supersaturation. At one week of leaching most columns were undersaturated with respect to the various minerals examined (Table 1). The column A1, during run 2 (N_2 at 12.7°C), was the only example of a run containing eluents that were supersaturated with respect to more than one mineral. The precipitation of basaluminite, jurbanite ($Al(OH)SO_4$), boehmite (γ-AlOOH), diaspore (χ-ALOOH), gibbsite ($Al(OH)_3$), goethite, hematite (χ-Fe_2O_3), lepidocrocite, magnetite (Fe_3O_4), and possibly an amorphous $Al(OH)_3$ may have occurred in this column due to high pH (6.0) conditions that were not achieved by any other run or column at this stage of the experiment. Hematite was also exhibited at or above saturation in many other columns and runs. Gypsum, anhydrite ($CaSO_4$) and jurbanite were close to equilibrium in several columns during various runs.

After four weeks of leaching, column A1 exhibited tremendous changes in its SI for all runs. In this column, most of the eluent solutions were either at equilibrium or supersaturated with respect to the Al and Fe oxide, hydroxide and hydroxysulfate minerals modeled (Table 2). This correlates well with the high pH conditions observed for column A1 at four weeks. With the exception of run 3 (N_2 at 29.4°C), however, SI's for column D1 did not change significantly after four weeks of leaching. Once again, this correlates well with the pH values at that point in time (Figure 5). The experimental conditions of run 3 resulted in pH levels that were sustained above 6.0 in column D1, thus allowing for the potential precipitation of various Fe and Al minerals. The episodic flushing of columns A2 and D2 resulted in lower pH values and higher element concentrations at four weeks than that observed in the constantly saturated columns. As such, the saturation indices did not change greatly through the course of the experiment. A slight enrichment in SI was observed at four weeks for the iron minerals goethite, lepidocrocite, magnetite and hematite for column D2 during run 1 (O_2 at 12.7°C). Saturation indices for pyrite remained highly unsaturated throughout the course of the

Table 1. Saturation indices (log Q/K)† of column effluents for selected minerals after one week of leaching.

Column§	A1				A2				D1				D2			
Treatment† Mineral	1	2	3	4	1	2	3	4	1	2	3	4	1	2	3	4
Al(OH)₃(a)	-8.98	-0.13	-6.65	-6.02	-9.19	-8.19	-8.94	-7.92	-7.96	-7.85	-5.78	-6.35	-8.40	-8.06	-6.44	-7.78
Basaluminite	-19.27	9.89	-15.38	-13.31	-20.31	-16.59	-21.96	-19.54	-14.96	-15.49	-11.80	-14.41	-16.48	-15.65	-13.54	-18.61
Jurbanite	-1.56	1.02	-1.53	-1.18	-1.93	-1.26	-1.32	-1.73	-0.26	-0.98	-0.35	-1.08	-0.27	-0.58	-0.27	-1.24
Alunite	-24.69	-24.10	-24.41	-23.40	-25.07	-23.14	-19.55	-23.25	-20.36	-21.76	-19.92	-21.25	-20.13	-20.86	-19.87	-21.11
Anhydrite	-0.64	-1.75	-0.57	-0.81	-2.59	-1.89	-0.36	-0.32	-1.34	-2.18	-1.77	-2.00	-0.70	-1.33	-1.33	-0.61
Boehmite	-6.81	2.03	-4.42	-3.79	-7.03	-6.02	-6.71	-5.68	-5.79	-5.68	-3.55	-4.12	-6.23	-5.89	-4.21	-5.55
Diaspore	-4.98	3.86	-2.73	-2.11	-5.20	-4.20	-5.02	-4.00	-3.97	-3.86	-1.87	-2.43	-4.41	-4.07	-2.52	-3.86
Ferrihydrite	-4.64	-1.42	-4.47	-4.00	-4.88	-3.81	-5.32	-4.69	-3.85	-3.45	-4.15	-3.41	-4.16	-3.60	-3.60	-3.70
Gibbsite	-5.87	2.97	-3.60	-2.97	-6.08	-5.08	-5.89	-4.87	-4.85	-4.74	-2.73	-3.30	-5.29	-4.95	-3.39	-4.73
Goethite	-1.83	1.38	-1.79	-1.32	-2.07	-1.01	-2.63	-2.01	-1.04	-0.66	-1.47	-0.72	-1.36	-0.79	-0.92	-1.02
Gypsum	-0.33	-1.44	-0.33	-0.57	-1.50	-1.57	-0.11	-0.09	-1.02	-1.87	-1.54	-1.76	-0.39	-1.02	-1.09	-0.37
Hematite	-1.33	5.10	-1.17	-0.23	-1.81	0.31	-2.85	-1.61	0.24	1.01	-0.53	0.95	-0.39	0.73	0.56	0.35
H-Jarosite	-7.40	-10.29	-9.34	-8.35	-8.42	-5.93	-6.91	-7.75	-4.43	-4.82	-7.59	-5.76	-4.40	-4.09	-4.58	-4.10
Lepidocrocite	-2.21	1.01	-2.73	-2.29	-2.45	-1.38	-3.56	-2.98	-1.42	-1.07	-2.45	-1.69	-1.78	-1.19	-1.87	-1.99
Maghemite	-8.06	-1.62	-9.11	-8.23	-8.55	-6.41	-10.76	-9.61	-6.50	-5.78	-8.56	-7.03	-7.20	-6.03	-7.36	-7.63
Magnetite	-4.80	6.38	-3.38	-2.70	-5.87	-2.91	-6.19	-4.55	-3.06	-2.57	-2.98	-1.77	-3.73	-14.92	-1.69	-2.86
FeS (ppt.)	-52.80	-42.06	-42.51	-48.77	-56.31	-57.69	-43.44	-46.17	-57.32	-63.93	-46.69	-55.44	-54.54	-59.40	-48.81	-56.25
Pyrite	-73.34	-59.66	-56.81	-67.65	-79.08	-82.19	-56.98	-62.40	-81.29	-92.84	-63.88	-78.97	-76.30	-85.00	-67.54	-79.74

†Positive values indicate supersaturation and negative values undersaturation; Q = ion activity product; K = solubility constant.

‡Treatment 1 = O_2-equilibrated at 12.7⁰ C; 2 = N_2-equilibrated at 12.7⁰ C; 3 = N_2-equilibrated at 29.4⁰ C; 4 = O_2-equilibrated at 29.4⁰ C.

§Columns A1 and D1 received continuous wetting; columns A2 and D2 received periodic wetting.

Table 2. Saturation indices (log Q/K)† of column effluents for selected minerals after four weeks of leaching.

Column§	A1				A2				D1				D2			
Treatment‡	1	2	3	4	1	2	3	4	1	2	3	4	1	2	3	4
Mineral																
Al(OH)₃ (a)	1.16	1.74	-0.76	0.42	-9.18	-7.44	-6.94	-7.49	-5.97	-6.38	-0.96	4.34	-7.16	-6.83	-6.30	-6.39
Basaluminite	12.70	14.86	1.85	7.24	-20.18	-14.85	-17.04	-18.22	-9.43	-10.80	-0.82	-7.83	-13.22	-12.52	-15.44	-14.52
Jurbanite	0.04	0.39	-1.89	-0.115	-1.79	-1.56	-2.30	-1.71	-0.67	-0.60	-3.91	-0.82	-0.86	-1.01	-2.57	-1.31
Alunite	-30.05	-29.93	-33.26	-26.49	-25.26	-26.3	-26.61	-22.94	-24.23	-22.00	-41.58	-22.97	-23.16	-23.67	-28.09	-21.87
Anhydrite	-2.66	-1.66	-1.76	-2.19	-0.61	-0.83	-0.86	-1.35	-1.63	-2.03	-1.78	-1.94	-1.99	-2.04	-1.88	-2.12
Boehmite	3.33	3.91	1.46	2.64	-7.02	-5.27	-4.71	-5.26	-3.80	-4.20	1.26	-2.11	-4.99	-4.66	-4.07	-4.16
Diaspore	5.15	5.73	3.15	4.33	-5.19	-3.45	-3.02	-3.58	-1.98	-2.39	2.94	-0.42	-3.17	-2.84	-2.38	-2.48
Ferrihydrite	2.17	0.36	-0.86	-0.46	-4.81	-4.83	-4.09	-3.78	-3.17	-3.34	-0.04	-1.94	-2.19	-3.58	-4.30	-3.39
Gibbsite	4.27	4.85	2.28	3.47	-6.07	-4.33	-3.89	-4.44	-2.86	-3.27	2.08	-1.29	-4.05	-3.73	-3.25	-3.34
Goethite	4.98	3.17	1.82	2.22	-2.00	-2.03	-1.41	-1.10	-0.36	-0.55	2.63	0.73	0.61	-0.78	-1.62	-0.71
Gypsum	-2.35	-1.35	-1.52	-1.95	-0.29	-0.51	-0.62	-1.12	-1.32	-1.72	-1.55	-1.70	-1.68	-1.72	-1.64	-1.88
Hematite	12.30	8.68	6.05	6.85	-1.68	-1.72	-0.41	0.19	1.60	1.23	7.68	3.88	3.56	0.76	-0.83	0.98
H-Jarosite	-3.97	-9.19	-10.93	-8.59	-7.93	-10.92	-9.14	-5.84	-7.17	-6.59	-12.07	-4.88	-2.20	-7.26	-11.55	-6.07
Lepidocrocite	4.59	2.79	0.86	1.27	-2.39	-2.44	-2.35	-2.07	-0.76	-0.98	1.66	-0.22	0.21	-1.20	-2.58	-1.68
Maghemite	5.54	1.95	-1.91	-1.09	-8.43	-8.53	-8.35	-7.79	-5.16	-5.61	-0.30	-4.09	-3.21	-6.05	-8.80	-7.01
Magnetite	13.87	11.79	9.25	9.62	-5.34	-4.40	-2.22	-2.66	-1.31	-1.53	12.48	2.32	0.24	-2.16	-2.66	-2.63
FeS (ppt.)	-70.41	-42.32	-28.89	-35.37	-53.16	-45.50	-42.43	-53.09	-61.63	-58.29	-22.75	-58.33	-72.05	-57.74	-41.72	-63.72
Pyrite	-110.07	-62.12	-38.59	-49.47	-73.82	-61.54	-57.10	-74.67	-89.74	-83.87	-29.51	-85.39	-97.42	-82.78	-56.06	-92.89

†Positive values indicate supersaturation and negative values undersaturation; Q = ion activity product; K = solubility constant.

‡Treatment 1 = O₂-equilibrated at 12.7⁰ C; 2 = N₂-equilibrated at 12.7⁰ C; 3 = N₂-equilibrated at 29.4⁰ C; 4 = O₂-equilibrated at 29.4⁰ C.

§Columns A1 and D1 received continuous wetting; columns A2 and D2 received periodic wetting.

137

experiment. Given the redox conditions of the experiment (Figure 6), the re-precipitation of pyrite or formation of metal sulfides was not anticipated and was not considered as a pathway for metal retention in the columns during the flow-through experiments.

5. CONCLUSIONS

In an effort to evaluate remedial options for an abandoned fly ash/ reject coal landfill, a laboratory column experiment was conducted to characterize the geochemistry of the waste material and associated lechate under differing environmental conditions. Columns were generated using materials from the surface of the landfill, primarily coal rubble, and those found deeper in the profile, primarily fly ash residue. The coal rubble zone contained large clasts of coal and detrital pyrite, secondary iron and aluminum precipitates, and evaporative iron-sulfate and calcium-sulfate salts, which formed from the weathering of pyrite and dissolution of carbonates. The iron-sulfate salts (coquimbite) represent a significant source of acidity on the landfill. The fly ash zone contained authigenic pyrite framboids that were likely formed through the weathering of the overlying detrital pyrite fragments, and secondary iron precipitates. Because of their dissimilar compositions, the two materials responded differently to the environmental conditions of the flow-through experiment.

Initial eluents from the coal rubble zone columns (zone A) were substantially lower in pH ($\sim \leq 1.0$) than that from the fly ash columns, regardless of temperature or flow regime. In continuous flowing (saturated) columns, the pH of fluids from the coal rubble zone established a dynamic equilibrium at a pH of 6 to 7 within the first week of the experiment, whereas eluent pH from the fly ash zone fluctuated more erratically from low to high. Eluent pH also increased consistently within columns that were episodically flushed; regardless of the temperature or flow regime, although the long-term increase in pH was significantly lower than that observed in the continuous flow experiments. A temperature effect on acidity was exhibited in the fly ash columns. At 12.7^0C, effluent chemistry was not affected greatly by gas composition or flow regime, and pH generally did not increase above 3.5. However, eluent pH increased both episodically and periodically at 29.4°C, suggesting that reactive surfaces were partially stabilized and inorganic salts were effectively mobilized from the system. This effect was more pronounced in columns that were exposed to nitrogen compared to those exposed to oxygen.

As with the pH response, initial concentrations of major and minor elements were substantially higher in the coal rubble effluents than the fly ash zone. However, concentration ranges were similar between the two zones once the initial flush had occurred. Elemental concentration patterns for eluents from the continuous flowing experiments were similar between columns from the two zones regardless of gas phase or temperature. In the episodically flushed columns, significant differences with respect to the eluent elemental concentrations were observed under the varying environmental conditions. These columns, which most closely resemble actual conditions on the basin, showed the dynamic nature of materials in response to alterations in saturation and oxidation state. In general, the column materials responded to the periodic wetting and drying cycles with great swings in elemental concentrations. The beginning of a cycle would generally correspond with high element concentrations in the eluents, followed by a decline through the wetting period, and a subsequent rebound to higher levels following a dry-down period. This is most likely a result of the deposition and resolubilization of reactive salts during drying and wetting cycles, respectively. The magnitude of this swing was influenced by both temperature and gas phase. The concentration gradient from beginning to end of a cycle was greatest at the higher temperatures for elements directly associated with pyrite oxidation (i.e. Fe and S), while species that may be influenced by subtle changes in redox (i.e. As and Se) responded more clearly to differences with respect to the gaseous composition.

These results suggest that the low pH fluids and high elemental concentrations emanating from materials in the 488-DAB are closely associated with environmental factors that control dissolved and/or gaseous oxygen and to periodic wetting and drying events. As such, a dry cover should reduce the production of acid lechate over time if it is designed to retard or eliminate fluid flow and oxygen transport to the subsurface. In addition, the use of a cover to minimize the occurrence of evaporative salts on the landfills' surface may significantly reduce the contaminant load given that they are a major contributor of acidity in this system.

ACKNOWLEDGEMENTS

The authors would like to express their gratitude to Miles Denham with the Westinghouse Savannah River Technology Center for his assistance during this project. This work was funded in part by the U.S. Department of Energy (Financial Assistance Award DE-FC09-96SR18546) to the Savannah River Ecology Laboratory through the University of Georgia Research Foundation, Inc.

REFERENCES

1. Kleinmann, R.L.P. Acidic mine drainage: U.S. Bureau of Mines researches and develops control methods for both coal and metal mines, 1989, 161.
2. Nordstrom, D.K., Aqueous pyrite oxidation and the consequent formation of secondary iron minerals. *In*: J.A. Kittrick et al. (ed.) Acid sulfate weathering. SSSA Spec. Pub. No. 17, SSSA, Madison, WI, 1986, 223.
3. Singer, D.C., and W. Stumm, Acidic mine drainage: the rate determining step. Science, 119, 1121, 1970.
4. Stumm, W., and J.J. Morgan. Aquatic Chemistry, 2nd ed.,J. Wiley & Sons, New York, 1981.
5. Pierce, W.G., B. Belzile, M.E. Wiseman, and K. Winterhalder, 1994. Proceedings of the International Land Reclamation and Mine Drainage Conference, US Dept. of the Interior. SP06B-94, 148, 1994.
6. Evangelou, V.P, Pyrite Oxidation and its Control. CRC Press. Boca Raton, 1995.
7. Dobson, M.C., and A.J. Moffat, The Potential for Woodland Establishment on Landfill Sites. HSMO Press, 1993.
8. Lalvani, S.B., B.A. DeNeve, and A. Weston, Passivation of pyrite due to surface treatment. Fuel, 69, 1567, 1990.
9. Belzile, N., S. Maki, Y, Chen, and D. Goldsack, Inhibition of pyrite oxidation by surface treatment. Sci. of the Total Environ., 196, 177, 1997.
10. Allison, J.D., D.S. Brown, and K.J. Novo-Gradac, MINTEQA2/PRODEFA2, a Geochemical Assessment Model for Environmental Systems, Version 3.0 User's Manual, USEPA-Environmental Research Laboratory, Athens, GA, 1990.
11. Gee, G.W., and J.W. Bauder, Particle size analysis. *In*: A. Klute (ed.) Methods of Soil Analysis, Part 1: Physical and Mineralogical Methods, 2nd Edition. SSSA, Madison, WI, 1986, 393.

12. Bingham, J.M., and D.K. Nordstrom, Iron and aluminum hydroxysulfates from acid sulfate waters. *In*: C.N. Alpers et al (eds.) Reviews in Mineralogy and Geochemistry, Vol 40: Sulfate Minerals, Mineralogical Society *of* America, Washington, DC, 2000, 351.

13. Jambor, J.L., D.K. Nordstrom and C.N. Alpers, Metal-sulfate salts from sulfide mineral oxidation. *In*: C.N. Alpers et al (eds.) Reviews in Mineralogy and Geochemistry, Vol 40: Sulfate Minerals, Mineralogical Society of America, Washington, DC, 2000, 305.

14. Nordstrom, D.K., and H.M. May, Aqueous equilibrium data for mononuclear aluminum species. *In*: G. Sposito (ed.) The Environmental Chemistry of Aluminum, 2nd Edition, CRC Press/Lewis Publishers, 1996, 39.

15. Karathanasis, A.D., and W.G. Harris, Quantitative thermal analysis of soil materials. *In*: J.E. Amonette and J.W. Stucki (eds.) Quantitative Methods in Soil Mineralogy, SSSA, Madison, WI, 1994, 360.

16. Paulik, J.F., and M. Arnold, Simultaneous TG, DTG, DTA, and EG techniques for the determination of carbonate, sulfate, pyrite and organic material in minerals, soils and rocks. J. Thermal Anal., 25, 327, 1982.

17. Almeida, C.M.V.B., and B.F. Giannetti, Comparative study of electrochemical and thermal oxidation of pyrite. J. Solid State Electrochemistry, 6, 111, 2002.

18. Schwertmann, U., and R.M. Taylor, Iron oxides. *In*: J.B. Dixon and S.B. Weed (eds.) Minerals in the Soil Environment, 2nd Edition, SSSA, Madison, WI, 1989, 379.

19. Karathanasis, A.D., and B.F. Hajek, Revised methods for rapid quantitative determination of minerals in soil clays. Soil Sci. Soc. Am. J., 46, 419, 1982.

20. Tudor, D.N., Thermal Analysis of Minerals. Abacus Press, Kent, England, 1976.

21. Blowes, D.W., E.J. Reardon, J.L. Jambor, and J.A. Cherry, The formulation and potential importance of cemented layers in inactive sulfide mine tailings. Geochim. Cosmochim. Acta., 55, 965, 1991.

22. Nordstrom, D.K., C.N. Alpers, C.J. Ptacek., and D.W. Blowes, Negative pH and extremely acidic mine waters from Iron Mountain, California. Environ. Sci. Technol., 34, 254, 2000.

23. Garrels, R.M., and C.L. Christ, Solutions, Minerals, and Equilibria. Jones and Bartlette Publishers, Inc., Boston, MA, 1990.

24. Faure, G., Principles and Applications of Inorganic Chemistry. Macmillan Pub. Co., New York, NY, 1991.

25. Lindsay, W.L., Chemical Equilibria in Soils. John Wiley & Sons, New York, NY, 1979.

26. Elrashidi, M.A., D.C. Adriano, S.M. Workman, and W.L. Lindsay, Chemical equilibria of selenium in soils: A theoretical development. Soil Science, 144, 141, 1987.

SELENIUM CONTENT AND OXIDATION STATES
IN FLY ASHES FROM WESTERN U.S. COALS

Shas V. Mattigod and Thomas R. Quinn

Pacific Northwest National Laboratory
Richland, WA 99352, USA.

ABSTRACT

A selective extraction scheme was developed for the determination of the oxidation states of Se species in coal ashes. As compared to HF dissolution, extractions with 70% $HClO_4$ mobilized 90 to 100% of all compound and redox forms of Se from four of the five fly ashes. Extractions with 16M HNO_3 did not mobilize all forms of Se as effectively as perchloric acid. Both oxidized forms of Se (IV and VI) were completely mobilized by 12M HCl extraction. Deionized-distilled water was not an effective extractant for mobilizing all compound forms of Se(IV) from fly ashes. Extraction data (70% $HClO_4$, 16M HNO_3, 12M HCl, DI water) indicated that the solid:solution ratio is a critical factor in Se extractability from fly ashes. Maximum extractions in all cases were obtained only with very high (1:500) solid:solution ratios. Extraction times from 1.5 to 25 hours did not significantly change Se extractability with any of the extractants except with 12M HCl, which required a minimum reaction time of 48 hours to attain maximum Se extractability. Reaction times shorter than the critical time and low solid:solution ratios significantly affected Se extractability from these fly ashes. Measurements of Se content and redox state in particle size and density fractions five western United States coal ashes indicated that typically, the Se content increased with decreasing particle size.. However, no consistent trend in Se concentration between the light and heavy density fractions of <2.7-μm size fraction was observed. Selenium redox state data indicated that only Se(0) and Se(IV) forms were present in these five coal ashes. The presence of Se(IV) is significant since it is much more easily mobilized than the elemental form. Examination of fly ashes by the proposed scheme to determine Se redox species could permit better estimation of the Se content of plants grown on fly ash amended soils.

INTRODUCTION

In the past two decades, coal production has doubled in the United States. In 1996, 1.06 billion tons of coal was produced[1] with electric utilities consuming approximately 55% of the production. Currently, coal combustion by these utilities annually results in the collection of about 63 million tons of ash residues[2]. The need to dispose of such large quantities of ash and questions of environmental effects prompted extensive research to determine alternative, beneficial uses for this material. Research into the chemical and physical properties of fly ash, which have been reviewed extensively[3-5] indicated that it might be suitable for use as an agricultural soil amendment. Some of the early studies showed that fly ashes could supply essential elements to crop species[6-11], favorably affect some soil physical properties[12], and also function as a liming material[13]. However, trace elements, such as Se, were also identified as potentially limiting the use of fly ash as a soil amendment[3,4,14].

Selenium, although essential to animal and possibly human nutrition is toxic to animals at relatively low concentrations. The range of Se concentrations between essential (>0.04 ppm[15]) and toxic (>3-5 ppm[16]) levels in animal forage is narrow and leaves little room for error. Potentially toxic levels of Se have been found in crop species grown on fly ash amended soils[20,21]. In one study, Furr et al.[17] found that only one of four fly ashes with similar total Se content (16 ppm) yielded crops containing potentially toxic levels of Se. The other ashes yielded Se levels only one third as high, which did not represent a significant hazard. Mbagwu[16] determined that fly ashes could effectively serve as a Se supplement to soils, which produce Se deficient forage. Since Se deficiency has been recognized as a significant factor in livestock losses in large areas of the United States[15, 18], identification of fly ash characteristics responsible for these observed plant uptake differences would be beneficial.

Earlier work on Se toxicity in the Great Plains would seem to explain some of the observed differences. Workers identified Se speciation as being of critical importance in plant uptake[19,20]. In a two-year field experiment, total Se uptake by mustard, as a percentage of added Se was determined to be 30% from K_2SeO_4, 4% from K_2SeO_3, and 0.10% from elemental Se[21]. These observations, which confirm the work of other researchers, are a result of the differing solubilities of the Se species. Selenates are very soluble while selenites are only sparingly soluble and their dissolved concentrations are further limited by adsorption on to hydrous oxide surfaces. Elemental Se is relatively insoluble. Based on extensive observations of Se uptake by terrestrial plants growing on a coal fly ash landfill, a number of investigators concluded that selenite and selenate species were taken up by plants by passive and active modes, respectively, whereas, elemental selenium was unavailable to plants until oxidized to more soluble forms[22-25]. These data clearly established that oxidation state of Se species in fly ashes would influence the mobilization of this element upon disposal of these ashes.

A selective extraction method that provides the oxidation state of Se species in fly ashes would be beneficial in several respects. The Se uptake of plant species grown on fly ash amended soils could be examined with respect to the oxidation states of Se originally present in fly ashes and its subsequent oxidation and mobilization. This could prove valuable in better understanding the behavior of Se in fly ash amended soils. In addition, the potential hazards of Se mobilization from fly ash disposal sites could be more readily evaluated.

Selenium in fly ashes can potentially exist in 0, IV, and VI oxidation states. Selenides that may exist in parent coals are oxidized during combustion (similar to sulfide oxidation) and, therefore, will not be found in fly ashes. In an earlier study, Andren et al.[26] reported the presence of only elemental Se and Se(IV) in a single sample of fly ash. However, as procedures for extraction of Se were not reported, methods used in this study to determine the oxidation state of Se species present in fly ashes are somewhat in doubt.

Recently, several methods have been proposed for Se extraction from and speciation in soils[27-32], water, and sodium hydroxide leachates from fly ashes[30,33,34]. However, Martens and Suarez[31] pointed out failure to use of standard reference compounds and spikes with known concentrations and oxidation states would limit the utility of schemes designed to determine Se speciation. Additionally, many of the studies were focused on determining Se speciation in only water, seawater, and hydroxide leachable fractions. Such extractions do not provide data on speciation of all Se (not an arbitrarily extractable fraction in whole ash) originally present whole fly ashes. Also, it has been clearly established that factors such as solid:solution ratio and length of time of extraction significantly affect extractions of trace elements from solids such as soils[35]. A review of published literature[36] listed wide variations in fractions of water and acid extractable Se in various fly ashes thus confirming the effects of factors such as solid:solution ratio and length of time of extraction. These data pointed out that a significant need exists to accurately assess the effects of the various types and strengths of extractants, extraction times, and solid:solution ratios on Se extractability from fly ashes.

At present, there are no systematic studies of speciation of total Se in fly ashes that includes evaluation of factors such as the extractant type, solid:solution ratio, and extraction times that includes appropriate use of standard reference compounds and Se spikes with known oxidation states. Therefore, our objectives were to 1) develop a scheme for determining speciation of total Se in fly ashes that included an evaluation of effectiveness of various extractants, effect of solid: solution ratio, and the length of extraction time on extractability of various Se redox species from fly ashes, and to 2) determine the Se content of size and density fractions and also Se speciation in of five western United States coal ashes.

MATERIALS AND METHODS

Fly Ash Samples

All fly ash samples were collected from the electrostatic precipitators of power generating stations. The fly ashes SB01, SB02, and SB03 were derived from combustion of subbituminous coals, and the fly ashes B01 and B02 resulted from burning of bituminous coals. After collection, the fly ashes were stored in glass jars under nitrogen to avoid oxidation of reduced Se species[31].

Size and Density Fractionation

The ash samples were initially dry sieved to separate the <45-μm size fraction from the coarser material. A Bahco microclassifier was then used to obtain <2.7-μm, 2.7-8-μm, 8-15-μm, and 15-45μm size fractions by means of air elutriation. Only fractions <15μm were used in these studies. In addition, the <2.7μm size fraction was subjected to sink-float separation in a 2.1-Mg m^{-3} solution. The solution was prepared from a mixture of 20% polyvinylpyrolidone (PVP) in absolute ethanol (EtOH), and 1,1,2,2,-tetrabromoethane (TBE) (Malinkrodt)[37]. A ratio of 39% v/v of PVP-EtOH and 61% TBE yielded the desired density. The PVP is included to prevent flocculation of ash particles. Approximately 150 mg of ash was placed on top of 20 ml of the density solution in a 40-ml glass centrifuge tube. After dispersing the ash in the solution by means of a vortex mixer, the tubes were centrifuged for about 4 hours at 750 rpm in an IEC Model K Centrifuge. After determining that separation had occurred, the float fraction and much of the solution were removed by means of a suction device. The float and the sink fractions were then centrifuge washed 5 times with absolute ethanol to remove traces of TBE and PVP. Excess ethanol was then removed from the samples in a vacuum desiccator. The total Se content was determined by HF decomposition[38], followed by Hydride Generation Atomic Absorption Spectroscopy (HGAA).

Effect of Time and Solid:Solution Ratio on Se Extraction

The effectiveness of different extractants used at different solid:solution ratios and leaching times on Se extractability was evaluated using the SB01 fly ash. The oxidation states of Se were determined by means of selective chemical extractions. Previous studies of soils have indicated that various redox species of Se are selectively dissolved in various acids and water[20,31,39]. It has been suggested that acids such as perchloric, nitric, and hydrofluoric will solubilize Se in the 0, IV, and VI redox states[26,39]. Based on these studies, several extractants were tested to ascertain the degree of selectivity for various redox species of Se that may be present in fly ash samples. The extractants tested were 70% HClO$_4$, 16M HNO$_3$, aqua regia + 48% HF, 12M HCl, distilled, and deionized (DO) water. All extractions were conducted in triplicate. The procedures for each of these was as follows:

1) Total Selenium
A. HClO$_4$ or HNO$_3$ Extraction: Fly ash sample and 70% HClO$_4$ or 16M HNO$_3$ were placed in 50-ml screw cap Erlenmeyer flasks to obtain solid:solution ratio ranging from 1:4 to 1:500. After shaking for 1.5 - 25 hours on a Burrell Model 75 wrist action shaker, the solutions were analyzed for Se by means of HGAA.
B. HF Decomposition[38,40]: 200 mg of coal ash was mixed with 1 ml of aqua regia (18/82 concentrated HNO3 and concentrated HCl v/v) before 6 ml of concentrated HF was added to Teflon-lined bombs (Parr 4745 general purpose bomb, Parr Instrument Co., Moline, Ill.). The bombs were heated at 105°C for 1.5 hours. Upon cooling to ambient temperature, the bomb contents were washed with water into 5.6 g H$_3$BO$_3$ in a polypropylene beaker. After dilution to approximately 80 ml, the samples were heated on a steam bath until the H$_3$BO$_3$ dissolved. The sample was cooled and transferred to a 100-ml volumetric flask. After the sample was made to volume with distilled water, the samples were filtered and stored in plastic containers. The Se content was then determined by means of HGAA.
2) Selenium (IV + VI)

HCl Extraction: Coal Ash and 12M HCl were placed in 50-ml or 100-ml screw cap Erlenmeyer flasks to obtain solid:solution ratios of 1:4 to 1:500. After shaking for 1.5 - 65 hours on a Burrell Model 75 wrist action shaker, the solutions were analyzed for Se by means of HGAA. As concentrated HCl reduces Se(VI) to Se(IV), this analysis cannot differentiate between these oxidation states[41].
3) Selenium (VI)

Distilled-deionized water extraction: Fly ash and DO water were placed in 50-ml screw cap Erlenmeyer flasks to obtain solid:solution ratios of 1:4 to 1:500. The samples were then shaken on a Burrell Model 75 wrist action shaker for time periods of 1.5 - 25 hours. The samples were then analyzed for Se(IV) and Se(VI) by means of HGAA. This was accomplished by analyzing the solutions before and after the reduction of Se(VI) to Se(IV). Since the HGAA procedure is specific for Se(IV), the difference between the Se concentration before and after reduction of Se(VI) with HCl (12M) would yield the quantity of Se(VI) present in the solutions.

Spike Recovery Experiments to Evaluate Extractant Specificity

The specificity of these procedures was tested through known additions (matrix spikes) of approximately 100 ppm of various compounds containing known oxidation states of Se to one of the (SB03) fly ashes. The reason for this choice was that this fly ash contained very low total Se concentrations (<1 ppm) and thus would not affect the extractability of compounds added as matrix spikes. The gray allotrope of elemental Se (Baker) was used to represent Se(0). Sodium selenate (Na_2SeO_4; Alfa) was used as a source of Se(VI). The Se(IV) compounds used were sodium selenite (Na_2SeO_3; Alfa) and a synthetic iron oxide-selenite compound. The iron oxide-selenite compound was prepared by adding selenous acid (0.1226 g SeO_2 in 40 ml distilled water) to 10 g Fe_2O_3 (Mallinkrodt). This mixture was centrifuge washed 5 times with distilled water to remove any excess Se. It was then dried in a vacuum oven at 60°C. Selenium content of this compound was found to be 730 ppm Se by HF decomposition method. These Se amended samples were then analyzed by the extraction procedures so that the accuracy of these procedures could be known with greater certainty.

The HGAA procedure used to determine the Se content of the samples utilized a MHS 10 Hydride Generator (Perkin-Elmer) coupled to a Perkin-Elmer Model 5000 Atomic Absorption Spectrometer which was set at the 196.0 nm wavelength for Se.

The MHS 10 contains a reservoir filled with 3% sodium borohydride (Aldrich) and 1% sodium hydroxide (Mallinkrodt) in water. Nitrogen gas (38 psi) was used to transfer several milliliters of this solution to the sample holder containing 0.1 - 4.0 ml of sample plus 10 ml of 4N HC1. The Se(IV) present in the sample was then reduced to H_2Se and swept from the sample holder by the N_2 gas into a quartz cell in the flame of the Perkin-Elmer 5000. The H_2Se decomposes in the flame allowing the Se absorption peak to be recorded on a strip chart recorder. The Se concentrations of the samples were then determined by comparison of the sample peak heights to a standard curve.

Selenium Concentrations in Size and Density Fractions and Oxidation States of Selenium in Bulk Fly Ashes.

The total Se concentrations in the bulk ashes, the size fractions, size-density fractions, and the NBS 1633a reference fly ash sample were determined by the HF dissolution method[38,40]. The redox states of Se in all bulk fly ash samples were determined using the optimum extraction scheme that was developed in this study.

RESULTS AND DISCUSSION

Particle size and Density Fractions

The total Se concentrations in the bulk ashes, the size fractions, size-density fractions, and the NBS 1633a reference fly ash sample determined by the HF dissolution method are presented in Table 1. The value of 10.8 ± 0.6 ppm found for the NBS 1633a compares favorably with the reported value of 10.3 ± 0.6 ppm. Nadkarni[41] reported similar accuracy when analyzing NBS reference materials by HGAA.

Table 1. Total Selenium Content (µg/g) of Size and Density Fractions for Five Western U. S. Fly Ashes Determined by HF Dissolution

Sample	Bulk*	8 - 15	2.7 - 8	<2.7	<2.7 Density (Mg m⁻³) >2.1	<2.1
SB01	21.6 ± 1.4	12.9 ± 0.5	24.7 ± 1.6	34.6 ± 2.2	36.3 ± 3.6	62.3 ± 6.1
SB02	5.2 ± 0.5	4.3 ± 0.3	8.5 ± 0.4	12.9 ± 1.2	14.7 ± 2.0	9.3 ± 1.0
SB03	0.9 ± 0.2	0.5 ± 0.1	1.2 ± 0.2	1.1 ± 0.2	1.3 ± 0.2	1.1 ± 0.2
B01	9.4 ± 0.4	15.1 ± 1.2	19.1 ± 1.3	19.7 ± 1.0	18.1 ± 1.3	28.4 ± 2.4
B02	7.9 ± 0.5	5.6 ± 0.6	17.1 ± 1.8	23.6 ± 1.5	28.5 ± 2.2	15.3 ± 0.8

*NBS 1633a fly ash standard analyzed at 10.8 ± 0.6 µg/g as compared to a reported value of 10.3 ± 0.6 µg/g.

These results demonstrated a particle-size dependence of Se in these fly ashes. Studies by a number of investigators have firmly established the size dependence concentration characteristic of Se in fly ashes[26,42-49]. Volatilization of Se with subsequent condensation on particle surfaces is an explanation for this observed behavior. With higher surface area per unit mass, the finer particles achieve significantly higher Se concentrations. Except for SB03, which contained the lowest concentration of Se, each of the ashes examined in this study exhibited greater Se concentrations in the <2.7 μm fraction than in the bulk ash. In another detailed study of four size and density fractionated fly ashes Mattigod et al.[50] noted similar particle-size dependence of Se concentration.

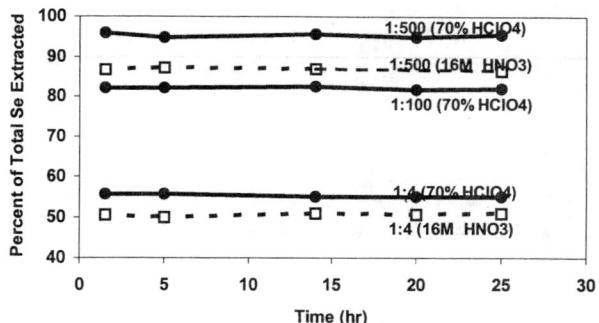

Figure 1. The Effect of Solid to Solution Ratio and Time on Selenium Extractability from SB01 Fly Ash Using 70% Perchloric and 16M Nitric Acids.

If particle size is the only criteria in Se concentration on coal ash particulates, one would then expect concentration to be independent of particle density and to observe similar concentrations of Se in the >2.1 Mg m-3 and the <2.1 Mg m-3 density fractions of the <2.7 μm size fraction. However, these data clearly demonstrated that in four of the five ashes studied, significant Se concentration differences existed between two density fractions of the same particle size fraction of (Table 1). The SB01 and B01 ashes have the greatest Se concentrations in the <2.1-Mg m-3 fraction while the SB02 and B02 ashes have higher Se concentrations in the >2.1 Mg m^{-3} fraction. These differences may be related to the differences in volatile and nonvolatile fractions of Se that may be present in the source coals or to differences in combustion techniques. It is known that coals contain both organic and inorganic Se sources that have significantly different volatility during coal combustion[51]. These differences would need to be examined further to reach definite conclusions about Se distribution among density fractions of fly ashes.

Comparison of Extraction Methods

Selenium extractability from bulk SB01 fly ash using various extractants as a function of different solid:solution ratios and leaching times are shown in Figures 1 - 3. In all cases, the extractability of Se was expressed as a percentage of total Se in bulk fly ash as measured by the HF decomposition method.

The data indicated (Figure 1) that extraction times ranging from 1.5 - 25 hours did not change Se extractability with either acids at all solid:solution ratio. In all cases 70% $HClO_4$ extracted significantly higher proportion of total Se than 16M HNO_3. The results of these extractions show that the solid:solution ratio significantly affects Se extractability. At low solid:solution ratio (1:4) both acids extracted only ~50 – 55% as compared to ~87 – 97% Se extraction at 1:500 solid to solution ratio. These data show that the solid:solution ratio is a major factor in Se extractability from fly ash. Also, these tests demonstrated that 70% $HClO_4$ acid extractions conducted for 1.5 to 25 hours at 1:500 solid:solution ratio provides an excellent indication of total Se (as determined by HF method) in fly ashes.

The 12M HCl extraction (Figure 2) did not yield constant Se extractability until 48 hours of reaction. As in the other extractions, the 1:500 solid:solution ratio yielded the high percentage of the added Se, but there was no significant difference between the 1:200 and 1:500 solid:solution ratios. Extending the period of extraction beyond 48 hours did not significantly increase the concentration of Se in solution. It is clear from these data that 12M HCl extractions conducted at much lower solid:solution ratios, and for shorter time periods resulted in significantly lower Se extractability. Thus, a 1:500 solid:solution ratio with a reaction time of 48 hours appeared essential for maximum Se extraction by 12M HC1.

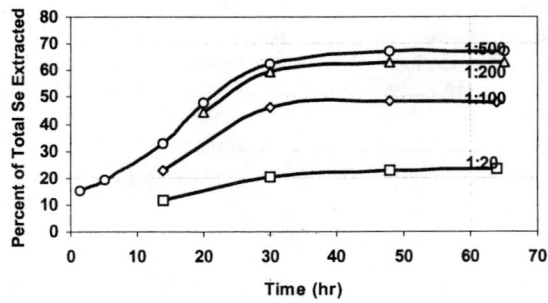

Figure 2. The Effect of Solid to Solution Ratio and Time on
Selenium Extractability from SB01 Fly Ash Using 12 M HCl

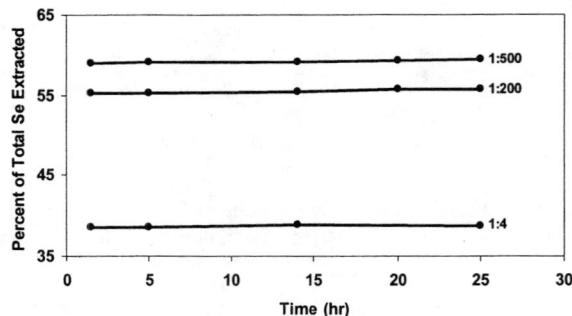

Figure 3. The Effect of Solid to Solution Ratio and Time on Selenium
Extractability from SB01 Fly Ash Using Deionized Distilled Water.

The distilled-deionized water extraction, like that of $HClO_4$ and HNO_3 extractions yielded constant Se extractability for reaction times ranging from 1.5 to 25 hours (Figure 3). A 1:500 solid- solution ratio yielded the best results, but, as in the case of HCl, the results were not significantly different than a 1:200 solid:solution ratio. At a 1:4 solid:solution ratio, however, Se(IV) was completely oxidized to Se(VI) during a period of approximately one week (data not shown). The pH values of the 1:4 extracts were between 11.4 and 12.3 and favor the formation of Se(VI) species under oxidizing conditions[52]. The pH of the 1:500 solid:solution ratio mixtures was approximately 10.0. While Se(VI) species are still favored under oxidizing conditions at pH 10 , These results showed that distilled-deionized water extraction was also significantly influenced by the solid:solution ratio.

All these extraction data (70% $HClO_4$, 16M HNO_3, 12M HCl, DI water) indicated that solid:solution ratio is an extremely important factor in Se extractability from fly ashes. Maximum extractions in all cases were obtained with the highest (1:500) solid:solution ratios. Additionally, DI water extractions of alkaline fly ashes at low solid:solution ratios (1:4) promoted Se(IV) oxidation. Also, for all extractants except 12M HCl, reaction times between 1.5 - 25 hours did not significantly change Se extractability. Extractions with 12M HCl indicated that a minimum of 48 hours of reaction time was required to maximize Se extractability.

These results suggested that Se extractability for fly ashes reported in literature that were generated at unoptimized (especially low) solid:solution ratios using extractants such as 70% $HClO_4$, 16M HNO_3, 12M HCl, and DI water are probably low estimates and do not represent potentially maximum Se extractability.

Extractant Specificity

The recovery of Se compounds added to SB03 fly ash is presented in Table 2. It can be seen that essentially complete recovery of elemental Se is attained from the 70% $HClO_4$ extraction. The 70%

$HClO_4$ extraction also yielded complete recovery of added Se(IV) and Se(VI) species. Therefore, 70% $HClO_4$ appears to be a suitable extractant for the determination of total Se in the coal ashes tested in this study. However, the validity of 70% $HClO_4$ extraction for determining total Se in acidic fly ashes and other coal combustion byproducts such as bottom ash need to be established by comparing the results with total Se values obtained with HF decomposition method.

The 12M HCl extraction did not solubilize detectable quantities of elemental Se (Table 2). It was, however, able to yield essentially complete recovery of added Se(IV) and Se(VI) compounds. Thus, the 12M HCl extraction did provide the sum of the Se(IV) and Se(VI) species present in this ash. For these fly ashes, the difference in values between the HF dissolution (or 70% $HClO_4$ extraction) and the 12M HCl extraction yields a measure of the elemental Se present in the ash.

Table 2. Extractant Recovery Efficiency (%) of Selenium Compounds added to SB03 Fly Ash

Se Species Added[a]	Extractant[b]		
	70% $HClO_4$	12M HCl	DD water
Se(0) Gray Allotrope	99.4 ± 2.8	BD	BD
Se(IV) Na_2SeO_3	100.2 ± 2.3	98.8 ± 3.1	99.1 ± 2.3
Se(IV) $Fe_2O_3 + SeO_3^{2-}$ Complex**	98.3 ± 2.9	100.2 ± 3.4	81.9 ± 3.0
Se(VI) Na_2SeO_4	99.1 ± 2.5	98.7 ± 3.2	100.1 ± 3.8

BD: Below Detection

[a]$100 \mu g/g$. ** $Fe_2O_3 + SeO_3^{2-}$ Complex contained $430 \mu g/g$ of Se.
[b]All extractions were conducted at 1:500 solid to solution ratio. Perchloric acid and water extractions were conducted for 24 hours, and HCL extractions were conducted for 48 hours.

The distilled-deionized water extraction was able to solubilize all the added sodium selenate and sodium selenite, but only 81.9% of the 730 ppm Se from the iron-selenium (IV) compound. Thus water cannot perform the function of the 12M HCl in obtaining the sum of the selenate and selenite species. Because Se(VI) compounds are very soluble in water, it is still possible to quantify Se(VI) and Se(IV) species present in these ashes. The HGAA method determines only Se(IV) species. Therefore, analysis of water extracts before and after reduction of Se(VI) to Se(IV) [reduction is accomplished by HCl^{41}] permits quantification of Se(VI). The Se(IV) value is obtained by subtracting the Se(VI) value from the combined Se(IV) and Se(VI) value obtained in the HCl extraction.

Oxidation States of Selenium in Fly Ashes.

Water extracts analyzed after reduction with HCl showed that there were no measurable concentrations of Se(VI) in these ash samples (Table 3). Therefore, Se in these ash samples existed mainly in two oxidation states [Se(0) and Se (IV)]. These data also revealed that between ~0 – 50% of the Se(IV) present was not extracted by DD water. This result suggests that significant fractions of Se(IV) in these ashes may be similar to the water inextractable phase of $Fe_2O_3 + SeO_3^2$ complex (Table 2).

Table 3. Selenium Redox Forms Extracted from Five Western U. S. Fly Ashes ($\mu g/g$)

Fly Ash	Total: Se(0)+Se(IV)[a]		Total Se(IV)[b]		Water soluble Se(IV)[c]	
	Bulk	<2.7 μm	Bulk	<2.7 μm	Bulk	<2.7 μm
SB01	21.6 ± 1.4	34.6 ± 2.2	14.7 ± 1.7	30.4 ± 1.3	12.6 ± 0.8	15.5 ± 0.7
SB02	5.2 ± 0.5	12.9 ± 1.2	4.3 ± 0.3	12.4 ± 0.7	2.8 ± 0.2	9.1 ± 0.4
SB03	0.9 ± 0.2	1.1 ± 0.2	0.2 ± 0.1	0.2 ± 0.1	0.2 ± 0.1	0.2 ± 0.1
B01	9.4 ± 0.4	19.7 ± 1.0	8.3 ± 0.5	16.0 ± 0.8	3.4 ± 0.3	7.8 ± 0.3
B02	7.9 ± 0.5	23.6 ± 1.5	6.8 ± 0.4	19.3 ± 0.8	4.8 ± 0.3	13.7 ± 0.4

[a]Determined by HF dissolution
[b]12M HCl extraction (48 hour extraction at 1:500 Solid to solution ratio).
[c]DD water extraction (24 hour extraction at 1:500 Solid to solution ratio)

The percentages of Se(0) and Se(IV) in the bulk and the <2.7-μm fractions of the ashes is shown in Figures 4 and 5. These figures show that except in fly ash SB03, Se in all other ashes exists principally in the oxidized form as Se(IV). The presence of significantly different levels of Se(IV) in fly ashes helps to explain the results of other investigators who found that ashes with similar total Se contents yielded significantly different Se uptake by cabbage[53,54].

Elemental Se and Se(IV) accounted for essentially all Se except in the case of the SB03 ash. This ash had a low Se content relative to the other ashes and appeared to contain Se in some unextractable phase. Therefore, when determining Se oxidation states in coal ashes, it is recommended that both the $HClO_4$ extraction and the HF decomposition be run on the samples. A comparison of these results would indicate if significant levels of Se were not extractable by $HClO_4$.

The presence of Se(0) and Se(IV) oxidation states in these fly ashes does not provide answers for the actual Se species present. From the results obtained, however, some inferences may be made. The elemental Se would exist as the gray allotrope as this form is the most stable at temperatures over 60°C. The Se(0) could have been formed from either the oxidation of selenides or the reduction of Se(IV) species present in the coal during combustion. The reduction of Se(IV) to Se(0) in the presence of SO_2 was used by Andren et al.[26] as a possible explanation for Se(0) being the only Se oxidation state found in their ash sample. However, it is apparent that this reduction mode is not a dominant factor in the coal fly ashes examined in this work.

The Se(IV) present could be in the form of a number of compounds. Lakin[15] suggested that SeO_2 was a possible Se compound in fly ashes. The results obtained in this work were not inconsistent with SeO_2 being a significant fraction of the total Se(IV) present in the ashes. However, the lack of total water solubility of the entire Se(IV) fraction could indicate Se(IV) species with reduced solubility (Table 3). Iron oxide-selenite species could account for the observed lack of complete water solubility (Table 2). The presence of less soluble species is important since they serve to reduce the availability of Se to plant species and the environment in general. Further work would be necessary to elucidate the nature of these less soluble Se(IV) species that exist in fly ashes.

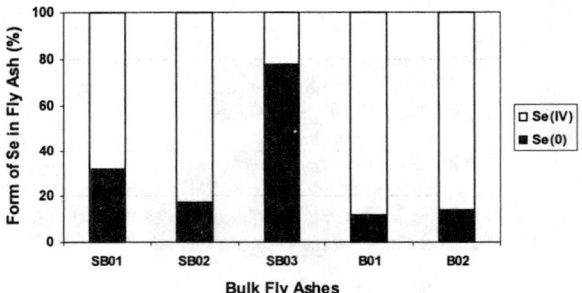

Figure 4. Redox Species Distribution in Bulk Fly Ashes.

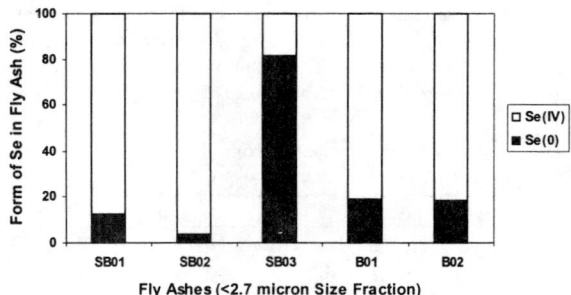

Figure 5. Redox Species Distribution in <2.7 µm Particle Size Fraction of Fly Ashes.

SUMMARY AND CONCLUSIONS

We developed a scheme for determining Se speciation in fly ashes by evaluating the effects of factors such as specificity various extractants, varying solid to solution ratio and reaction time on extractability of Se redox species. We also determined the Se content of bulk, size and density fractions, and also Se speciation in fly ashes derived from three subbituminous and two bituminous coals from the Western United States.

Based on extractions of a fly ash sample (B02) spiked with various compounds that contained various redox forms of Se, extractions with 70% $HClO_4$ completely mobilized all compound and redox forms of Se. However, this extractant mobilized 90 – 100% of all Se from only four of the five fly ashes. Extractions with 16M HNO_3 did not mobilize all forms of Se as effectively as perchloric acid. Both oxidized forms of Se (IV and VI) were completely mobilized by 12M HCl extraction. Deionized-distilled water was not an effective extractant for mobilizing all compound forms of Se(IV) from fly ashes.

Extraction data (70% $HClO_4$, 16M HNO_3, 12M HCl, DI water) indicated that solid:solution ratio is a critical factor in Se extractability from fly ashes. Maximum extractions in all cases were obtained with the highest (1:500) solid:solution ratios. Extractions of these alkaline fly ashes with deionized-distilled water at low solid:solution ratios (1:4) promoted Se(IV) oxidation. Also, for all extractants except 12M HCl, reaction times between 1.5 to 25 hours did not significantly change Se extractability. However, a minimum reaction time of 48 hours was needed to attain maximum Se extractability with 12M HCl. Reactions times shorter than the critical time and low solid:solution ratios significantly affected Se extractability from these fly ashes.

The five western United States coal ashes were examined for Se content and Se oxidation states. The Se content varied inversely with particle size. Selenium concentrations in density fractions of the <2.7-pm size fraction did not show any consistent trends. Further work is necessary to explain the observed Se concentration characteristics in these density fractions.

Only Se(0) and Se(IV) were shown to be present in these five fly ashes. Se(VI) was not detected in water-soluble fractions of any of these ashes. The presence of Se(IV) is significant since it is much more easily mobilized than the elemental form. In addition, Se(IV) may be oxidized to Se(VI) at high pH values. The presence of these Se(IV) and Se(VI) species has been shown to be important to plant uptake of Se. Examination of fly ashes by the method presented could permit better estimation of the Se content of plants grown on fly ash amended soils.

ACKNOWLEDGEMENT

This research was partially funded through United States-Israel Binational Agricultural Research Development Fund (BARD) Project 1-97-80. We thank Ms Pat Hays for editorial assistance, and Dr. J. E. Amonette for peer review.

REFERENCES

1. National Mining Association. Mining Facts/About Coal. http://www.nma.org/coalfacts.html, 2001.
2. EPA, http://www.epa.gov/epaoswer/non-w/recycle/jtr/comm/cfa.htm, 2001
3. Adriano, D.C., A.L. Page, A.A. Elseewi, A.C. Chang, and I. Straughan. Utilization and disposal of fly ash and other coal residues in terrestrial ecosystems: a review. *J. Environ. Qual.* 9, 333, 1980.
4. Page, A. L., Elseewi, A. A., and I. Straughan, I., Physical and chemical properties of fly ash from coal-fired power plants with reference to environment impacts. *Residue Rev.* 71, 88, 1979.
5. Mattigod, S. V., Dhanpat Rai, Eary, E., and Ainsworth, C. C., Geochemical factors controlling the mobilization of selected inorganic constituents from fossil fuel combustion residues. Part 1: Review of the major elements. *J. Env. Qual.* 19, 187, 1990.
6. Martens, D.C., Availability of plant nutrients in fly ash. *Compost Sci.* 12, 15, 1971.
7. Adams, L.M., Capp, J. P., and Gillmore, D. W., Coal mine spoil and refuse bank reclamation with power plant fly ash. *Compost Sci.* 13, 20, 1972.
8. Doran, J.W., and Martens, D. C., Molybdenum availability as influenced by application of fly ash to soil. *J. Environ. Qual.,* 1:186, 1972.
9. Plank, C.O., and Martens, D. C., Boron availability as influenced by application of fly ash to soil. *Soil Sci. Soc. Am. Proc.* 38, 974, 1974.
10. Elseewi, A.A., Bingham, F. T., and A.L. Page, A. L., Growth and mineral composition of lettuce and Swiss chard grown on fly ash amended soils. In D.C. Adriano and I.L. Brisbin, Jr. (Eds.): *Environmental chemistry and cycling processes.* Proc. Symp. Augusta, GA April 28-May 1 (1976). DOE Symposium Series 45, CONF-760429, pp. 568, 1976.
11. Elseewi, A.A., Bingham, F. T., and Page, A. L., Availability of sulfur in fly ash to plants. *J. Environ. Qual.* 7, 69, 1978.
12. Chang, A.C., Lund, L. J., A.L. Page, A. L., and Warneke, J. E., Physical properties of fly ash-amended soils. *J. Environ. Qual.* 6, 167, 1977.
13. Phung, H.T., Lund, L. J. and Page, A. L., Potential use of fly ash as a liming material. p. 504-515. In D.C. Adriano and I.l. Brisben (ed.) in *Environmental chemistry and cycling processes.* CONF-760429. U.S. Dep. Commerce, Springfield, Va., 1978.

14. Straughan, I.R., Elseewi, A. A., and Page, A. L., Mobilization of selected trace elements in residues from coal combustion with special reference to fly ash. pp. 389-402. In: 0.0. Hemphill (ed.), *Trace Substances in Environmental Health XII Symposium.* 1978.

15. Lakin, H.W., Selenium in our Environment. In *Trace elements in the environment.* Ed. E. L. Kothny. Adv. Chern. Ser. 123. Am. Chem. Soc. Washington D. C. pp. 96, 1973.

16. Mbagwu, J., Selenium concentrations in crops grown on low-selenium soils as affected by fly-ash amendment. *Plant and Soil.* 74, 75, 1983.

17. Furr, A.K., Kelly, W. C., Bauche, C. A., Gutenmann, W. H., and Lisk, D. J., Multielement uptake by vegetables and millet grown in pots on fly ash amended soil. *J. Agric. Food Chem.* 24, 885, 1976.

18. National Academy of Sciences. *Selenium in Nutrition.* Revised Edition. National Academy of Sciences Press. Washington, D.C. 174, 1983.

19. Hurd-Karrer, A.M., Factors affecting the absorption of selenium from soils by plants. *Jour. Agr. Res.* 50, 413, 1935.

20. Williams, K. T., and Byers, H. G., Selenium Compounds in Soils. *Ind. Eng. Chem.* 28, 914, 1936

21. Gissel-Nielson, G. and Bisbjerg. B., The uptake of applied selenium by agricultural plants 2. The utilization of various selenium compounds. *Plant Soil* 32, 382, 1970.

22. Arthur, M. A., Rubin, G., Schneider, R. E.,and Weinstein, L. H., Uptake and accumulation of selenium by terrestrial plants growing on a coal fly ash landfill: Part 1. Corn, *Environ. Toxiol. Chem.,* 11, 541, 1992.

23. Arthur, M. A., Rubin, G., Schneider, R. E.,and Weinstein, L. H., Uptake and accumulation of selenium by terrestrial plants growing on a coal fly ash landfill: Part 2. Forage and Root Crops, *Environ. Toxiol. Chem.,* 11, 1289, 1992.

24. Arthur, M. A., Rubin, G., Schneider, R. E., and Weinstein, L. H., Uptake and accumulation of selenium by terrestrial plants growing on a coal fly ash landfill: Part 3. Forbs and grasses, *Environ. Toxiol. Chem.,* 11, 1301, 1992.

25. Woodbury, P. B., McCune, D. C., and Qweinstein, L. H., A Review of Selenium Uptake, Transformation, and Accumulation by Plants with Particular Reference to Coal Fly Ash Land Fills. in *Biogeochemistry of Trace Elements in Coal and Coal Combustion Byproducts,* Sajwan, K.E, Alva, A. K., and Keefer, R.F. Kluwer Academic/Plenum Publishers, New York, U. S. A., 309, 1999.

26. Andren, H.W., D.H. Klein, and V. Talmi. 1975. Selenium in coal-fired steam plant emissions. *Environ. Sci. Technol.* 9, 856, 1975.

27. Cutter, G. A. *Anal. Chem.* 57, 2951, 1985.

28. Chao, T. T., and Sanzolone, R. F., Fractionation of soil selenium by sequential partial dissolution, *Soil Sci. Soc. Am.,* 53, 385, 1989.

29. Kang, Y., Yamada, H., Kyuma, K, and Hattori, T. *Soil Sci. Plant Nutr.* 39, 331, 1993.

30. Niss, N. D., Schabron, J. F., and Brown, T. H., determination of selenium species in coal fly ash extracts., *Env. Sci. Technol.* 27, 827, 1993.

31. Martens, D. A., and Suarez, D. L., Selenium speciation of soil/sediment determined with sequential extractios and hydride generation atomic absorption spectrophotometry, *Environ. Sci. Technol,* 31, 133, 1996.

32. Jackson, B. P., and Miller, W. P., Soluble arsenic and selenium species in flyash/organic waste-amended soils using ion chromatography-inductively coupled plasma mass spectrometry, *Environ. Sci. Technol,* 33, 270, 1999.

33. van der Hoek, E. E., van Elteren, J. T., and Comans, R. N. J., Determination of As, Sb, and Se speciation in fly ash leachates, *Intern. J. Environ. Anal. Chem,*63, 67, 1996.

34. Jackson, B. P., and Miller, W. P., Arsenic and selenium speciation in coal flyash extracts by ion chromatography-inductively coupled plasma mass spectrometry, *J. Anal. At. Spectrom,* 13, 1107, 1998.

35. Soltanpour, P. N., A. Khan, and W. L. Lindsay., Factors affecting DTPA extractable zinc, iron, manganese and copper from soils. *Comm. Soil Sci. Plant Anal.* 7, 791, 1976.

36. Eary, L. E., Dhanpat Rai, S. V. Mattigod, S. V., and Ainsworth, C. C., Geochemical factors controlling the mobilization of selected inorganic constituents from fossil fuel combustion residues. Part 1: Review of the major elements. *J. Env. Qual.* 19, 202, 1990.

37. Mattigod, S. V. and Ervin, J., Scheme for density separation and identification of compound forms in size-fractionate fly ash. *Fuel* 16, 927, 1983.

38. Bernas, B., A new method for decomposition and comprehensive analysis of silicates by atomic absorption spectrometry. *Anal. Chem.* 40, 1682, 1968.

39. Cary, E. E., Wieczorek, G. A., and Allaway, W. W., Reactions of selenite-selenium added to soils that produce low-selenium forages. *Soil Sci. Soc. Am. Proc.* 31, 21, 1967.

40. Pierce, F. J., Dowdy, R. H., and O. F. Grigal, O. F., Concentrations of six trace elements in some major Minnesota soil series. *J. Environ. Qual.* 11, 416, 1982.

41. Nadkarni, R., Applications of hydride generation-atomic absorption spectrometry to coal analysis. *Anal. Chem. Acta.* 135, 363, 1982.

42. Davison, R.L., Natusch, D. F. S., J.R. Wallace, J. R., and Evans Jr., C. A., Trace elements in fly ash: Dependence of concentration on particle size. *Environ. Sci. Technol.* 8, 1107, 1974.

43. Natusch, D.F.S., Wallace, J. R., and Evans, Jr., C. A., Toxic trace elements: Preferential concentration in respirable particles. *Science* 183, 202, 1974.

44. Lee, R. E., Jr., Crist, H. L., Riley, A. E., and MacLeod., K. E., Concentration and size of trace metal emissions from a power plant, a steel plant, and a cotton gin. *Environ. Sci. Technol* 9, 646, 1975.

45. Linton, R.W., Loh, A. Natusch, D.F.S., Evans, C. A. Jr., and P. Williams, P., Surface predominance of trace elements in air-borne particles: *Science* 191, 853, 1975.

46. Klein, D.H., A.W. Andren, J.A. Carter, J.F. Emery, C. Feldman, W. Fulkerson, W.S. Lyon, J.C. Ogle, Y. Talmi, R.I. Van Hook, and N. Bolton., Pathways of thirty-seven trace elements through coal-fired power plants. *Environ. Sci. Technol.* 9, 973, 1975.

47. Campbell, J. A., Laul, J. C., Nielson, K. K., and Smith, R. D., Separation and chemical characterization of finely sized fly ash particles. *Anal. Chem.* 50, 1032, 1978.

48. Smith, R., Campbell, J., K. Nielson, K., Concentration dependence upon particle size of volatilized elements in fly ash. *Environ. Sci. Technol.* 13, 553, 1979.

49. Coles, D., Ragaini, R., Ondov, J., Fisher, G., Silberman, D., and Prentice, B., Chemical studies of stack fly ash from a coal-fired power plant. *Environ. Sci. Technol.* 13, 455, 1979.

50. Mattigod, S. V. Dhanpat Rai, and Amonette, J. E. Concentrations and distributions of major and and selected trace elements in size-density fractionated fly ashes, in *Biogeochemistry of Trace Elements in Coal and Coal Combustion Byproducts,* Sajwan, K.E, Alva, A. K., and Keefer, R.F. Kluwer Academic/Plenum Publishers, New York, U. S. A., 115, 1999.

51. Finkelman, R., *The modes of occurence of trace elements in coals.* Ph.D. dissertation. University of Maryland. 1980.

52. Leutwein, F., Solubilities of Compounds which control Selenium concentrations in Natural Waters; Valence States in Natural Environments; Adsorption. In. *Handbook of Geochemistry* Ed. K. H. Wedephol. Vol 11/3,34- H-3, Springer-Verlag, New York. 1978.

53. Dreesen, D.R., Gladney, E. S., Owens, J. W., Perkins, B.L., Wienke, C. L., and Hansen L. E., Comparison of levels of trace elements extracted from fly ash and levels found in effluent waters from a coal-fired power plant. *Environ. Sci. Technol.* 11, 1017, 1977.

54. Furr, A.K., Parkinsen, T. F., Hinrichs, R. A. van Camper, D. R., Bauche, C. A., Gutenmann, W. H., St. John, Jr., L. E., Pakkala, I. S., and Lisk, D. J., National survey of elements and radioactivity in fly ashes. Absorption of elements by cabbage grown in fly ash-soil mixtures. *Environ. Sci. Technol.* 11,1104, 1977.

RARE EARTH ELEMENTS IN FLY ASHES AS POTENTIAL INDICATORS OF ANTHROPOGENIC SOIL CONTAMINATION

Shas V. Mattigod

Pacific Northwest National Laboratory
Richland, WA 99352, USA

ABSTRACT

Studies of rare earth element (REE) content of disposed fly ashes and their potential mobility were neglected for decades because these elements were believed to be environmentally benign. A number of recent studies have now shown that REE may pose a long-term risk to the biosphere. Therefore, there is a critical need to study the REE concentrations in fly ash and their potential mobilization and dispersal upon disposal in the environment. We analyzed the REE content of bulk, size fractionated, and density separated fractions of three fly ash samples derived from combustion of sub bituminous coals from the western United States and found that the concentrations of these elements in bulk ashes were within the range typical of fly ashes derived from coals from the North American continent. The concentrations of light rare earth elements (LREE) such as La, Ce, and Nd, however, tended towards the higher end of the concentration range whereas, the concentrations of middle rare earth elements (MREE) (Sm and Eu) and heavy rare earth elements (HREE) (Lu) were closer to the lower end of the observed range for North American fly ashes. The concentrations of REE did not show any significant enrichment with decreasing particle size, this is typical of nonvolatile lithophilic element behavior during the combustion process. The lithophilic nature of REE was also confirmed by their concentrations in heavy density fractions of these fly ashes being on average about two times more enriched than the concentrations in the light density fractions. Shale normalized average of REE concentrations of fly ashes and coals revealed significant positive anomalies for Eu and Dy. Because of these distinctive positive anomalies of Eu and Dy, we believe that fly ash contamination of soils can be fingerprinted and distinguished from other sources of anthropogenic REE inputs in to the environment.

INTRODUCTION

Environmental aspects of coal resource utilization have come under increasing scrutiny in recent years. This is in part due to the recognition that coal combustion represents a large-scale mobilization of all naturally occurring elements in a magnitude that is comparable to the rates of mobilization and transport that occur as part of natural geochemical cycling of elements.[1,2] Fly ash from coal combustion contains significant quantities of trace elements that may be potentially beneficial or toxic to biological systems. As a result, extensive investigations have been conducted to examine elemental associations in coal fly ash so that the environmental effects could be assessed.

Fly ash utilization as a soil amendment has also generated some concern. Among the trace elements in fly ash, As, B, Cd, Hg, Mo, Pb, and Se are considered to be the elements of greatest concern, whereas, V, Cr, Ni, Cu, and Zn are considered to be elements of moderate concern.[3,4] Although previously the REE in fly ashes were considered to be on no particular concern, a number of recent studies have indicated that these elements may pose a long-term risk to the biosphere.[5,6,7,8,9,10] Also, REE concentrations have been used as indicators of anthropogenic contamination in soils, sediments, and water.[11,12,13,14] Because large volumes of fly ash are disposed off in the environment, there is a significant need to understand their REE composition of fly ashes and assess their effects on the biosphere. Therefore, measuring REE content of fly ash can help to critically assess both the beneficial and harmful effects on biota, as well as, assessing the applicability of the REE content as an indicator of anthropogenic contamination. Although a number of investigators have measured the concentrations of REE in bulk fly ashes,[15,16,17,18,19,20,21] only a few studies have examined particle size[22,23] and the density dependence of REE concentrations.[23] Therefore, the objective of our study was to examine the elemental characteristics of three fly ashes including the REE contents within fractions of differing size and density. We assessed the applicability of REE concentration anomalies as potential environmental tracers. Such characterization would add to the current body of available information on REE contents of fly ashes and provide an improved basis for assessing its potential environmental effects.

MATERIALS AND METHODS

Fly ash samples (SB01, SB02, and SB03) were obtained from the electrostatic precipitators of three power stations that used sub bituminous coals from the western United States. The fly ashes were stored under nitrogen in glass jars to avoid hydration of species such as anhydrite and oxides of Ca and Mg and oxidation of reduced trace element species such as Se (0) and Se (IV). The ash samples were initially dry sieved to separate the <45 μm fraction from the coarser material. A Bahco microclassifier was then used to obtain <2.7 μm, 2.7-8 μm, 8-15 μm size fractions by means of air elutriation.

To determine the variation of elemental concentrations with density, the <2.7 μm size fraction was subjected to sink-float separation in a 2.1 g/cm^3 solution. The solution was prepared from a mixture of 20% polyvinylpyrrolidone (PVP) (Eastman 15420) in absolute ethanol and 1, 1, 2, 2-Tetrabromoethane (TBE) (Mallinckrodt).[24] A ratio of 39% v/v of PVP-EtOH and 61% TBE yielded the desired density. The PVP was included to prevent flocculation of fly ash particles. Approximately 150 mg of ash was placed on top of 20 ml of density solution in a 40 ml glass centrifuge tube. After dispersing the ash in the solution by means of a vortex mixer, the tubes were centrifuged for about four hours at 750 rpm in an IEC Model K centrifuge. After determining that separation had occurred, the float fraction and much of the solution were removed by means of a suction device. The float and the sink fractions were then centrifuge-washed five times with absolute ethanol to remove traces of TBE. Excess ethanol was then removed from the samples in a vacuum desiccator.

The bulk ashes and the size and density fractions were then analyzed by means of instrumental neutron activation analysis (INAA), inductively coupled argon plasma (ICAP), and hydride generation atomic absorption spectroscopy (HGAA). For INAA, samples (200-700mg) were weighed and sealed into 0.5 dram polyvials (Olympic Plastics). The fly ash samples were subjected to two irradiations in the thermal neutron flux of the UC Irvine TRIGA reactor. The first irradiation (30 sec @ 1x10^{10} neutron cm^{-2} sec^{-1}) was followed a minute later by a two minute count on an 18% efficiency Ge-Li detector. This yielded data for Al and V. The remaining elements were determined after a five-hour irradiation (1.5 ×10^{12} neutron cm^{-2} sec^{-1}). After a two-day decay interval, K, Na, As, and La were determined while the remaining elements (Ba, Ce, Co, Cr, Cs, Eu, Fe, Hf, Lu, Nd, Rb, Sb, Sc, Sm, Sr, Ta, Th, Zn, and Zr) were determined after a 14 day decay period. Gamma-ray spectrometry was performed by means of an 18% efficiency Ge-Li detector coupled to a 4096-channel spectrometer (Norland Industries, ULTIMA II). Spectra were stored on disk and the peaks analyzed using MACROGAM software; the elemental concentrations were determined by comparison of selected peak areas to those of standards.

The ICAP samples were prepared by means of the bomb digestion method.[25] Samples of 100-150 mg were placed in Teflon Parr Bombs with 2 ml of aqua regia and 6 ml of 48% hydrofluoric acid (Baker). After being heated at 105°C in an oven for 1.5 hours, the samples were allowed to cool. The contents were then transferred into polypropylene beakers containing 5.6 g boric acid (Mallinckrodt). After diluting to 80 ml with distilled deionized water, the samples were heated on a steam bath until the boric acid dissolved. The solutions were cooled and transferred to 100 ml volumetric flasks. The samples were then filtered (Whatman #2) into 125 ml plastic containers. The samples and standards were then analyzed by means of a Jarrel-Ash ICP coupled with an Atom Comp Series 800 multi-channel spectrometer to determine the concentrations of Ca, Mg, Mn, Ni, and Si. Hydride Generation Atomic Absorption Spectroscopy was used to obtain Se values used for ICP. An MHS-I0 hydride generator coupled with a Perkin-Elmer 5000 was utilized. The MHS-I0 contains a reservoir filled with 3% sodium borohydride (Aldrich) and 1% sodium hydroxide in water. Nitrogen gas pressure (38 psi) was used to transfer several milliliters of this solution to the sample reservoir, which contained 0.01-5 ml of sample plus 10 ml 6N HCl. Selenium (IV) was reduced to H_2Se and was swept from the sample reservoir by N_2 gas into a quartz cell in the flame of the Perkin-Elmer 5000. The H_2Se decomposes as a result of the heat and the Se absorption peak was then recorded on a strip chart recorder. The Se concentrations of samples were determined by comparison of the unknown and standard sample peak height.

RESULTS

Major and Trace Elements

The results of the major and trace elemental analyses are presented in Tables 1 – 3. Generally, the composition of bulk ashes was within the ranges for fly ashes reported in the literature.[22,26,27] The elemental concentrations in these fly ashes increased with decreasing particle size, which is typical of elemental distribution in fly ashes.[28] Due to higher volatility, the chalcophilic elements (associated mainly with sulfides and organics in coal), such as As, Se, Sb, and Zn, in these fly ashes exhibited enhanced enrichment in finer particles, which is also one of the typical characteristics of fly ashes.[27] The density distribution of major and minor elements in these fly ashes exhibited similar trends as those in other fly ashes described in several review articles.[26,27,28]

Rare Earth Elements

The concentrations of the REE in bulk fly ashes were within the range of values observed for fly ashes derived from North American coals (Table 4). In these fly ashes, the concentrations of LREE, such as La, Ce, and Nd, tended towards the higher end of the concentration range, whereas the concentrations of MREE (Sm and Eu) and HREE (Lu) were closer to the lower end of the observed range for North American fly ashes.

Table 1. Elemental Concentrations in Size and Density Fractions of SB01 Fly Ash

Major Elements	Bulk Ash	Density Fractions (g/cm³)		Particle Size Fractions (µm)		
		<2.1	>2.1	<2.7	2.7 - 8	8 - 15
			%			
Al	11.10	16.70	24.70	10.40	10.40	10.30
Si	27.30	34.70	15.60	25.80	17.70	18.70
Na	1.43	2.80	2.40	1.91	1.65	1.30
K	0.74	0.97	0.95	0.76	0.70	0.74
Mg	0.86	0.83	1.28	1.13	0.95	0.77
Ca	5.64	3.94	7.61	7.30	7.15	5.15
Fe	3.72	1.98	4.62	3.87	3.53	3.38
Trace Elements			mg/kg			
As	14.7	25.6	43.1	29.9	16.3	7.8
Se	21.6	66.0	32.0	34.0	25.0	13.5
Sb	78.0	66.0	202.0	174.0	76.0	37.0
Rb	62.0	79.0	74.0	73.0	65.0	54.0
Cs	35.0	76.0	58.0	45.0	33.0	41.0
Sr	2079.0	581.0	2916.0	2612.0	2468.0	2205.0
Ba	4596.0	2505.0	7066.0	6260.0	4758.0	4239.0
Sc	18.2	12.5	28.3	23.0	20.1	18.5
V	81.0	--	180.0	113.0	110.0	69.0
Co	20.0	24.4	35.8	31.2	22.1	16.4
Cr	31.4	45.9	118.0	102.0	60.0	48.0
Mn	163.0	212.0	232.0	206.0	148.0	126.0
Ni	36.0	54.0	45.0	40.0	33.0	42.0
Zn	101.0	73.0	183.0	96.0	82.0	52.0
Zr	91.0	--	--	--	--	500.0
Hf	18.5	6.0	18.3	16.3	16.5	21.2
Ta	1.8	--	2.1	2.1	1.9	1.3
Th	24.1	17.8	33.0	28.7	24.2	24.7

The concentrations of REE (Table 4, Figure 1 a,b,c) indicated that within the ranges of particle sizes examined in this study, there was no significant enrichment of REE concentrations with decreasing particle size. In fly ashes, similar particle size independent concentrations for lithophilic elements (e.g., REE) have also been observed by other investigators.[15,16,17,22,23,29,30,31,32,33] However, concentrations of REE in heavy density fraction of these fly ashes (Table 4, Figure 1 c,d,f) were on average about two times more concentrated as compared to the light density fractions. Such enhancement in heavier density fractions can be attributed to the fact that bulk of REE are lithophilic elements which occur in coals principally in inorganic mineral forms such as phosphates, silicates, and carbonates.[34] The REE bearing

Table 2. Elemental Concentrations in Size and Density Fractions of SB02 Fly Ash

Major Elements	Bulk Ash	Density Fractions (g/cm³)		Particle Size Fractions (µm)		
		<2.1	>2.1	<2.7	2.7 - 8	8 - 15
		%				
Al	12.8	17.5	12.2	15.7	12.6	12.5
Si	20.4	20.8	20.4	20.5	18.5	19.6
Na	1.3	2.2	2.0	2.3	1.8	1.6
K	0.6	0.8	0.7	0.7	0.6	0.6
Mg	0.3	0.4	0.6	0.5	0.5	0.5
Ca	1.9	2.5	8.0	3.4	2.8	2.2
Fe	2.4	1.5	2.8	2.6	2.3	2.2
Trace Elements		mg/kg				
As	12.2	39.2	58.5	59.4	31.2	14.8
Se	5.2	9.3	14.7	12.9	8.5	4.3
Sb	41.0	104.0	162.0	157.0	102.0	61.0
Rb	42.0	52.0	40.0	42.0	46.0	36.0
Cs	29.0	39.0	31.0	37.0	32.0	27.0
Sr	384.0	351.0	559.0	565.0	258.0	316.0
Ba	647.0	475.0	919.0	861.0	704.0	628.0
Sc	10.9	10.3	15.8	14.7	13.6	12.1
V	70.0	67.0	92.0	118.0	93.0	64.0
Co	8.6	13.3	20.7	19.6	14.1	10.2
Cr	25.1	37.0	53.0	50.0	38.0	27.0
Mn	102.0	66.0	152.0	140.0	137.0	118.0
Ni	14.0	24.0	48.0	42.0	32.0	68.0
Zn	68.0	107.0	146.0	148.0	108.0	78.0
Zr	236.0	--	494.0	335.0	373.0	321.0
Hf	17.4	11.2	21.8	20.6	18.2	16.9
Ta	1.5	1.4	1.9	2.0	2.0	1.7
Th	23.2	22.7	29.4	28.4	26.8	24.7

Table 3. Elemental Concentrations in Size and Density Fractions of SB03 Fly Ash

Major Elements	Bulk Ash	Density Fractions (g/cm³)		Particle Size Fractions (µm)		
		<2.1	>2.1	<2.7	2.7 - 8	8 - 15
		%				
Al	13.7	19.6	12.6	13.1	12.6	13.1
Si	18.1	18.5	18.6	17.2	18.4	15.6
Na	1.4	2.1	1.5	1.8	1.1	1.4
K	0.8	0.9	0.6	0.6	0.6	0.7
Mg	0.5	0.4	0.6	0.5	0.5	0.5
Ca	4.3	1.2	5.1	5.3	7.0	4.9
Fe	2.1	1.3	2.2	1.9	1.8	1.8
Trace Elements		mg/kg				
As	15.3	12.0	60.9	35.5	18.7	9.3
Se	0.9	1.1	1.3	1.1	1.2	0.5
Sb	50.0	45.0	113.0	95.0	55.0	29.0
Rb	45.0	53.0	51.0	41.0	39.0	47.0
Cs	22.0	34.0	27.0	26.0	23.0	25.0
Sr	776.0	464.0	805.0	802.0	915.0	865.0
Ba	3192.0	2104.0	4342.0	3840.0	2566.0	2141.0
Sc	13.0	11.6	15.6	14.1	14.1	14.1
V	52.0	62.0	93.0	87.0	60.0	71.0
Co	9.2	10.7	14.6	14.3	10.0	7.8
Cr	29.6	28.0	50.0	46.0	36.0	28.0
Mn	156.0	90.0	222.0	202.0	217.0	178.0
Ni	24.0	38.0	30.0	32.0	37.0	26.0
Zn	79.0	82.0	123.0	107.0	89.0	63.0
Zr	428.0	329.0	574.0	324.0	--	--
Hf	24.4	14.4	21.9	19.5	21.1	23.2
Ta	2.3	1.9	2.5	2.1	2.1	2.3
Th	31.6	29.9	34.5	31.8	31.9	33.3

Table 4. REE Distribution in Fly Ashes derived from Sub bituminous Coals from Western U.S.

REE	Range in Fly Ashes*	Bulk Ash	Density Fractions (g/cm³)		Particle Size Fractions (µm)		
			<2.1	>2.1	<2.7	2.7 - 8	8 - 15
			SB01				
La	8 - 94	71.2	51.9	105.1	100.9	76.4	73.6
Ce	5 - 180	101.0	84.0	147.0	125.0	108.0	105.0
Nd	10 -300	49.0	--	--	76.0	57.0	58.0
Sm	<2 - 70	9.0	7.0	14.7	11.4	10.0	9.1
Eu	0.5 - 20	2.1	1.2	2.5	2.2	1.8	2.0
Lu	0.5 -1.7	0.8	0.4	1.1	0.9	0.9	0.9
			SB02				
La	8 - 94	61.3	58.4	78.4	73.9	72.3	63.5
Ce	5 - 180	94.0	94.0	122.0	119.0	109.0	100.0
Nd	10 -300	41.0	33.0	69.0	54.0	52.0	42.0
Sm	<2 - 70	9.7	9.0	13.0	10.0	9.0	7.8
Eu	0.5 - 20	1.5	1.3	1.9	2.0	1.7	1.6
Lu	0.5 -1.7	0.5	0.4	0.8	0.7	0.6	0.6
			SB03				
La	8 - 94	78.4	63.1	91.4	86.9	81.6	79.6
Ce	5 - 180	117.0	101.0	133.0	120.0	117.0	119.0
Nd	10 -300	46.0	33.0	64.0	53.0	57.0	37.0
Sm	<2 - 70	13.0	7.7	14.2	10.4	10.0	12.3
Eu	0.5 - 20	1.8	1.5	2.1	1.9	1.9	1.9
Lu	0.5 -1.7	0.7	0.5	0.7	0.7	0.7	0.7

*North American and Chinese fly ashes[18,29]

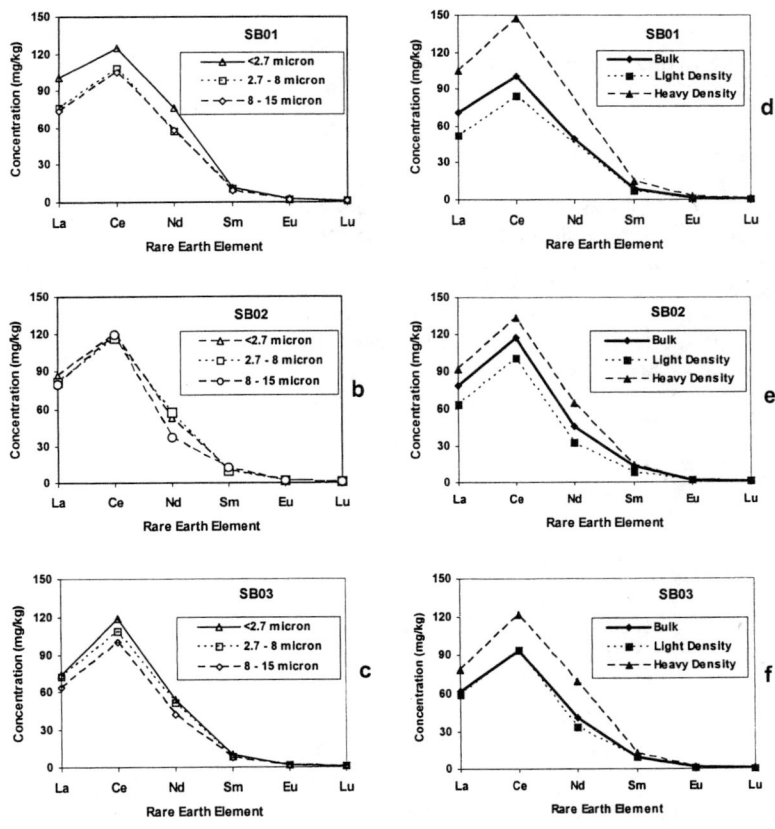

Figure 1. (a,b,c) Distribution of REE among Different Particle Size Fractions, and (d,e,f) Density Distribution of REE for three Fly Ashes.

heavier density minerals (e.g., zircon) are not affected by coal combustion and, therefore, end up relatively intact in fly ashes.[24,35] Minor fractions of organically associated REE would be converted to nonvolatile oxides upon combustion and be retained in fly and bottom ashes.[36] Therefore, combustion would result in significant enrichment in the REE content of fly ashes. Such enrichment is evident from a comparison of average REE contents of North American coals and fly ashes and REE content of bulk fly ashes from this study (Figure 2). The magnitude of enrichment ratios (defined as the concentration ratio of REE content in fly ash to that in source coals) obtained from literature and this study indicated that the REE originally present in coal would be retained in solid byproducts upon combustion (Table 5). The average REE enhancement ratios for fly ashes in this study (calculated using the average North American coal composition) were similar to the typical North American and Chinese fly ashes (Table 5); thus confirming that REE in coals are predominantly lithophilic. The magnitude of the overall average REE enrichment ratio for typical North American and Chinese fly ashes (6.2 ± 0.5) indicated that fly ashes are a rich source of anthropogenically mobilizable REE. A comparison between the REE content of fly ashes and soils indicated that fly ashes contain, on average, two to three fold higher concentrations than world average REE content of soils.[37]

Table 5. Typical Enrichment Ratios for REE in Fly Ashes

REE	N. American FA *	Chinese FA *	SB FA
La	6.0	6.3	6.7
Ce	6.5	5.8	5.8
Sm	6.8	6.3	6.3
Eu	6.3	5.7	4.5
Tb	5.2	5.7	--
Dy	6.2	5.6	--
Yb	6.4	6.1	--
Lu	7.0	6.8	5.5

*North American and Chinese fly ashes[29], Av. SB fly ashes: This Study.

Rare Earth Element Anomaly in Fly Ashes

To examine the REE anomaly in fly ashes, we calculated the weighted average of REE concentrations of fly ashes, coals, and soils from various published sources.[15,16,17,18,23,29,33] The REE profile for these materials show the typical saw tooth variation in absolute concentrations conforming to the Oddo-Harkins effect/rule[38] (Fig 3). According to this effect/rule, increased nuclear stability of even atomic numbered REE show relatively higher abundances in chondritic and natural terrestrial materials. To compensate for this effect, and to observe the REE anomalies, the measured concentrations are typically normalized to chondrite or shale composition.[39]

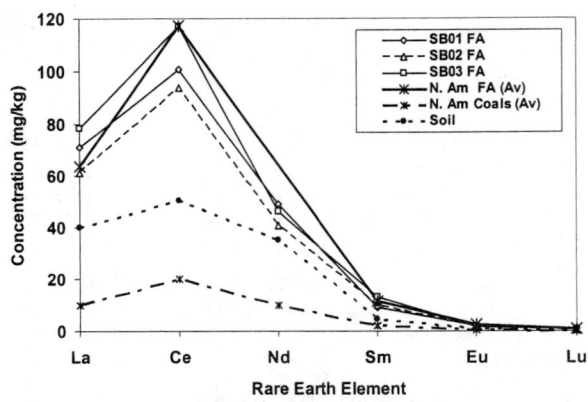

Figure 2. Comparison of REE Concentrations of SB Fly Ashes from this study with Average REE Composition of NA Fly Ashes, Coals, and Soils.

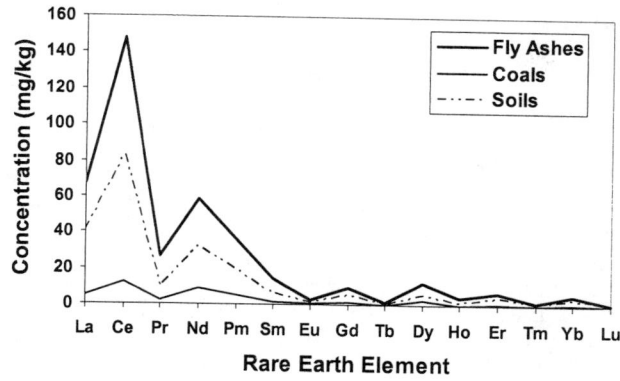

Figure 3. REE Concentrations in Fly Ashes, Coals, and Soils.

To observe any potential REE anomalies we normalized the REE weighted average compositions of fly ashes[15,16,17,18,23,32,36], coals, and soils.[37] The normalized data revealed significant anomalies for Eu and Dy in both fly ashes and coals (Figure 4), whereas, no such anomalies were evident for soils. Clearly, no soil REE anomalies are expected because soils are typically derived from crustal rocks including shales and, therefore, would have similar relative abundances of REE. Comparatively, coalification of organic material with minor amount of primary and secondary mineral matter would result in noticeable REE fractionation relative to normal soil forming processes. Not too surprisingly, the Eu and Dy anomalies for fly ashes and coals are similar because all REE in coals being non-volatile are retained during combustion process, thus preserving the coal's original anomalous REE imprint.

It is clear from these observations, that an REE anomaly in fly ashes relative to soils can be used as a tracer to identify the presence of fly ash (inadvertent or deliberate additions) in soils. Recent observations[11,12] indicate that anthropogenic REE inputs into soils and sediments are relatively easy to extract as compared to the native REE in contaminated soils. Our study suggested that anthropogenic fly ash additions could be detected by extracting soils with acids and by observing the Eu and Dy REE anomalous signatures.

To understand the observed identical REE anomalies of fly ashes and parent coals, it is essential to examine the source of REE and their probable transformation during the combustion process. Two review articles[26,27] that examined the elemental and mineral distributions in coals and fly ashes suggested that REE, being mainly lithophilic in nature, are typically associated with the inorganic mineral component of coals. Upon combustion, mineral-associated REE in coals are transformed into oxides of low volatility and retained in fly ashes, thus preserving the REE anomalous signature of parent coals. For instance, minerals such as feldspars, micas, and carbonate minerals occur most frequently and constitute the major inorganic fraction of coals.[26,34] A number of observations of normalized REE profiles for

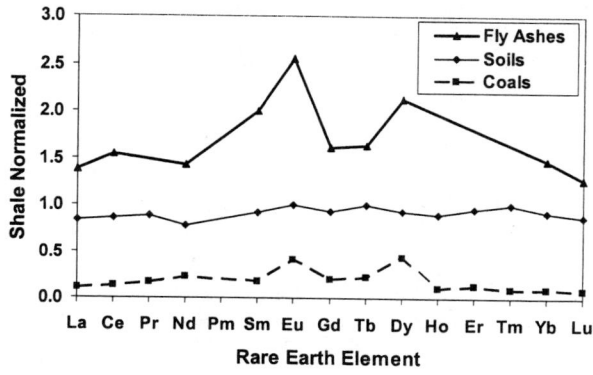

Figure 4. Europium and Dysprosium REE Anomalies in Fly Ashes and Coals.

plagioclase feldspar and carbonates show consistent positive Eu anomaly; whereas, mica minerals show a distinct positive Eu and Dy anomalies.[40,41,42] The reason for this anomaly is that Eu being a redox species, in its divalent state, substitutes for alkaline earth elements in plagioclase feldspars and carbonates.[43] Similarly, Eu (II) can substitute for Ca in interlayer positions and Dy(III), due to its ionic radius being similar to Al(III), can substitute in the octahedral position in micaceous minerals.[44]

Therefore, the significant positive Eu and Dy anomalies in fly ashes can be used as a marker for detecting anthropogenic fly ash input into soils. This is feasible in view of recent studies which show that anthropogenic input of sewage sludges, and medical wastes in to sediments, soils, and water can be detected by positive anomalies of several REE such as La, Ce, Nd, Sm, and Gd[11,13,14] and negative anomalies for Ce in surface soils due to its low leachability.[12] These studies clearly suggest that the anthropogenic perturbations of natural geochemical cycle of REE are reflected as anomalies in environment near the source of discharge. Because of the distinctive positive anomalies of Eu and Dy, we believe that fly ash contamination of soils can be fingerprinted and distinguished from other sources of anthropogenic REE inputs in to the environment.

SUMMARY AND CONCLUSIONS

The results of the major and trace elemental analyses indicated that the composition of bulk ashes was within the ranges for fly ashes reported in the literature. The elemental concentrations in these fly ashes increased with decreasing particle size, which is typical of elemental distribution in fly ashes. Due to higher volatility, the chalcophilic elements, associated mainly with sulfides and organics in coal, (e.g., As, Se, Sb, and Zn) in these fly ashes exhibited enhanced enrichment in finer particles, which is also a typical characteristic of fly ashes. The density distribution of major and minor elements in these fly ashes exhibited similar trends observed for a number of fly ash samples.

The concentrations of the REE in bulk fly ashes were within the range of values observed for fly ashes derived from North American coals. The concentrations of LREE (e.g., La, Ce, and Nd) tended towards the higher end of the concentration range; whereas, the concentrations of MREE (Sm and Eu) and HREE (Lu) were closer to the lower end of the observed range for North American fly ashes. The concentrations of REE indicated that within the ranges of particle sizes examined, there was no significant enrichment of REE concentrations with decreasing particle size. In fly ashes, similar particle size independent concentrations for lithophilic elements, such as REE, have also been observed by other investigators. Concentrations of REE in heavy density fractions, however, were on average two times more concentrated as compared to the light density fractions. Such enhancement in heavier density fractions can be attributed to the fact that the REE are lithophilic elements.

The average REE enhancement ratios for the three fly ashes (calculated as a ratio of concentration in fly ashes to the average concentration in North American coals) were similar to the enrichment observed for typical North American and Chinese fly ashes. The magnitude of the overall average REE enrichment ratio for fly ashes (6.2 ± 0.5) indicated that fly ashes are a rich source of anthropogenically mobilizable REE.

Shale normalized REE data for typical fly ashes, coals, and soils revealed significant anomalies for Eu and Dy in both fly ashes and coals; whereas, as expected, no such anomalies were evident for soils. Recent studies have shown that anthropogenic input of sewage sludges and medical wastes into sediments, soils, and water can be detected by positive anomalies of several REE, such as La, Ce, Nd, Sm, and Gd, and negative anomalies for Ce in surface soils. These studies clearly suggest that the anthropogenic perturbations of natural geochemical cycle of REE are reflected as anomalies in environment near the source of discharge. Because of the distinctive positive anomalies of Eu and Dy, we suggest that fly ash contamination of soils can be fingerprinted and distinguished from other sources of anthropogenic REE inputs into the environment.

ACKNOWLEDGEMENT

This research was partially funded through United States-Israel Binational Agricultural Research Development Fund (BARD) Project 1-97-80. We thank Dr. Kenneth Sajwan of Savannah State University for reviewing the manuscripts and providing valuable comments.

REFERENCES

1. Bertine, K. K. and E. D. Goldberg. Fossil fuel combustion and major sedimentary cycle. *Science* 173, 223, 1971.

2. Hedrick, J. B. The global rare-earth cycle. *J. Alloys and Compounds* 225: 609-618, 1995.
3. National Research Council, Panel on the Trace Element Geochemistry of Coal Resource Development Related to Health. *Trace-Element Geochemistry of Coal Resource Development Related to Environmental Quality and Health.* NAS. Washington, D.C. 1980.
4. Valkovic, V. *Trace elements in coal.* Vol II, CRC Press 1983.
5. Evans, C. "Toxicology and pharmacology of the lanthanides" Chap 8 in *Biogeochemistry of the Elements*, Plenum New York. 1990.
6. De Boer, J. L. M., W. Verweij, T. van der Velde-Koerts, and W. Mennes. Level of rare earth elements in Dutch drinking water and its sources. Determination by ICP-MS and toxicological implications. A Pilot Study. *Wat. Res.* 30, 190, 1996.
7. Wen, B., D. Yuan, X. Shan The influence of rare earth element fertilizer application on the distribution and bioaccumulation of rare earth elements in plants under field conditions. *Chem. Speciation & Bioavail.* 13, 39, 2001.
8. Li, F., X. Shan, S. Zhang. Evaluation of single extractants for assessing plant availability of rare earth elements in soils. *Comm. Soil Sc and Plant Anal* 32, 2577, 2001.
9. Wang, Z., X. Shan, S. Zhang. Comparison of speciation and bioavailability of rare earth elements between wet rhizosphere soil and air-dried bulk soil. *Ana. Chim. Acta* 441, 147, 2001.
10. Liang, T, B. Yan, S. Zhang. Contents and the biogeochemical characteristics of rare earth elements in wheat seeds. *Biogeochemistry* 54, 41, 2001.
11. Olmez, I., E. R. Sholkovitz, D. Hermann, and R.P. Eganhouse. Rare earth elements in sediments off Southern California: A new anthropogenic indicator. *Environ. Sci. Technol.* 25, 310, 1991.
12. Steinmann, M. and P. Stille. Rare earth element behavior and Pb, Sr, Nd isotope systematics in a heavy metal contaminated soil. *App. Geochem.* 12, 607, 1997.
13. Nozaki, Y., D. Lerche, D. S. Alibo, and M. Tsutsumi. Dissolved indium and rare earth elements in three Japanese rivers and Tokyo Bay: Evidence for anthropogenic Gd and In. *Geochim. Cosmochim. Acta*, 64, 3975, 2000.
14. Elbaz-Poulichet, F., J.L. Seidel, and C. Othoniel. Occurrence of an anthropogenic gadolinium anomaly in river and coastal water in southern France. *Wat. Res.* 36, 1102, 2002.
15. Klein, D.H., A.W. Andren, J.A. Carter, J.F. Emery, C. Feldman, W. Fulkerson, W.S. Lyon, J.C. Ogle, Y. Talmi, R.I. Van Hook, and N. Bolton., Pathways of thirty-seven trace elements through coal-fired power plants. *Environ. Sci. Technol.* 9, 973, 1975.
16. Block, C. and R. Dams. Study of fly ash emission during coal combustion. *Environ. Sci.Technol.* 10, 1011, 1976.
17. Smith, R., Campbell, J., K. Nielson, K., Concentration dependence upon particle size of volatilized elements in fly ash. *Environ. Sci. Technol.* 13, 553, 1979.
18. Kronberg, B. I. J. R. Brown, W. S. Fyfe, M. Pierce, and C. G. Winder. Distribution of trace elements in western Canadian coal ashes. *Fuel*, 60, 59, 1981.
19. Warren C. J., and M. J. Dudas. Leaching behavior of selected trace elements in chemically weathered alkaline fly ash. *Sci Tot. Environ.* 76, 229, 1988.
20. Vassilev, S. V., and C. G. Vassileva. Geochemistry of coals, coal ashes and combustion wastes from coal-fired power stations. *Fuel Proc. Tech.* 51, 19, 1997.
21. Karayagit, A. I., T. Onacak, R. A. Gayer, and S. Goldsmith. Mineralogy and geochemistry of feed coals and their combustion residues from Cayirhan power plant, Ankara, Turkey. *App. Geochem.* 16, 911, 2001.
22. Page, A.L., A.A. Elseewi, and I. Straughan. Physical and chemical properties of fly ash from coal-fired power plants with reference to environmental impacts. *Residue Rev.* 71, 83, 1979.
23. Querol, X, J. L. Fernandez-Turiel, and A. Lopez-Soler. Trace elements in coal and their behavior during combustion in a large power station. Fuel. 74, 331, 1995.
24. Mattigod, S. and J. Ervin. Scheme for density separation and identification of compound forms in size-fractioned fly ash. *Fuel* 16, 927, 1983.
25. Bernas, B. A new method for decomposition and comprehensive analysis of silicates by atomic absorption spectrometry. *Anal. Chem.* 40, 1682, 1968.
26. Mattigod S. V., Dhanpat Rai, L. E. Eary, and Ainsworth, C. C., Geochemical factors controlling the mobilization of selected inorganic constituents from fossil fuel combustion residues. Part 1: Review of the major elements. *J. Env. Qual.* 19, 188, 1990.
27. Eary, L. E., Dhanpat Rai, Mattigod, S. V., and Ainsworth, C. C., Geochemical factors controlling the mobilization of selected inorganic constituents from fossil fuel combustion residues. Part 1: Review of the major elements. *J. Env. Qual.* 19, 202, 1990.

28. Mattigod, S. V. Dhanpat Rai, and Amonette, J. E. Concentrations and distributions of major and and selected trace elements in size-density fractionated fly ashes, in *Biogeochemistry of Trace Elements in Coal and Coal Combustion Byproducts,* Sajwan, K.E, Alva, A. K., and Keefer, R.F. Kluwer Academic/Plenum Publishers, New York, U. S. A., 115, 1999.

29. Jingxin, S., and R. E. Jervis. Concentrations and distributions of trace and minor elements in Chinese and Canadian coals and ashes. *J. Radioanal. Nucl. Chem. Art.* 114, 89, 1987.

30. Campbell, J.A., J.C. Laul, K.K. Nielson, R.D. Smith. Separation and chemical characterization of finely-sized fly-ash particles. *Anal. Chem.* 50, 1032, 1978.

31. Coles, D., R. Ragaini, J. Ondov, S. Fisher, D. Siberman, and B. Prentice. Chemical studies of stack fly ash from a coal-fired power plant. *Env. Sci. Technol.* 13, 455, 1979.

32. Davison, R.L., D.F.S. Natusch, J.R. Wallace, and C.A. Evans, Jr. Trace elements in fly ash: dependence of concentration on particle size. *Env. Sci. Technol.* 8, 1107, 1974.

33. Klein, D.H., A.W. Andren, J.A. Carter, J.F. Emery, C. Feldman, W. Fulkerson, W.S. Lyon, J.C. Ogle, V. Talmi, R.I. Van Hook, and N. Bolton. Pathways of thirty-seven trace elements through coal-fired power plants. *Env. Sci. Technol.* 9, 973, 1975.

34. Finkelman, R. *The modes of occurrence of trace elements in coals.* Ph.D. Dissertation, University of Maryland. 1980.

35. Mattigod, S. V., Characterization of fly ash particles. *Scan. Elec. Micro. II,* 611, 1982.

36. Smith, R. The trace element chemistry of coal during combustion and the emissions from coal-fired plants. *Prog. Ener. Combust. Sci.* 6, 53, 1980.

37. Bowen, H. J. M. *Environmental Chemistry of Elements.* Academic Press, London 1979.

38. Nozaki,Y., D. Lerche, D. S. Alibo, and A. Snidvongsi. The estuarine geochemistry of rare earth elements and indium in the Chao Phraya river, Thailand. *Geochim. Cosmochim. Acta,* 64, 3983, 2000.

39. McLennan, S. M. Rare earth elements in sedimentary rock: Influence of provenance and sedimentary processes. In *Reviews in Mineralogy, Vol 21 Geochemistry and mineralogy of rare earth elements* (eds B. R. Lipin and G. A. McKay) 169-200. The Mineralogical Soc. of America. 1989.

40. McKay, G. A. Partitioning of rare earth elements between major silicate minerals and basaltic melts. In *Reviews in Mineralogy, Vol 21 Geochemistry and mineralogy of rare earth elements* (eds B. R. Lipin and G. A. McKay) 45-74. The Mineralogical Soc. of America. 1989.

41. Hanson, G.N. an approach to trace element modeling using a simple igneous system as an example In *Reviews in Mineralogy, Vol 21 Geochemistry and mineralogy of rare earth elements* (eds B. R. Lipin and G. A. McKay) 79-97. The Mineralogical Soc. of America. 1989.

42. Grauch R. I. Rare earth elements in metamorphic rocks, In *Reviews in Mineralogy, Vol 21 Geochemistry and mineralogy of rare earth elements* (eds B. R. Lipin and G. A. McKay) 147-167. The Mineralogical Soc. of America. 1989.

43. Burt, D. M. Compositional and phase relations among rare earth element minerals. In *Reviews in Mineralogy, Vol 21 Geochemistry and mineralogy of rare earth elements* (eds B. R. Lipin and G. A. McKay) 259-307. The Mineralogical Soc. of America. 1989.

44. Roaldset, E. rare earth element distributions in some precambrian rocks and their phyllosilicates, Numedal, Norway. *Geochim. Cosmochim. Acta* 39, 455, 1975.

MOBILITY AND ADSORPTION OF TRACE ELEMENTS IN A COAL RESIDUES-AFFECTED SWAMP

G.S. Ghuman[1], K.S. Sajwan[1], S. Paramasivam[1], D.C. Adriano[2], and G.L. Mills[2]

[1]Department of Natural Sciences and Mathematics
Savannah State University
Savannah, GA 31404, U.S.A.

[2]Savannah River Ecology Laboratory
University of Georgia
Aiken, SC, 29802, U.S.A.

ABSTRACT

A study was conducted to determine the mode of transport of Cd, Cu and Ni through the Steel Creek sediments at the Savannah River Site and to find the adsorption of these metals to the sediments. The sediments collected from three adjacent sites showed considerable heterogeneity but had a uniform pH approximating 6.0. Three periodic collections of sediment cores (0-20 cm) from two sites and surface sediments (0-10 cm) of 3rd site were analyzed for three metals in the pore water and DTPA extract of wet sediments. The concentrations of dissolved metals in the pore water were quite low in the ranges of 0.02-0.5 μg L^{-1} Cd, 0.0-6.0 μg L^{-1} Cu, and 2.3-11.9 μg L^{-1} Ni. DTPA-extractable metals in the sediments ranged 0.0-34.67, 0.0-491.4, and 44.6-676.4 μg L^{-1} for Cd, Cu and Ni, respectively. Surface water enclosed in two polyethylene chambers imbedded in the stream path was spiked with 100 μg L^{-1} of each metal. The metals moved downward and outside the chambers through the sediments and equilibrated with the ambient within two weeks. Adsorption study with seven sediment samples using Cd and Cu solutions in the concentration range of 50-110 μg L^{-1} provided essentially linear adsorption isotherms at 25° C with significant correlation coefficients. The amounts of Cd adsorption were greater as compared to Cu adsorption. Study revealed low concentrations of dissolved metals in the surface and pore water and but relatively high adsorption by the sediments and particulate matter which forms the mobility pattern of these metals as a mechanism of transport in the swamp system.

INTRODUCTION

Increased use of coal for electrical generation has resulted in its greater outdoor storage. This is a potential source for the release of large amounts of toxic metals into the surrounding environment. At the Savannah River Site (U. S. Department of Energy Center in Aiken, South Carolina) several creeks running through the area are affected by discharges from coal-fired power plants. Creek waters initially enter the swamp area before discharging into the Savannah River. These waters provide the essential mechanism for the transport and deposition of heavy metals in the swamp area. The

Chemistry of Trace Elements in Fly Ash, edited by Sajwan *et al.*
Kluwer Academic/Plenum Publishers, 2003

waters carrying the runoff from the surroundings of several coal-fired power plants are generally contaminated with iron, aluminum, sulfate, chromium and nickel with lesser amounts of trichloroethylene, cadmium and lead. Aquatic sediment particles containing organic matter have a direct bearing on the accumulation and transport of heavy metals released from coal and ambient ecosystems[1]. Present study was conducted to determine the distribution and mobility of Cd, Cu and Ni in the ambient water and the sediment column in the swamp zone of Steel Creek at the Savannah River Site. To determine the metal retention capacity of swamp sediments adsorption studies were conducted with Cd and Cu using sediments from three locations in the swamp. Several earlier studies were concerned with the adsorption or sorption of heavy metals by sediments and sewage sludge [2,3,4]

MATERIALS AND METHODS

Study Area

Steel Creek is one of the five main creeks that traverse the forested area of the Savannah River Site near Aiken, South Carolina. These creeks flow from north to south and discharge into the Savannah River via the swamp along the riverbank. Steel Creek receives the seepage and runoff water from the surrounding area as well as the effluent water from the L-Reactor located in the center of the site complex. Narrow creek channels expand into a broad swamp area, which is interspersed with small streams and extends over a distance of 1.5 km up to the bank of the river. Creek water is mostly turbid due to suspended particles and flows at the rate of sixty cm/second near the entry point of swamp area where the sampling sites were selected (Figure 1). Since the swamp area is quite loose soil, a wooden board walk erected from the stream edge to the middle of swamp, enabled to locate the sampling sites in the interior of the swamp. Three sampling sites I, II, and III were located adjacent to the board walk at distances of 30 m, 75 m and 150 m, respectively away from the edge of the stream. This swamp area is covered with a thick vegetation of pine and cypress trees, button bush plants and low surface grass in open patches. Due to irregular deposition of water transported materials, the physical and chemical characteristics of the study area are quite heterogeneous.

Sampling Procedure

Sediment samples were collected from three selected sites during the months of June and July. Using a method employed by Menon et al.[5] sediment core samples were collected from sites I and II with a core sampler (a stainless steel tube with a dia. 2.8 cm and length 50 cm) which was lined inside with a polyethylene tube. Upon withdrawing each sample core, it was laid on a plastic board and was cut into three sections as indicated by the distinct color and texture changes of horizons with depth. A total of three cores were withdrawn at each site and composite soil sections were transferred into 125 mL wide- mouthed polyethylene bottles. At site III, only the surface grab samples were collected because the soil at this site was merely a fluid-like sediment-water suspension.

In addition, surface water samples were collected at one-week intervals for a period of three weeks from two experimental chambers set up at sites I and II. Separate core samples were collected from the same chambers at the termination of the experiment. A separate field trip was made to collect relatively large amounts of sediment samples for adsorption experiments and other basic analyses. For this purpose a stainless steel core sampler with internal dia. 16 cm and length 50 cm, previously used by Ghuman and Menon [6] was employed at sites I and II. Upon penetration of the core sampler, surface water was removed with a plastic cup. Three sediment core sections

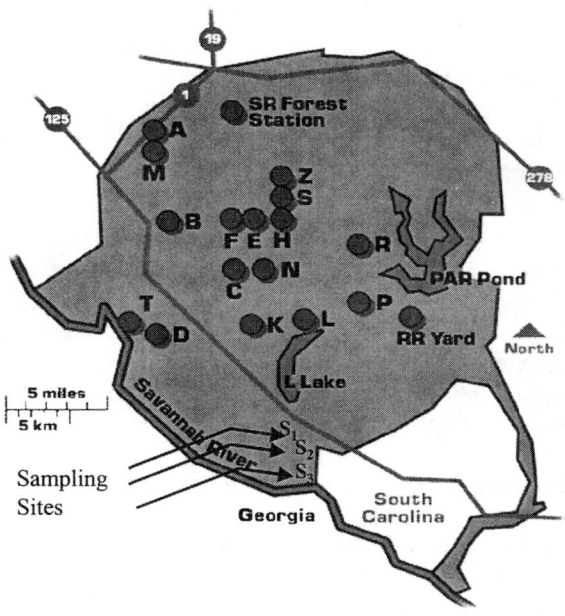

Figure1. Sampling Sites at the Savannah River Site near Aiken, South Carolina.

were scooped and stored in polyethylene bags. At site III, four one-Liter wide- mouthed polyethylene bottles were filled with sediment suspension to obtain the required amount of sample.

Analyses of Cd, Cu and Ni in Pore Water and Sediments

All bottles, centrifuge tubes, filters and sample processing materials were pre-cleaned by soaking in 2% nitric acid. Wet sediment samples were thoroughly mixed with a teflon spatula, pH was measured and pore water was extracted by filtration through 0.45 μm NUCLEOPORE polycarbonate, 47 mm dia. Filter. Pore water was acidified to pH 2.0 with a few drops of 1N HCl and stored for analysis. Duplicate samples of filtered wet sediments (10-15 g each equivalent to 8-12 g dry samples) were weighed into 50-mL centrifuge tubes. Twenty mL of DTPA reagent (0.005 M DTPA (Diethylene Triamine Penta-acetic acid) + 0.01 M $CaCl_2$ + 0.1 M TEA (Triethanol Amine) was added to each sample and the mixture was shaken for 24 hours on a flat-bed shaker at 100 oscillations /min.[7] The suspension was centrifuged at 2000 rpm for 8 minutes and clear supernatant was transferred to a 50-mL volumetric flask. The residue was washed with additional 10 mL DTPA solution and the extract was combined with the previous one in the flask. The combined extract was acidified to pH 2.0 with 0.5 mL (1:1) nitric acid. Volume was made up to the mark with de-ionized water and the extract was filtered through Whatman No. 42 filter to remove the suspended particles. Duplicate blank samples of DTPA solution were also prepared in the same way as the test samples. Cadmium, copper and nickel were analyzed in the acidified pore-water and DTPA extract of sediments by flameless atomic absorption spectroscopy.

Field Chamber Experiment

For the study of transport pattern of metals through the water and sediment column, two circular polyethylene chambers (dia. 37.6 cm) were implanted at sites I and

II in the paths of flowing streams. To design the chambers, the bases of polyethylene containers were removed and the lids were cut to obtain wide openings for sampling. Each chamber was penetrated to a depth of 25 cm in the sediment column, which enclosed a surface water layer of 4 cm depth, thus providing a volume of four liters of enclosed water. Surface water in the chamber was spiked with 0.4 mL each of 1000 ppm solutions of Cd, Cu and Ni, which resulted in 100 μg L^{-1} additional concentration of each metal.

Added metal solutions were thoroughly mixed with the surface water using a Teflon rod. After allowing a settling time of 15 minutes, duplicate aliquots of water samples were taken from the chambers for analysis. Subsequently, water samples were taken from the chambers for two additional weeks at one-week intervals after which the experiment was terminated. At the time of termination, sediment core samples were also taken from each chamber, splitting the cores into three sections as was done for the contemporary cores taken from the spots adjacent to the chambers.

Analysis of Chamber Water and Sediments

After each collection, the volumes and pH of chamber water samples were measured. For the estimation of total metal concentrations in water, one of the duplicate samples was acidified to pH 2.0 with a few drops of 1N HCl and then all samples were stored in a refrigerator. Next day, both types of water samples were filtered through 0.45 μm NUCLEOPORE polycarbonate filters. Non-acidified water was then acidified to pH 2.0 and both types of samples were analyzed for Cd, Cu and Ni. Pore water from the wet sediment samples was separated by filtration and portions of filtered sediments were extracted with DTPA solution as described earlier. All three metals were analyzed in the pore water and the DTPA extract of sediments.

Determination of Total Concentrations of Cd, Cu and Ni in Sediments

Remaining portions of wet sediments and the other bulk amounts of collected sediments were dried in an air oven at 45°C. Dried samples were ground in an agate pestle and mortar to pass through a 35-mesh sieve (0.5 mm esd.) and the coarse sand fraction was discarded. Two-grams sample of each sieved and homogeneous sediment fraction was weighed into a 140-mL Pyrex beaker. Sample was treated with 8 mL of reagent grade conc. HNO_3 and heated at 60° C for one hour with occasional shaking. Digested mixture was diluted with deionized water and transferred to a 40-mL Pyrex centrifuge tube and centrifuged for eight minutes at 2000 rpm. Clear supernatant was decanted into a 50-mL volumetric flask. Sediment residue was washed with 5 mL water by centrifuging and combining the decantate into the volumetric flask. Volume was made up and the digest was filtered through Whatman No. 42 filter paper and the total concentrations of three metals were determined by atomic absorption spectroscopy.

Adsorption Equilibrium Study of Sediments for Cd and Cu

Absorption study with Cd and Cu was conducted on seven sediment samples collected from sites I, II and III in the Steel Creek swamp. Sediments dried at 45° C and ground to pass through a 35-mesh sieve were used for this study. Seven solutions ranging from 10 to 110 μg L^{-1} concentrations of Cd and Cu were separately prepared from atomic absorption standards in the nitrate form. Each of these solutions contained 0.01M sodium nitrate to provide background electrolyte. Sodium nitrate was chosen as against calcium chloride because it was thought that at very low equilibrium solution concentrations, the adsorption of Cd might be inhibited in this study. Four-gram duplicates of prepared sediments were added to 40 mL volumes of varying metal

Table 1. Selected properties of sediments collected from the Steel Creek Swamp.

Site No.	Sample depth (cm)	pH	Mechanical Composition		Chemical Composition		
			Coarse Sand 2.0 – 0.5 mm %	Fine Sand, Silt, & Clay (< 0.5 mm) %	Cd	Cu	Ni
					(μg kg^{-1} Dry Sediment)		
I	0-8	6.66	4.2	95.8	91.49	5775.0	2206.3
	8-14	6.60	9.7	90.3	19.34	487.5	362.5
	14-20	6.61	14.2	85.8	23.00	1087.5	478.3
II	0-8	6.44	32.0	68.0	16.10	562.5	340.8
	8-15	6.12	38.7	61.3	45.54	2437.5	793.8
	15-20	6.34	65.7	34.3	11.54	375.0	56.3
III	0-10	6.72	0.0	100.0	126.59	12337.5	3956.3

Table 2. Analyses of periodic collections of pore water and sediments from the Steel Creek Swamp.

Site No.	Sample depth (cm)	pH	Pore Water			Sediment(DTPA extract)		
			Cd	Cu	Ni	Cd	Cu	Ni
			μg L^{-1}			μg kg^{-1} Dry Sediment		
				June 21 Collection				
I	0-10	6.45	0.27	0.0	3.83	3.97	88.9	314.9
	10-20	6.49	0.12	0.1	4.04	0.35	54.0	302.2
II	0-10	6.62	0.50	0.4	5.08	4.51	88.2	195.3
	10-18	6.61	0.30	0.4	4.48	7.81	142.9	370.3
	18-25	6.83	1.10	0.7	9.36	24.50	327.4	610.6
III	0-10	6.49	0.08	0.0	3.96	0.76	183.2	676.4
				July 4 Collection				
I	0-8	6.47	0.05	3.0	8.70	1.67	25.4	85.1
	8-14	6.57	0.24	3.5	8.70	3.88	141.8	162.5
	14-20	6.55	0.13	24.4	4.50	6.59	130.8	175.8
II	0-8	6.76	0.07	5.0	9.50	6.94	87.3	58.7
	8-15	6.33	0.03	3.0	4.50	0.73	64.1	121.0
	15-20	6.70	0.40	6.0	11.90	1.06	36.9	194.4
III	0-10	6.24	0.02	2.0	4.50	0.0	0.0	544.9
				July 18 Collection				
I	0-8	6.67	0.13	2.5	3.00	34.67	333.3	499.0
	8-14	6.60	0.22	2.5	3.90	27.00	491.4	140.2
	14-20	6.62	0.41	3.7	5.70	60.53	961.1	337.3
II	0-8	6.45	0.10	2.5	3.40	14.01	305.2	157.3
	8-15	6.14	0.09	2.5	2.30	3.59	59.3	448.8
	15-20	6.33	0.10	2.5	3.40	5.31	143.9	44.6
III	0-10	6.80	0.07	2.5	3.00	7.01	95.6	397.5

concentration solutions contained in 50-mL polyethylene tubes. Initial pH of each solution was 6.0, but the pH of suspension was again adjusted to 6.0 with 0.1N high-purity nitric acid or NaOH. The tubes were shaken for four hours in a shaker with constant temperature bath at 25° C. Preliminary study indicated that four hours shaking time was sufficient to reach equilibrium. At the end of shaking time, the tubes were centrifuged at 2000 rpm for eight minutes and the supernatant was analyzed for Cd and Cu using atomic absorption spectroscopy. The adsorbed Cd and Cu were calculated from the difference between each metal present in the initial and that determined in the equilibrium solutions. The pH of the suspensions after shaking time of the tubes and that of the supernatant solutions were measured in all samples. The metal uptake (adsorption) concentrations were plotted against the initial solution concentrations for all the seven sediment samples.

RESULTS AND DISCUSSION

The selected properties of sediments collected from the steel creek Swamp are presented through Table 1.Three periodic collections of sediment samples from the swamp exhibited nearly constant pH values in the range of 6.14-6.83 (Table 2). This indicated that all pH-dependent reactions in the Steel Creek swamp might be relatively constant. Kanungo and Mohapatra[8] reported that trace metal ions release from fly ash decreases sharply with increase of pH from 3.0 to 8.0. Filtered and acidified pore water samples had very low concentrations of dissolved Cd, Cu and Ni, but showed appreciable variations between the sampling periods, sampling sites and the depths of sediments. Dissolved Ni concentrations of pore water ranged from 2.3 to 11.9 μg L^{-1}, with respect to other samples, the maximum value was found in the bottom sample (15-20 cm depth) of July collection at site II. Dissolved Cu concentrations of pore water varied between 0.0 and 6.0 μg L^{-1}, with one exceptionally high value of 24.4 μg L^{-1} Cu in the bottom sample (14-20 cm depth) of July collection at site I. All Cd concentrations of pore water samples fell within the range of 0.02 – 0.50 μg L^{-1}, with one exceptional value of 1.1 μgL^{-1} Cd in the bottom sample (18-25 cm depth) of June collection at site II. Surprisingly, the consistent depth values of all three metals in the pore water from site III were lower than in pore water from sites I and II. The concentrations of three metals ranged in the decreasing order as : Ni > Cu > Cd.

The patchiness of sediment deposition in the swamp resulted in the heterogeneous concentrations of DTPA-extractable metals (Table 2). Cadmium concentrations of most samples fell between 0.0 and 34.67 μg kg^{-1} with one exceptional value of 60.53 μg kg^{-1}. Copper concentrations were in the range of 0.0 – 491.4 μg kg^{-1} with one exceptional value of 961.1 μg kg^{-1} and the Ni concentrations were in the range of 44.6 – 676.4 μg kg^{-1}. Just like the pore water metal concentrations, the DTPA-extractable metals of sediments also range in the decreasing order as: Ni > Cu > Cd. The DTPA-extractable concentrations of Cu and Ni were much lower than those reported by Tobin et al.[9] for the flood plain sediments of Idaho. None of the samples exceeded the threshold contamination concentration of 170 μg g^{-1} Cu and 100 μg g^{-1} Ni, concentrations considered as critical threshold levels for contamination (New Jersey Department of Environmental Protection[10]). The average concentration of copper in American soils is 25 μg g^{-1} and the natural range is 1 - 300 μg g^{-1} (Shacklette et al.)[11]

Results of Chamber Experiment

Periodic analysis of surface water samples from the two chambers clearly distinguished the changes in metal concentrations (Table 3). Analysis of unfiltered acidified water (B) provided the total metal concentrations and the analysis of filtered acidified water (A) gave the dissolved metal concentrations. The difference between these two types of concentrations is the particulate metal concentration. The samples collected just after spiking the surface water of two chambers with 100 μg L^{-1} metal solutions showed significant difference between the total and dissolved concentrations of Cd, Cu and Ni.This indicates that the applied metals were immediately taken up by the particulate matter (organic carbon, clay and other mineral particles) in the surface water. Besides, at the prevailing pH of 7.0 ± the applied metals could be rapidly precipitated as insoluble colloidal forms. The differences between the total and dissolved decreased with time and by the end of 3rd week, no significant difference was observed. This may be due to the complete settling of the metal carrying particulate matter or its transport outside the chamber via the moving stream below the open chamber bottom and dilution by the diffusion of fresh water into the chamber. A comparison between the metal

Table 3. Periodic analysis of surface water of the experimental chambers spiked with 100 µg L^{-1} each of Cd, Cu and Ni.

Site No.	Sample Date	A: Filtered acidified water				B: Unfiltered acidified water			
		pH	Cd	Cu	Ni	pH	Cd	Cu	Ni
			---------- µg L^{-1} --------				----------- µg L^{-1} ----------		
I	* July 4	6.81	9.62	6.50	15.0	6.95	62.04	44.25	50.4
	July 11	7.27	0.20	2.50	5.0	7.57	0.82	9.00	9.5
	July 18	7.42	0.11	3.70	4.5	7.30	0.15	3.70	3.7
II	July 4	7.02	5.56	7.00	11.9	7.15	60.00	40.23	37.4
	July 11	7.60	0.17	2.50	8.7	7.79	0.62	3.10	22.5
	July 18	7.56	0.08	2.50	5.0	7.72	0.15	2.50	4.0

* Chamber waters were spiked with metals on July 4 and sample aliquots were taken for A and B type analysis after 15 minutes of settling time.

Table 4. Analysis of pore water and sediments collected from the experimental chambers after the termination of the experiment on July 18.

Site No.	Sample depth (cm)	pH	Pore water			Sediment (DTPA extract)		
			Cd	Cu	Ni	Cd	Cu	Ni
			----------- µg L^{-1} --------			----- µg kg^{-1} dry sediment ----		
I	0-8	6.75	0.08	2.2	5.5	38.14	313.0	291.6
	8-14	6.64	0.09	1.9	3.8	6.83	108.3	64.0
	14-20	6.76	0.35	2.5	3.6	38.90	775.1	404.2
II	0-8	6.70	0.10	2.5	3.7	24.42	397.6	197.9
	8-15	6.63	0.17	2.5	3.9	10.32	146.5	58.0
	15-20	6.56	0.17	2.5	2.5	12.67	272.5	94.8

Table 5. Results of total concentrations of metals in the HNO_3 digest of sediments.

Site No.	Sample depth (cm)	Metal concentration (µg kg^{-1})		
		Cd	Cu	Ni
I	0 - 8	28.75	1387.5	412.5
	8 - 14	15.05	187.5	312.5
	14 - 20	38.09	2150.0	706.3
II	0 - 8	39.82	1925.0	696.8
	8 - 15	17.73	650.0	365.8
	15 - 20	15.08	700.0	437.5

concentrations of water samples collected after two weeks (Table 3) with the metal concentrations of pore water (Table 4) indicates that an equilibrium had been reached between the two phases of water in both the chambers.

Results of DTPA-extractable and concentrated HNO_3 – extractable total concentrations of Cd, Cu and Ni (Tables 4 and 5) in the sediments suggest a downward movement of spiked metals, which partially enriched the bottom sediments of both chambers. When metal concentrations in the chamber sediment columns are compared with those in the adjacent open-surface sediments (Table 1), an appreciable deposition of fresh particulate matter is seen in the top layer of site I.

Results of Adsorption Equilibrium Study

Cadmium and copper adsorption isotherms (Figures 2 to 6) for the seven sediment samples from the three sites are essentially linear. The linear regression equations showed significant correlation coefficients ranging 0.986 – 1.0 with small variations indicating a good fit to the Langmuir Equation.

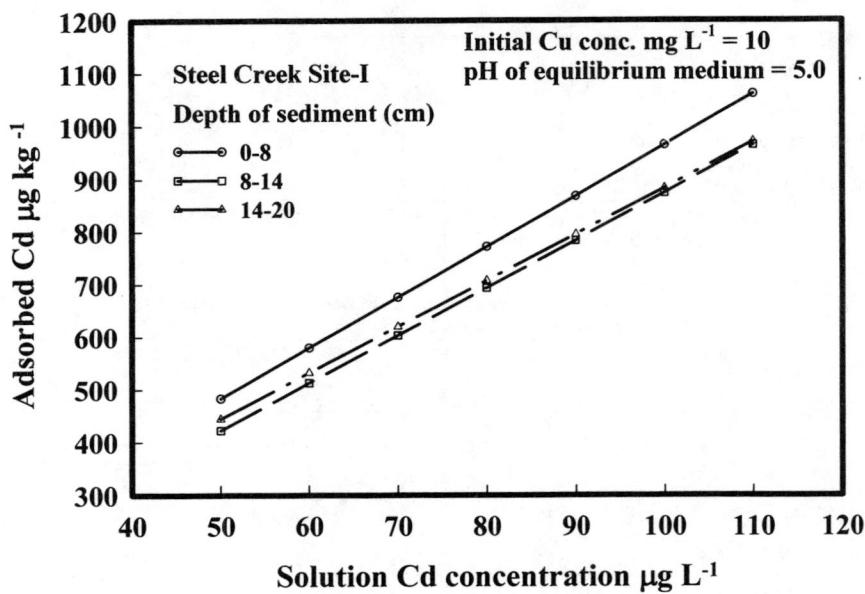

Figure 2. Adsorption Isotherms for Cd for Steel Creek Sediments of Site-I.

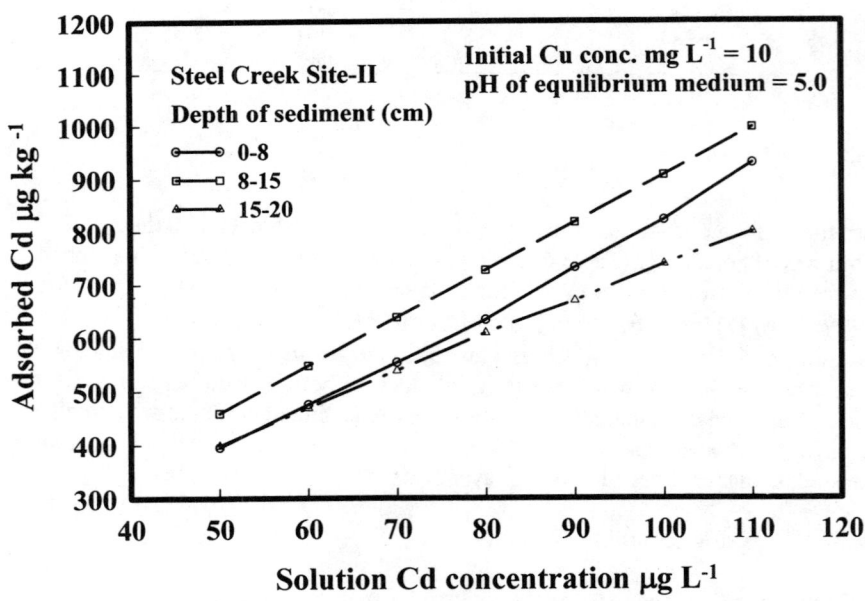

Figure 3. Adsorption Isotherms for Cd for Steel Creek Sediments of Site-II.

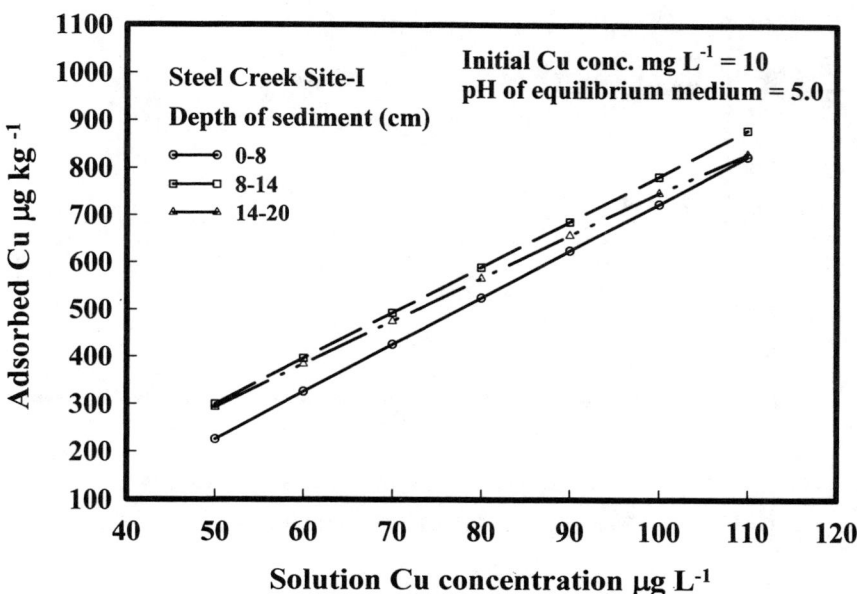

Figure 4. Adsorption Isotherms for Cu for Steel Creek Sediments of Site-I.

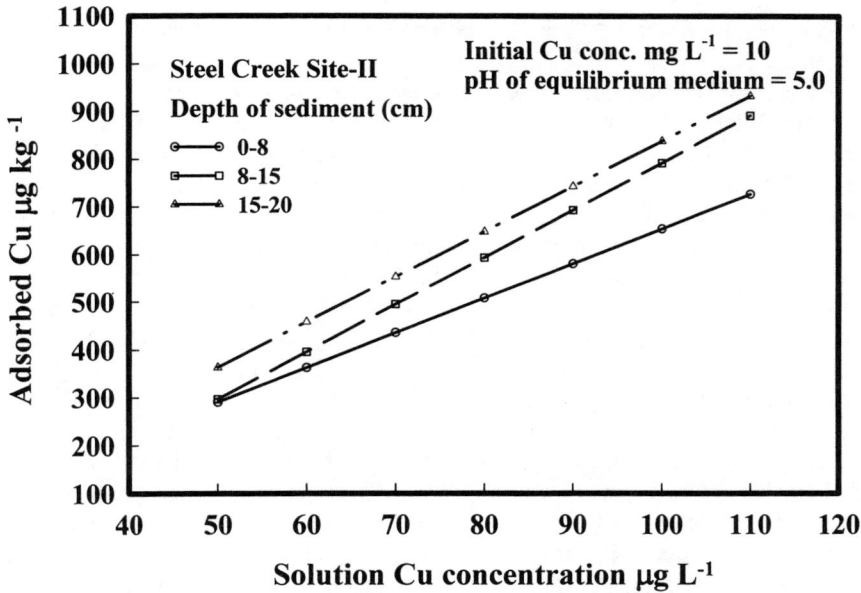

Figure 5. Adsorption Isotherms for Cu for Steel Creek Sediments of Site-II.

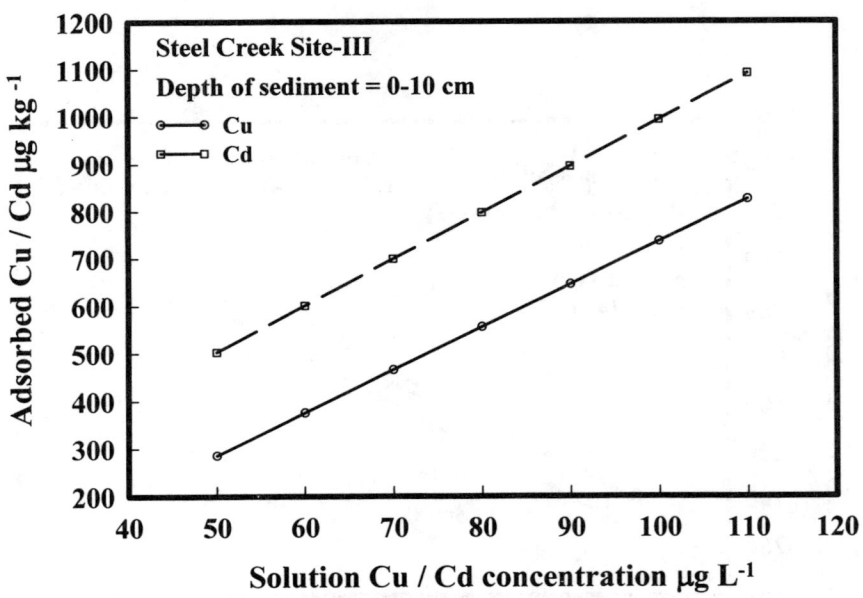

Figure 6. Adsorption Isotherms for Cu/Cd for Steel Creek Sediments of Site-III.

Slopes of Cd adsorption curves for the sediments of sites I and II (Figs. 2 and 3) are in the range of 7.013 – 9.650, the low values were for the site II sediments. At equilibrium, only small proportions of Cd (3.2 – 26.8%) remained unadsorbed. Cadmium adsorption curves are perfectly linear for all sediments of site I and of medium layer of site II. Top and bottom layer sediments of site II (Fig. 4) are curvilinear at high Cd concentrations, the top-layer ones curved upwards, while the bottom-layer ones curved downward.

Slopes of Cu adsorption curves for the sediment core samples of sites I and II range from 7.255 to 9.992. Appreciable proportions of added Cu (15.1 – 52.8%) remained unadsorbed at equilibrium, and the maximum adsorption occurred from the solutions of initial high concentrations. The unadsorbed fraction of Cu at equilibrium in site III sediments was in the range of 21.9-45.1%, while Cd adsorption for this site was 99-100%. Relatively greater Cd adsorption than Cu adsorption may be attributed to the low initial Cd content of these sediments (Table 1). Copper adsorption depicted by these isotherms indicates only a small faction of maximum adsorption possible by these sediments. Additional Cu concentrations for maximum adsorption would have produced departures from linearity, giving curvilinear isotherms. Such a curving tendency is apparent in the Cu adsorption isotherm for the bottom sediments of site I (Fig. 4) when solution concentration used was 90 $\mu g\ L^{-1}$ of Cu. Adsorption solutions prepared in 0.01M $NaNO_3$ represent specific adsorption of Cu in this study. Elliott and Denney[12] reported that Cd adsorption was the least at low pH, but increased sharply to a maximum uptake as pH approached neutrality (6.0-7.0), and leveled off or decreased slightly at more alkaline conditions. McBride et al.[13] observed that organic matter and clay in soil were responsible for strong adsorption of Cd and thus were able to limit Cd uptake by plants. Gerritse and Driel[14] found that the exchangeable fractions were in the range of 10 – 50% of total Cd, Zn and Cu in temperate soils. In view of these reports it is conceivable that organic carbon, exchange sites on clay particles and oxides of iron and manganese are responsible for the linear adsorption of Cd and Cu by the sediments of Steel Creek Swamp.

ACKNOWLEDGEMENTS

This study was funded, in part, by the contract number DE-FG09-96SR-18558 U.S. Department of Energy and Environmental Protection Agency.

REFERENCES

1. Ekpo. B. O., and Ibok, U. J., Seasonal variation and partition of trace metals (Fe, Zn, Cu, Mn, Cr, Cd and Pb) in surface sediments; relationship with physico-chemical variables of water from Calabar River, S-E Nigeria. *Environ. Geochem and Health*, 20,113, 1998.

2. Bowman, R. S., Essington, M. E., and O'Connor, G. A., Soil sorption of nickel: Influence of solution composition, *Soil Sci. Soc. Am. J.* 45, 860, 1981.

3. Harter, R.D., Effect of soil pH on adsorption of lead, copper, zinc, and nickel, *Soil Sci. Soc. Am. J.*, 47, 47, 1983.

4. Riffaldi, R.Levi-Minzi, Saviozzi, A. and Tropea, M., Sorption and release of Cadmium by some sewage-sludges, *J. Environ. Qual.*, 12, 253, 1983.

5. Menon, M.P., Ghuman, G.S., and Emeh, C.Obi., Trace element release from estuarine sediments of South Mosquito Lagoon near Kennedy Space Center, *Water, Air and Soil Pollution*, 12, 295, 1979.

6. Ghuman, G.S., and Menon, M.P., Distribution of heavy metals in a marshland ecosystem of the southeastern Atlantic Coast, *in Trace substances in environmental health*, XVIII Proc. Ann. Conf., Univ. Missouri, Columbia, D.D. Hemphill, (ed.) 237, 1984.

7. Bingham, F. T., Page, A. L., Mahler, R. J., and Ganje, T. J., Yield and cadmium accumulation of forage species in relation to cadmium content of sludge-amended soil, *J. Environ. Qual.* 5, 57, 1976.

8. Kanungo, S.B., and Mohapatra, R., Leaching behavior of various trace metals in aqueous medium from two fly ash samples, *J. Environ. Qual.* 29, 188, 2000.

9. Tobin, G.A., Brinkmann, R., and Montz, B.E. Flooding and the distribution of selected metals in floodplain sediments in St. Maries, Idaho. *Environ. Geochem. And Health*, 22, 219, 2000.

10. New Jersey Department of Environmental Protection, Summary of approaches to Cleanup levels, Division of Waste Management, Trenton, New Jersey, 32p., 1987.

11. Shacklette, H.T., Hamilton, J.G., Boerngen, J.G. and Bolwes, J.M., Elemental composition of surfacial in the coterminous United States. U.S. Geological Survey Paper 574-D, Government Printing Office: Washington, D.C., 71., 1971.

12. Elliott, H.A., and Denny, C.M., Soil adsorption of cadmium from solutions containing organic ligands, *J. Environ. Qual.*, 11, 658, 1982.

13. McBride, M.B., Tyler, L.D., and Hovde, D.A., Cadmium adsorption by soils and by plants as affected by soil chemical properties, *Soil Sci. Soc. Am. J.* 45, 739, 1981

14. Gerritse, R.G., and Driel, W.Van. The relationship between adsorption of trace metals, organic matter and pH in temperate soils, *J. Environ. Qual.*, 13, 197, 1984.

POTENTIAL USE OF FLY ASH TO REMOVE CADMIUM FROM AQUEOUS SYSTEMS

M. Hajarnavis[1], K.S. Sajwan[2], S. Paramasivam[2], C.S. Chetty[2], and G.R. Reddy[2]

[1]National Environmental Engineering Research Institute, Nagpur, MS 440020
India
[2]Marine, Environmental Sciences and Biotechnology Research Center
Savannah State University, Savannah, GA 31404, U.S.A.

ABSTRACT

Removal of pollutants from water via adsorption on to activated carbon is a promising remediation technique. However, due to its high cost and limited availability, it is necessary to investigate alternate adsorbent sources. Fly ash is, an inexpensive and abundantly available by-product from thermal power plants that utilizes coal for the production of energy. A study was conducted to (i) identify the optimum conditions for Cd adsorption by fly ash, (ii) evaluate the potential use of fly ash to remove cadmium from a mixed metal solution of Cu, Cd, Mn, Ni, and Zn at room temperature mimicking industrial and municipal effluents and (iii) study the kinetics of Cd adsorption. Preliminary results of the study indicated that a pH of 5 was optimum for Cd removal. Results of the kinetics studies indicated that removal of Cd by fly ash increased with increasing contact time while Cd removal also marginally increased with increasing amount of adsorbent used. Fitting of Cd adsorption data for the full range of metal concentrations was described by a Freundlich model with a moderate correlation coefficient (r = 0.63) while the adsorption phenomena was described well by Langmuir isotherms at moderate metal concentration levels (5 to 100 mg L^{-1}) with high correlation coefficient (r = 0.85). This study revealed that fly ash could be used as an adsorbent to remove Cd from wastewater containing a mixture of various inorganic pollutants.

INTRODUCTION

The increase in industrialization results in a concomitant increase of wastewater production and contamination of water that require solutions for purification, recovery and reuse of wastewater. Potentially toxic trace metals (specific gravity > 5 and atomic number > 23) [1] are unlike other pollutants in that they occur naturally in the environment in the form of oxides and or sulfates. These metals are non-biodegradable, persistent and toxic to living organisms at relatively low concentrations. Discharge of effluents containing heavy and trace metals from industries and wastewater treatment plants poses a major threat to aquatic fauna and flora and to human health through food chain biomagnification. Higher concentrations of heavy and trace metals in treated wastewater from conventional treatment process are difficult to remove and prevent efficient reuse and disposal.

The chemical composition of processed wastewater produced by the wastewater treatment plants in cities around the globe varies with location and source of influent

Chemistry of Trace Elements in Fly Ash, edited by Sajwan *et al.*
Kluwer Academic/Plenum Publishers, 2003

wastewater. Unlike fly ash, influent wastewater collected from various locations for purification characteristically contains high levels of major plant nutrients, trace and heavy metals and organic compounds. Among the various metals (Zn, Cu, Fe, Mn, Ni, Cd, Cr, Pb, and Se) commonly present in influent wastewater, Cd is widely recognized as the most hazardous element [2, 3].

Though conventional adsorbents are available and used for treatment by wastewater treatment plants, activated carbon is commonly used and considered the most promising technique for removal the majority of organics contaminants and selected metals from aqueous systems [4]. However, due to its high cost and limited availability, it becomes necessary to identify and investigate alternate adsorbent sources. Fly ash is, an inexpensive and abundantly available byproduct, from thermal power plants, and consists of finely dispersed particulate matter [5, 6]. In India, the great majority of plants are coal-fired [6]. The quantity of coal burnt is at present very large and predicted to increase in the near future, resulting in a huge and potentially unmanageable quantity of fly ash. The composition of fly ash is similar to clay in that it contains large amounts of silica, aluminum and some unburnt carbon [7]. A potential use of fly ash, often overlooked, is as an adsorbent, capable of removing many organic contaminants due to its large surface area per unit volume and its high residual carbon content. The concentration of residual carbon contained within the fly ash determines its effectiveness as an adsorbent.

Several scientists have studied the use of fly ash in the treatment of low molecular weight organics compounds and industrial effluent [8, 9]. Khanna and Malhotra [10] studied the kinetics and mechanism of adsorption of phenol on to fly ash particles and designed an economical and flexible system of phenol removal. Gupta et al., [11, 12] treated aqueous solutions with chrome dye and hazardous dye-house wastewater using fly ash, coal, wallostonite, and china clay and concluded from the isotherm studies that the removal efficiency of the fly ash was higher than that of any other adsorbents tested. Vandenbusch and Sell [9] have tried six different fly ashes for removal of color, fluorescence, and reduction in the chemical oxygen demand (COD) from a municipal treatment facility effluent. Singh et al., [8] briefly explained sources and behavior of organic pollutants, which can be adsorbed on to fly ash as well as the conditions, and mechanisms that are involved in water and wastewater treatment using fly ash for removal of organic contaminants.

In early 90s, interest arose in the investigation of materials that could be used in non-conventional methods for scavenging heavy metal ions from industrial wastewaters [13]. Singh and Rawat [14] studied the adsorption of Cu from an aqueous system on bituminous coal and found that adsorption was dependent on concentration, pH and temperature. They also studied the kinetics and the mechanism of Cu sorption in this system. Several studies have indicated that the efficacy of trace metal and metalloid (such as Cd, Pb, Mn, Zn, As, and Se) and organic compound adsorption on fly ash could be enhanced by a CO_2 infusion process [15, 16, 17, 18]. It was also demonstrated [19, 20] that the sorptive behavior of trace metals on fly ash in aqueous systems is partly determined by the pH of the system, the trace metals As, Cd, Cr, Cu, Pb, Ni, and Zn attained the lowest solubility in solutions with a pH between 8 and 9. They attributed this observation to increased adsorption and precipitation processes. In addition, significant increase in solubility of these trace metals were observed when the pH dropped from 12 to 9 and again below 6 in the presence of alkaline fly ash in aqueous system [19].

Therefore, there is considerable interest in investigating the ability of fly ash to remove heavy metals from wastewater and contaminated aquatic systems. The objectives of this study were to (i) determine the feasibility of using fly ash to remove cadmium

Table 1. Selected characteristics of fly ash

Characteristic	Magnitude	Unit
Particle size	0.1 – 0.25	mm
Loss on Ignition (Carbon content)	3.71	% by weight
pH	8.43	
SiO_2	40.45	% by weight
Al_2O_3	16.75	% by weight
Fe_2O_3	4.13	% by weight
CaO	3.67	% by weight
MgO	2.58	% by weight
Cd	1.00	$mg\ kg^{-1}$
Mn	539.00	$mg\ kg^{-1}$
Specific gravity	1.27	$g\ mL^{-1}$
Specific surface area	3520.00	$cm^2\ g^{-1}$

from aquatic systems by using mixed of metals solution and (ii) to understand the kinetics and mechanisms associated with the adsorptive behavior of cadmium by fly ash.

MATERIALS AND METHODS

Adsorbent

Untreated fly ash (adsorbent) from the Koradi thermal power plant, Nagpur, India was used in this study. Basic characteristics of the fly ash are presented in Table 1.

Adsorbate

The combined salts on common inorganic contaminants of wastewater and aqueous systems (e.g. Cu, Cd, Zn, Ni, and Mn) were dissolved to prepare mixed metal solutions. Removal of metals is difficult when effluents contain a mixture of metals due to differences in the solubilities of those present [21].

Preliminary Study: pH Effect

The effect of pH ranging from 1 to 7 was investigated for an initial metal concentration of 10 mg L^{-1}. The removal of pollutants from wastewater by sorption was highly dependent on pH of the solution, which affects the surface charge of the adsorbent and speciation of sorbate. When a mixed metal solution was adsorbed on fly ash, cadmium adsorption was optimal at pH 4.5 to 5.0. This observation was in contrast to

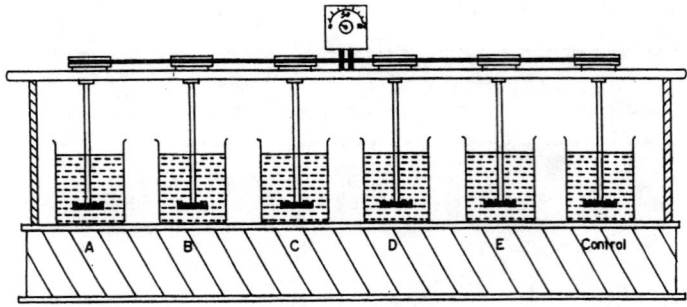

Figure 1. Mixing arrangement for batch experiment with fly ash.

observations by other scientists [19, 20] and may be caused by certain characteristics of the fly ash (Table 1) used in this study. Thus, mixed metal solutions were adjusted to a pH of 5 as an appropriate pH to fulfill the objective.

Adsorption Experiments

Adsorption studies were performed at room temperature ($25° \pm 2° C$) as batch experiments. Mixtures of predetermined quantities of metal solution (5, 25, 50, 75, and 100 μg g^{-1}) and adsorbent (25, 50, 100, 200, 300, 400, and 500 g L^{-1}) were agitated continuously for 0.25, 0.5, 1, 2, 4, and 8 hours of contact in batch reactors (Fig. 1). A blank control and treated samples were filtered through Whatman filter paper No. 42, removed from the reactors at the specified contact periods, and analyzed for pH, electrical conductivity, and metal concentration. Metals were analyzed by an 8000-polarized Zeeman atomic adsorption spectrophotometer.

Cadmium Kinetics Studies

Kinetic studies of the metal-fly ash systems were performed to study the effect of metal concentration in solution, quantity of fly ash mixed, and time of contact between the adsorbate and adsorbent. These studies were: (a) kinetics of metal adsorption for fixed initial metal concentrations in solution and varying fly ash concentrations, (b) kinetics of metal adsorption for varying initial metal concentrations and a fixed fly ash concentration, and (c) fly ash-metal adsorption isotherm. The kinetics studies were conducted with mixing times of 0.25, 0.5, 1, 2, 3, 4, 5, 6, 7, and 8 hours at room temperature ($25° \pm 2°$) and at pH 5.

RESULTS AND DISCUSSION

The adsorption of solute onto the adsorbent presumably occurred in three consecutive steps: (a) solute from solution moved to the exterior surface of the adsorbent, (b) the solute moved through the pores of the adsorbent by intraparticle diffusion, and (c) solute was adsorbed at the specific site on the surface of the adsorbent particle.

Effect of Adsorbent Concentration

Figure 2 shows the removal of Cd from mixed metal solution (10 mg L^{-1}) by varying quantities of fly ash ranging from 25 to 500 g L^{-1}. The removal curves can be expressed by equations 1 and 2:

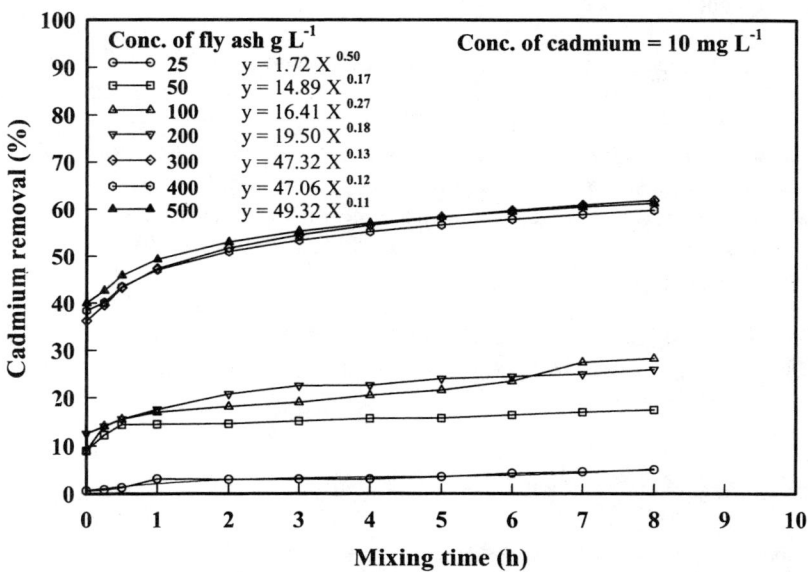

Figure 2. Kinetics of cadmium adsorption from various fly ash concentrations for fixed initial Cadmium concentration with time (h).

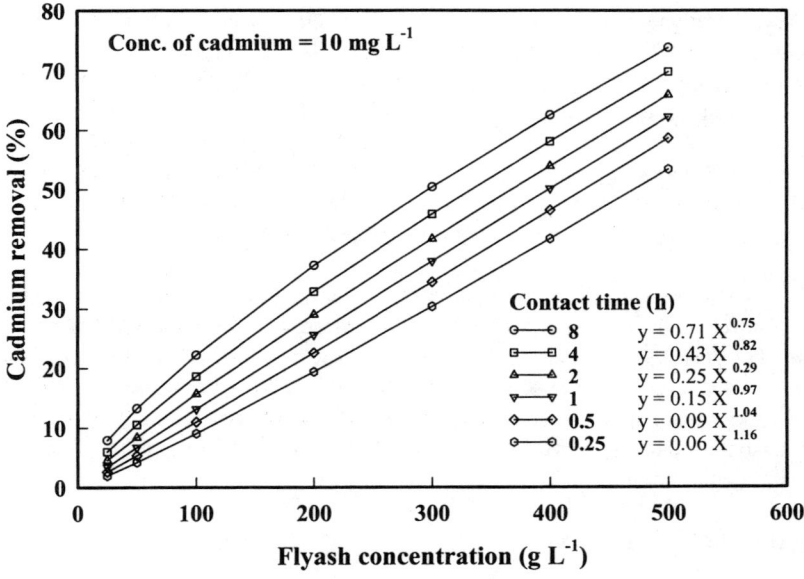

Figure 3. Relationship between percent cadmium removal and with varying concentrations of fly ash with time (h) at fixed initial cadmium concentration.

$$A_s = K_t \; x \; t^m \tag{1}$$
$$\text{or} \; \log A_s = m \log t + \log K_t \tag{2}$$

(see Appendix for characters used)

Equation 2 gives a straight line on a log-log plot with slope "-m-" varying from 0.23 to 1.693 and K_t value from 0.491 to 0.894 for cadmium. The rate of adsorption has been reported to be functions of m and K_t [21]. The amount of solute adsorbed per unit weight (K_t) increased with increasing fly ash concentrations indicating an increase in the rate of metal removed (Fig. 2). The increase in the percentage of metal removed fell substantially after a contact time of 2 hours. Equilibrium was attained after 7 to 8 hours of contact between adsorbate and adsorbent. Maximum adsorption mechanisms for high fly ash concentrations gave high values of "-m-".

The relationship between percent Cd removed and the concentration of fly ash is shown in Figure 3. The metal removed was a function of fly ash concentration as shown in equation 3 and 4:

$$A_s = K_f \; x \; C_f^n \tag{3}$$
$$\text{or} \; \log A_s = n \log C_f + \log K_f \tag{4}$$

(See appendix for characters used)

The rate of Cd removal at a particular contact time is the value for "-n-" (Fig. 3). The "-n-" value increased up to a contact time of 2 hours and then an increase in "-n-" appeared to become dependent on contact time. The percentage of metal removed by 100 g L^{-1} of fly ash showed that an increase in contact time reduced further beyond 2 hours. Thus, 2 hours of contact time appeared to be considered optimum.

One hundred gram of fly ash per liter removed 20% of the cadmium from solution after a contact time of 2 hours. For each subsequent 100 g L^{-1} increase in fly ash concentration, the percent removal of Cd was 14, 13, 12, and 12 (Fig. 3). Therefore, larger quantities of fly ash become comparatively less efficient per unit weight, notwithstanding higher total percent metal removal for high fly ash concentrations. Thus, taking into account handling problems for large quantities of fly ash quantities and the considerably small percent removal for low fly ash quantities, a fly ash concentration of 100 g L^{-1} can be considered optimum for further experimentation.

The present investigation revealed that the average value of "-m-" was 1.10 for Cd, which controlled the overall rate of adsorption because of the greater transport rate in the fly ash. The surface of fly ash develops positive and negative electrical charges in the presence of a water dipole. The potential determining ions of water molecules (H^+/OH^- ions) that become associated with the metal oxides of the adsorbent and their subsequent acid-base dissociations can produce comparatively different kinds of surface charge in the acidic and basic mediums.

Effect of Adsorbate Concentration

Figure 4 shows the Cd adsorption from systems where fly ash concentrations were fixed at 100g L^{-1} and concentration of metals varied from 5 to 100 mg L^{-1}. The relationship can be expressed by equation 2. An increase in concentration of Cd resulted in a decrease in slope "-m-" and reduction in "K_t." This reduction in immediate solute adsorption is probably due to the lack of available sites on the fly ash surface compared

182

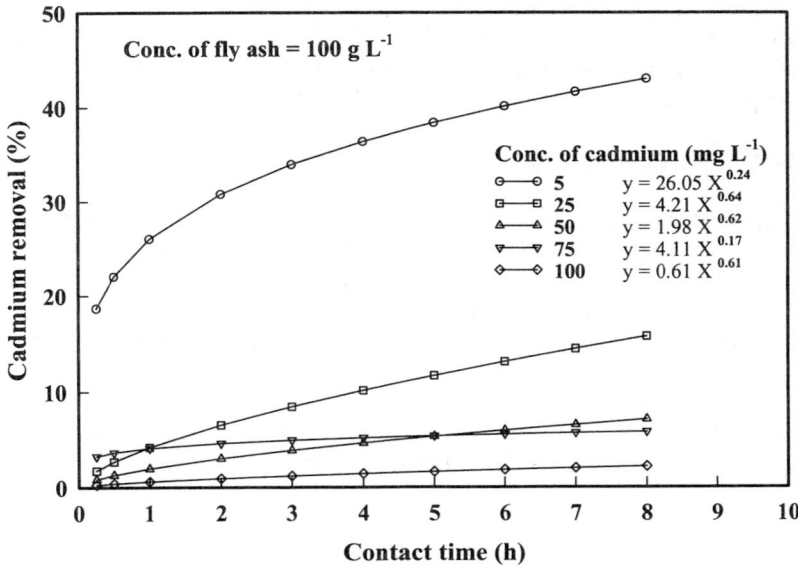

Figure 4. Kinetics of cadmium adsorption by fixed fly ash concentration for variable initial cadmium concentrations.

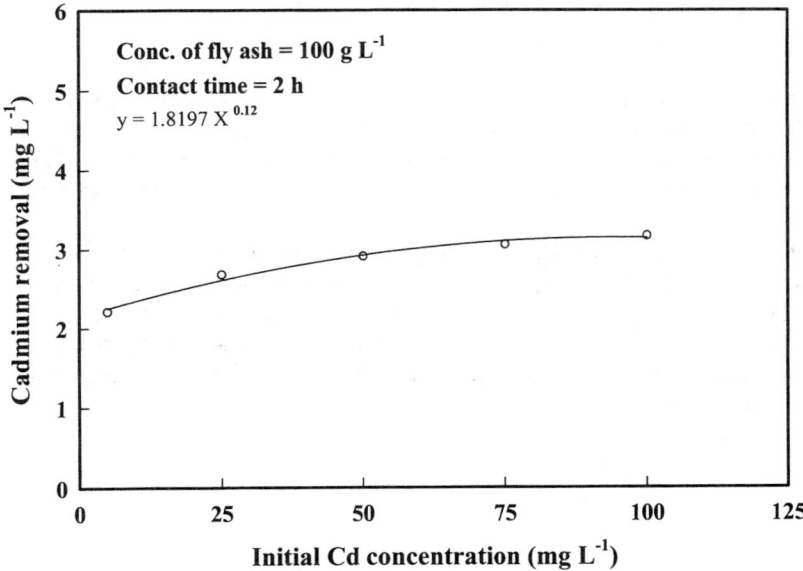

Figure 5. Relationship between cadmium adsorption by fixed fly ash concentration and contact time and variable initial cadmium concentration.

to the relatively large number of sites required for high metal concentrations. A large value of "-m-" implies strong bonds between adsorbate and adsorbent.

The relationship between Cd removed by a fixed fly ash concentration and varying initial Cd concentrations were plotted (Fig. 5) to reveal an almost decreasing rate of metal removal for high initial metal concentrations. The ultimate capacity of fly ash used in the experiment was estimated as 0.127 mg of Cd g^{-1} of fly ash.

Fly Ash-Metal Adsorption Isotherm

The degree of adsorption and resulting equilibrium relationship has been correlated according to the empirical relationship of Freundlich and the theoretically derived Langmuir relationship. An adsorption isotherm for a metal defines a functional expression for the distribution of the metal in liquid and solid phases, i.e., fly ash at constant temperature has been plotted and discussed for an efficient utilization of fly ash adsorption capacity.

The equilibrium data have been processed in accordance with Freundlich isotherm, defined by equations 5 and 6 (plot is not shown) for the full range of metal concentrations investigated:

$$qe \quad = \quad K_F \quad x \quad C^{1/n} \tag{5}$$
$$or \ \log \ qe \quad = \quad \log \ K_F + \tfrac{1}{2} \log \ C \tag{6}$$

(see appendix for characters used)

This is a special case of heterogeneous surface energies and its validity in the system is probably correct as fly ash consists of carbon and oxides of silica, alumina, and iron possessing different surface energies. A coefficient of correlation of this relationship was 0.63.

Metals become adsorbed on silica, alumina, and functional groups present in the fly ash. Adsorption of ions may be due to the desolvation of the sorbing species, change in size of the pores, and the enhanced rate of intraparticle diffusion of sorbate. The value of K_F represents the adsorption capacity of fly ash for a metal concentration of 1 mg L^{-1}.

Similarly, the data for moderate metal concentrations (5 to 100 mg L^{-1}) have been correlated with the Langmuir isotherms. The extent of steady-state adsorption is a function of the relative equilibrium concentration of the solute constituent according to a monolayer adsorption mode. Equation 7 defines the Langmuir isotherm and data were plotted (Fig. 5) for cadmium with a coefficient of correlation of 0.85, proving the strong validity for Cd.

$$\frac{1}{qe} \quad = \quad \frac{1}{Q^o} \quad + \quad \frac{1}{bQ^oC} \tag{7}$$

(see appendix for characters used)

The value of Q^o (1.63×10^{-5} mg Cd mg^{-1} of fly ash) represents the limiting amount of adsorbed metal to form a complete monolayer on the surface of the fly ash.

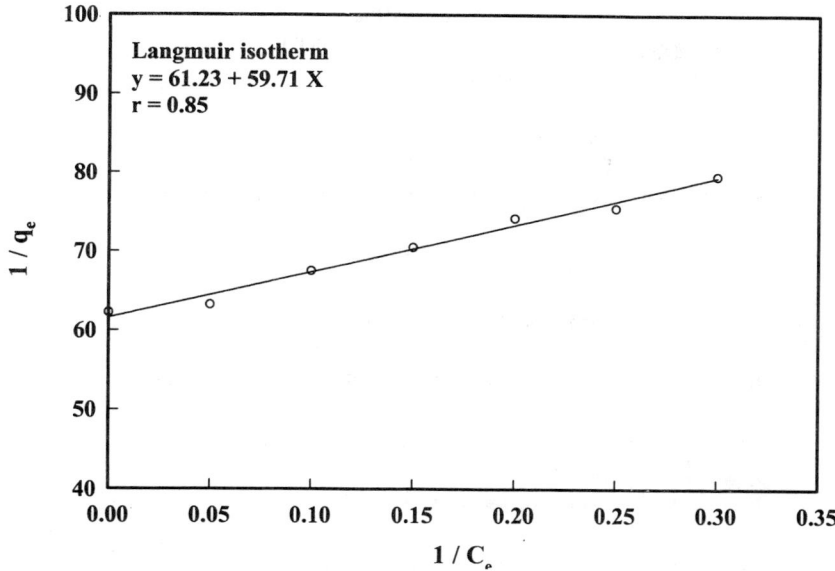

Figure 6. Langmuir isotherm for moderate (5 to 100 mg L^{-1}) cadmium concentration.

Cadmium data was tested with the BET (Brunauer, Emmett, and Teller) isotherm at low metal concentrations up to 5 mg L^{-1}. The BET isotherm can be represented by Equation 8:

$$\frac{C}{(C_s - C)qe} = \frac{1}{BQ^0} + \frac{C\,(B-1)}{C_s\,BQ^0} \qquad (8)$$

(see appendix for characters used)

The value of Q^0 calculated from the slope and intercept of the linear plot was 1.28 x 10^{-5} mg Cd mg^{-1} fly ash. However, the relationship was very weak (r = 0.002)

CONCLUSIONS

The time of contact between adsorbent and adsorbate can be ascertained by the rate of adsorption. The equilibrium of adsorption indicates the approximate ultimate capacity of adsorbent that affects such design parameters. This study further indicated that Cd adsorption phenomena by fly ash could be well described by a monolayer adsorption mode (Langmuir isotherm) with strong correlation coefficient (r = 0.85). These studies suggest that this experimental work can be applicable in a rational design of a fly ash adsorption unit for removal of metals, such as cadmium, under slightly acidic conditions so that the wastewater can be reused.

ACKNOWLEDGEMENTS

The senior author is thankful to the Director, National Environmental Engineering Institute (NEERI) Nagpur, India for providing facilities at the Institute and for his permission to publish this paper. The manuscript preparation was funded in part, by the contract number DE-FG09-96SR-18558 U.S. Department of Energy and Environmental Protection Agency.

APPENDIX

A_s = Percent metal removed

t = Mixing time

K_t = Rate factor (percent metal removed for $t = 1$ hr)

m = Slope of linear plot

C_f = Concentration of fly ash

K_f = Constant denoting extent of solute removal

K_F = Constant denoting adsorption capacity

n = Slope of linear plot

qe = Amount of metal adsorbed per unit weight of fly ash

C = Concentration of metal in solution at equilibrium

b, B = Constants related to energy of adsorption

C_s = Saturation Concentration of metal

REFERENCES

1. Phipps, D.A., Chemistry and biochemistry of trace metals in biological systems. *in Effects of heavy metals in plants,* N.W. Lepp (ed.), Applied Science Publishers.1, 1981.

2. Alloway, B.J., Heavy metals in soils. Blackie Publishers, London, 1995.

3. Jackson, A.P., and Alloway, B.J., The transfer of cadmium from agricultural soils to the human food chain. *in Biogeochemistry of Trace Metals,* D.C. Adriano (ed.) Lewis Publishers, Chelsea, Michigan, 109, 1992

4. Weber, W. J., Jr. and Morries, J. C., Kinetics of adsorption on carbon from solutions, *J. Sanitary Engineering, Div. Am. Civil Engr.* 89, (2), 31, 1963.

5. Sajwan, K.S., Alva, A.K., and Keefer, R.F., Biogeochemistry of trace elements in coal and coal combustion by-products, Kluwer academic / Plenum Publishers. New York, 359, 1999.

6. Khandekar, M.P., Bhide, A.D., and Sajwan, K.S., Trace Elements in Indian coal and coal fly ash. *in Biogeochemistry of trace elements in coal and coal combustion by-products,* K.S. Sajwan et al. (eds.). Kluwer Academic / Plenum Publishers, New York, 99, 1999.

7. Adriano, D.C., Page, A.L., Elseewi, A.A., Chang, A.C., and Straughan, I., Utilization and disposal of fly ash and other coal residues in terrestrial ecosystems: A Review, *J. Environ. Qual.* 9, 333, 1980.

8. Singh, B.K., Misra, N.M., and Rawt, N.S., Fly ash as adsorbent for toxic organic: A Review, Minetech, 14, (4), 35, 1993.

9. Vandenbusch, M. B., and Sell, N. J., Fly Ash, a Sorbent for the Removal of Biologically Resistant Organic Matter, *Res. Conserv. and Recycling*, 6, 95, 1992.

10. Khanna, P., and Malhotra, S. K., Kinetics and Mechanism of Phenol Adsorption on Fly Ash, *Indian J. Environ. Health,* 19, (3), 224, 1977.

11. Gupta, G. S., Prasad, G., and Singh, V. N., Removal of Chrome Dye from Aqueous Solutions by Mixed Adsorbents Fly Ash and Coal, *Water Res.* 24, (1), 45, 1990.

12. Gupta, G. S., Prasad, G., and Singh, V. N., Treatment of Hazardous Dye-House Wastewater by Low Cost Materials, *J. IAEM,* 18, 107, 1991.

13. Rao, A. J., Verma, N., and Kaur, A., Bottom Ash for Adsorption of Nickel Metal Ion from Industrial Wastewater, *Indian J. Environ. Health,* 32, (3), 280, 1990.

14. Singh, D., and Rawat, N. S., Removal of Copper from Aqueous System by Sorption on Bituminous Coal, *Asian Environ.*, 56, 1992.

15. Frye, G.C., and Thomas, M.M., Adsorption of organic compounds on carbonate minerals. 2: Extraction of carboxylic acids from recent and ancient carbonates, *Chem. Geol.,* 109, 215, 1993.

16. van, Proosdij, E.M.H., and Reddy, K.J., Immobilization of contaminants with in-situ calcite precipitation, A preliminary evaluation, *in. Contaminated Soils. 3rd International Conference on Biogeochemistry of Trace Elements*, R. Prost (ed.), Paris, France, 1997.

17. Zavarin, M., and Doner, H.E., Selenium, Nickel, and Manganese interactions with calcite, *in Sorption and Desorption of Trace Elements, Proceedings of 4th International Conference on the Biogeochemistry of Trace Elements,* I.K. Iskandar et al. (eds.), U.S. Army Cold Regions Research and Engineering Laboratory, Hanover, New Hapshire, 1997.

18. Reddy, K.J., Coal fly ash chemistry and carbon dioxide infusion process to enhance its utilization, *in Biogeochemistry of trace elements in coal and coal combustion by-products,* K.S. Sajwan et al. (eds.), Kluwer Academic / Plenum Publishers, New York, 133, 1999.

19. Theis, T.L., and Wirth, J.L., Sorptive behavior of trace elements on fly ash in aqueous systems, *Environ. Sci. Technol,* 11,1096, 1997.

20. Rai, D., Ainsworth, C.C., Eary, L.E., Mattigod, S.V., and Jackson, D.R., Inorganic and organic constituents in fossil fuel combustion residues, Electric Power Research Institute, EA-5176, 1987.

21. Weber, W. J., Physico-Chemical Processes for Water Quality Control. Wiley-Interscience, New York, NY, 1972.

COMPARATIVE STUDY OF ELEMENTAL TRANSPORT AND DISTRIBUTION IN SOILS AMENDED WITH FLY ASH AND SEWAGE SLUDGE ASH

S. Paramasivam[1], K.S. Sajwan[1], A.K. Alva[2], D.C. Adriano[3], T. Punshon[3], D. van Clief[4], and K.H. Hostler[4]

[1] Department of Natural Sciences and Mathematics
Savannah State University, Savannah, GA 31404, U.S.A.
[2] USDA-ARS PacificWest Area, Vegetable and Forage Crops Research Unit, 24106 N Bunn Rd., Prosser, WA 99350, U.S.A.
[3] Savannah River Ecology Laboratory, University of Georgia Aiken, SC 29802, U.S.A.
[4] Citrus Research and Education Center, University of Florida Lake Alfred, FL 33850, U.S.A.

ABSTRACT

Disposal of various coal combustion and municipal by-products is a serious and challenging problem due to strict environmental regulations. This is due to the threat of accumulation of certain heavy metals in soils, plants and groundwater when these products are used as soil amendments or disposed of in landfills. This chapter describes a study that was undertaken on a coarse and medium textured soils amended with single rate of (74.1 Mg ha^{-1}) fly ash (FA), sewage sludge (SS) and sewage sludge ash (SSA) to compare the transport and leaching potential of various elements and their distribution within 30-cm soil columns. Transport and leaching potential of macro nutrient elements were highest in SSA amended soils followed by SS and FA. Leaching potential of Cr, Cd and Ni were the lowest in SSA followed by SS and FA. In addition, the total quantity of metals in the leachate (total of 2640 mL) were generally greater when the amendments were applied to coarse textured soil compared to medium textured soil with the exception of Pb. The results also suggest that binding sites created by the interaction between soil and amendments may modify metal adsorption-desorption and solubility, with a consequential effect on leaching and downward transport of various elements.

INTRODUCTION

Despite accelerated development and use of alternate sources of energy, coal remains an important source of power in the global economy. To achieve the national goal of energy independence, coal combustion is expected to increase. This in turn results in increased release of potentially toxic organic and inorganic contaminants into the environment[1]. Fly ash (FA) traditionally forms the bulk of coal combustion products (CCP) at present. Since FA and other coal residues contain a variety of potentially toxic metals, their irrational disposal and management could cause considerable environmental impacts. Despite this fact, public interest in recycling and reuse of CCP continues to increase.

Sewage sludge (SS) is another major solid products released from wastewater treatment plants in cities around the globe[2, 3], which is generally disposed of in landfills [2, 4, 5, 6]. The composition of SS varies considerably depending on the locality from which the sewer and waste are derived.

Extensive research on the use of SS during the past three decades has helped to realize the beneficial effects of land application in terms of increased crop production and improved soil quality [4, 5, 7, 8, 9, 10, 11, 6]. This has resulted in renewed interest in reusing and recycling SS from wastewater treatment plants. Land application of SS dramatically increased from 20 % to 54 % in the USA from 1972 to 1995 respectively[11, 2].

Wastewater treatment plants, however, encounter public acceptance problems in disposing of SS because of unpleasant odors, high acidity and levels of some heavy metals in excess of maximum critical limits. The exploration of alternate disposal methods has resulted in the production of ash from the incineration of dewatered activated sewage sludge and weathered ash. The incinerated products are dissolved and stored in ash ponds adjacent to wastewater treatment plants. These products are termed as sewage sludge ash (SSA) and weathered sewage sludge ash (WSSA), respectively. The same terminology is used to denote these by-products throughout the rest of the manuscript.

Unlike CCPs, however, these relatively new wastewater treatment products are still recognized as hazardous waste by the U.S. Environmental Protection Agency (USEPA) and this presents regulatory obstacles to their effective use. In addition, products from both municipal (SS) and industrial (FA) sources are enriched with trace and heavy metals (e.g. Zn, Cd, Cr, Pb, and Ni), and this may also limits its application to land [12, 13]. Unlike SS and FA, information available on these new byproducts from SS is meager. To our knowledge there have been no studies that report data from a comparison of the leaching potential of various elements from FA and SSA amended soils with contrasting textural types. This leaching column study was designed to evaluate the leaching pattern, potential and distribution of various elements within soil columns amended with single application rate of FA, SSA, and SS.

MATERIALS AND METHODS

Soils and Amendments

An unweathered alkaline FA (pH 12.1), SS (pH 5.7) and SSA (8.0) were used in this study. FA was collected from a coal-fired power plant near Beach Island, South Carolina. SS and SSA were collected from President street wastewater treatment plant located in Savannah, GA. Selected chemical properties of the products are presented in Table 1.

The study was conducted on Ap horizon soil samples of a Candler fine sandy soil (sandy, hyperthermic, uncoated, Typic Quartzipsamments) collected from a citrus grove Polk County, FL, and an Orangeburg sandy loam (fine loamy, silicious, thermic Typic Paleudult) from Aiken, South Carolina. Selected properties of these soils are presented in Table 1.

Leaching Column Study

Plexiglass columns, 32-cm long and 7-cm inner diameter, were used to study the transport and distribution of various elements (including heavy metals) in soils amended

Table 1. Selected properties of fly ash, sewage sludge (SS) sewage sludge ash (SSA) and soils used in this study.

Properties[†]	Units	FA[‡]	SS[§]	SSA[§]	Candler[¶]	Orangeburg[#]
pH (1:1 water: soil)		12.1	5.7	8.0	7.0	5.5
P	mg kg⁻¹	1,268.0	3,597.0	1,625.0	913.5	175.0
K	mg kg⁻¹	1,447.0	6,020.0	3,578.0	285.0	3,270.0
Ca	mg kg⁻¹	5,654.0	2,308.0	1,890.0	285.0	3,270.0
Mg	mg kg⁻¹	214.0	872.0	417.5	562.0	21.6
Mn	mg kg⁻¹	93.0	110.0	27.5	13.7	14.1
Fe	mg kg⁻¹	121.0	619.0	505.0	335.0	330.0
B	mg kg⁻¹	57.0	0.5	nd	nd	nd
Cu	mg kg⁻¹	82.0	71.0	23.6	11.1	0.6
Zn	mg kg⁻¹	84.0	81.0	44.5	72.6	4.4
Pb	mg kg⁻¹	199.0	29.0	2.3	2.6	5.2
Cd	mg kg⁻¹	0.9	3.5	0.2	nd	nd
Ni	mg kg⁻¹	26.0	25.0	nd	nd	nd
Cr	mg kg⁻¹	106.0	23.0	0.4	0.2	0.6
Sand	g kg⁻¹				967.0	860.0
Silt	g kg⁻¹				8.0	100.0
Clay	g kg⁻¹				25.0	40.0
Organic matter	g kg⁻¹				13.0	18.5
CEC	cmol kg⁻¹				2.2	3.9
Texture					fine sand	sandy loam

[†] Elemental compositions were determined by ICP-OES on Mehlich-3 extract.
[‡] Collected from Augusta, GA.
[§] Collected from President street wastewater treatment plant in Savannah, GA.
[¶] Collected from Florida.
[#] Collected from Georgia.

with single rate (74.1 Mg ha⁻¹) of FA, SS or SSA. Soils were dried, and sieved to pass a 2 mm sieve. Whatman No. 42 filter paper was placed at the bottom of the leaching column and the soils were packed to a height of 30 cm to attain a bulk density of 1.5 g cm⁻³. Three replicate columns were used for each treatment. Appropriate quantities (32.7 g per column) of amendments were mixed with the top 2.5 cm soil and repacked to attain the same bulk density as above. Three more columns were included to accommodate an un-amended control. All soil columns were saturated with distilled water and allowed to drain overnight.

Whatman No. 42 filter paper was placed on the top of the soil column and de-ionized water was applied at 1.5 mL min⁻¹ using a peristaltic pump to facilitate leaching. Leachate was collected at one half-pore volume fractions (220 mL) for a total of 12 fractions. Leaching events were repeated at 6d intervals. Upon completion of each leaching event, soil columns were allowed to dry until the next leaching event (6d) at room temperature (30°C). When the ionic strength leachate of all treatments approached to that of non-treated control, addition of water was terminated. The total amount of leachate generated was equivalent to 60 cm rainfall, which is about 45% of the mean annual rainfall of southeastern coast of Georgia.

Concentrations of various nutrient elements (K, Ca, Mg, and P) and metals (Cr, Zn, Cd, Cu, Ni, and Pb) in the leachate were determined using inductively coupled plasma optical emission spectroscopy (ICP-OES Plasma RL 3300, Perkin Elmer Inc., Norwalk, CT). The quantity of metals leached was calculated using the concentrations of each metal and the volume of leachate fraction. The leaching potential for individual metals represents the cumulative amount of metal leached in all the leachate fractions. In addition, electrical conductivity (EC) and pH of leachate samples were measured using HI 8733 conductivity meter (Hanna Instruments, Singapore) and an Accumet Model 15 pH meter (Fisher Scientific, Pittsburgh, PA) respectively.

Distribution of Elements in Soil Column

After 6 pore volumes of water were leached through the soil columns, soil from each was divided into 3 sections at 10 cm increments (0-10, 10-20, and 20-30 cm depth sections). The soil was air-dried, homogenized and a sub sample was taken for pH analysis (1:1 suspension of soil: water). Another sub sample was used for Mehlich 3 (M-3) extraction using soil: extractant ratio of 1:10. The concentrations of various elements and heavy metals in the M-3 extract were measured using ICP-OES as described above.

Statistical Analysis

Experimental data were analyzed using the SAS (version. 8.1) completely randomized design (CRD) with four treatments and three replicates per treatment [14]. Duncan's Multiple Range Test (DMRT) at 0.05 probability-level was used to compare means of cumulative amounts of each individual element leached. The same test was used to evaluate the significance of means of various elemental concentrations at different depth sections of the soil columns with various treatments.

RESULTS AND DISCUSSION

The leaching and drying conditions adapted in this study were to mimic the soil conditions in areas Florida and southeastern Georgia during the summer months characterized by frequent intense rainfall and rapid drying because of high temperature.

Table 2. Comparison of mean leachate pH and total quantities of metals (μg) in 2,640 mL of leachate from Candler fine sand, or an Orangeburg loamy sand amended with single rate of fly ash (FA), sewage sludge (SS) or sewage sludge ash (SSA) materials and from unamended soils.

Treatments Mg ha^{-1}	μg metals[†] in 2,640 mL leachate						
	pH	Cr	Zn	Cd	Cu	Ni	Pb
Candler fine sand							
Unamended	7.04 a	nd	176 c	nd	444 c	nd	283 d
74.1 (FA)	7.19 a	235 a	190 c	62 a	492 c	269 a	671 b
74.1 (SS)	7.11 a	271 a	346 b	76 a	1,349 b	258 a	909 a
74.1 (SSA)	7.09 a	54 b	2699 a	22 b	5,229 a	92 b	436 c
Orangeburg loamy sand							
Unamended	5.13 b	nd	65 d	nd	7 c	39 c	837 c
74.1 (FA)	7.12 a	163 b	92 c	85 a	39 b	216 b	988 b
74.1 (SS)	7.09 a	200 a	126 b	101 a	60 a	277 a	1,070 b
74.1 (SSA)	7.03 a	24 c	694 a	53 b	24 b	57 c	2,107 a

[†] Elemental compositions in leachate were determined by ICP-OES. Similar letters after a mean indicate no significant difference ($p < 0.05$) among treatment means within a column for each soil according to Duncan's Multiple Range Test. nd: non detectable. ND: not determined.

193

pH of Leachate

Candler soil (fine sand) amended with various amendments had no significant effects on leachate pH value (mean) irrespective of the original pH of amendments (Tables 1 and 2). In contrast, the mean leachate pH values were greater in Orangeburg sandy loam regardless of source of amendments, compared to leachate from unamended Orangeburg sandy loam (Table 2). It is pertinent to note that the control leachate pH from both soils indicated that soil type did not influence leachate pH (Table 2). However, when soils were amended with material having varying pH, mean leachate pH approached near neutral. During the beginning of leaching events, pH values varied slightly around that of unamended soil. Elevated leachate pH arising from Orangeburg soil may have been due to the creation of binding sites as a result of interaction between the soil and the various amendments. These interactions are expected to modify metal solubility, with consequential effect on leaching and leachate pH.

Concentration of Major Nutrient Elements in Leachate

Irrespective of soil types, the highest concentration of nutrient elements measured in the leachate came from soils amended with SSA. Leachate from the fine sand amended with SSA contained the highest concentrations of Ca, K, Mg and P (Figs.1- 4) in contrast to sandy loam soil with same amendment. However, this finding was less pronounced for Mg (Fig. 3) in sandy loam soil amended with SSA. This is a clear indication of the greater leaching potential of the majority of elements in coarse textured soils compared to that from medium textured sand. Peak concentration of nutrient elements occur between the 4th and 6th leachate fractions in both soils amended with SSA. Thereafter, concentrations of nutrient elements declined rapidly. However, in all other treatments, peak nutrient element concentrations were observed much earlier at the 2nd or 3rd leachate fraction before declining to that of unamended soils.

The concentrations of Ca in the majority of leachate fractions from SSA amended Candler fine sand were consistently above 10 mg L^{-1}. However, Ca concentrations were generally below 10 mg L^{-1} in SSA amended Orangeburg sandy loam (Fig. 1). In addition, leachate Ca concentrations fell below 10 mg L^{-1} beyond the 4th leachate fraction in both soils amended with all treatments with the exception of SSA amendment.

The leaching patterns of K from both soils amended with various amendments were very similar; approximately 10-12 mg L^{-1}. The total concentration of K in various leachate fractions differed very little between Candler fine sand and Orangeburg sandy loam although a comparatively greater volume of water was required to leach K from sandy loam soil amended with SSA (Fig. 2).

The concentration of Mg in leachate remained below 12 mg L^{-1} (Fig. 3) in all soil-amendment combinations, except SSA-amended Candler fine sand. The peak Mg concentration was ~ 15 mg L^{-1} in fine sand amended with SSA observed at 6th leachate fraction and did not fall below 3 mg L^{-1} throughout the study.

The concentration of P in various leachate fractions was below 0.8 mg L^{-1} in both soils amended with FA and SS. In addition, the concentrations of P in various leachate fractions collected from both Orangeburg sandy loam and Candler fine sand were similar. When these soils were amended with SSA, peak concentrations of P from Candler fine sand was ~ 2.4 mg L^{-1} where as it was ~1.2 mg L^{-1} in Orangeburg sandy loam (Fig. 4). Concentrations of Ca and Mg in leachate fractions from both soils also followed a similar pattern. It is important to note that Mehlic extractable P content in Candler fine sand was

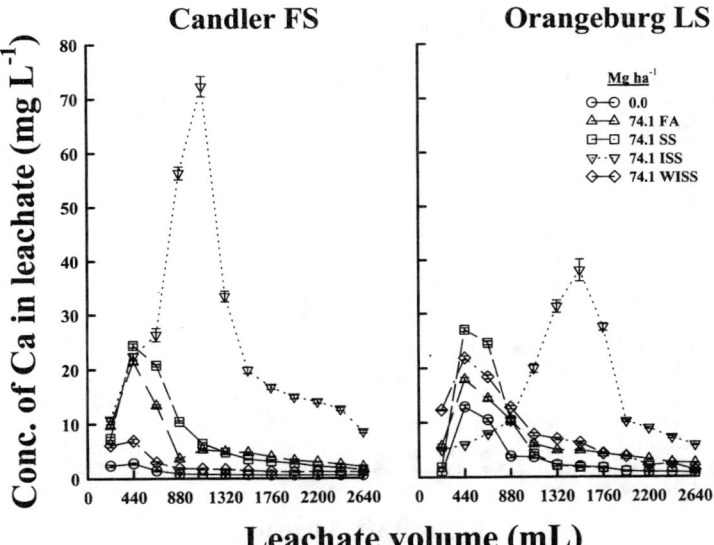

Fig. 1. Comparison of Ca leaching in Candler fine sand and Orangeburg sandy loam amended with single rate of various amendments. Error bar on each point represents standard error of the mean.

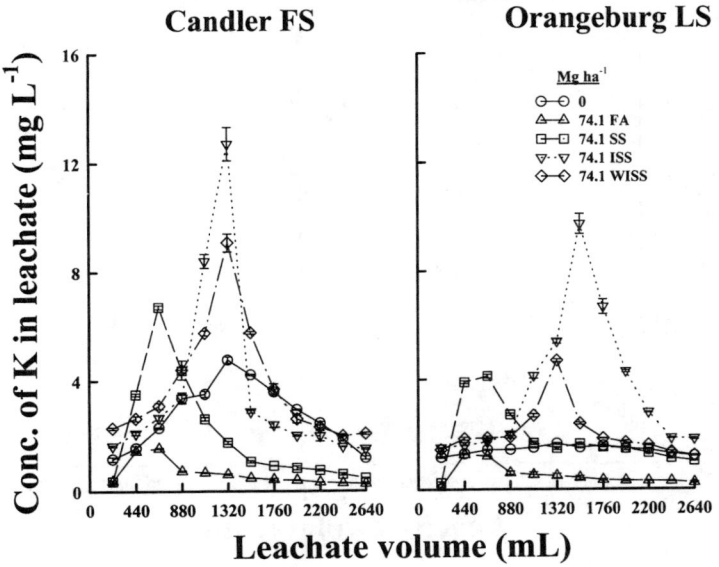

Fig. 2. Comparison of K leaching in Candler fine sand and Orangeburg sandy loam amended with single rate of various amendments. The error bars as in Fig. 1.

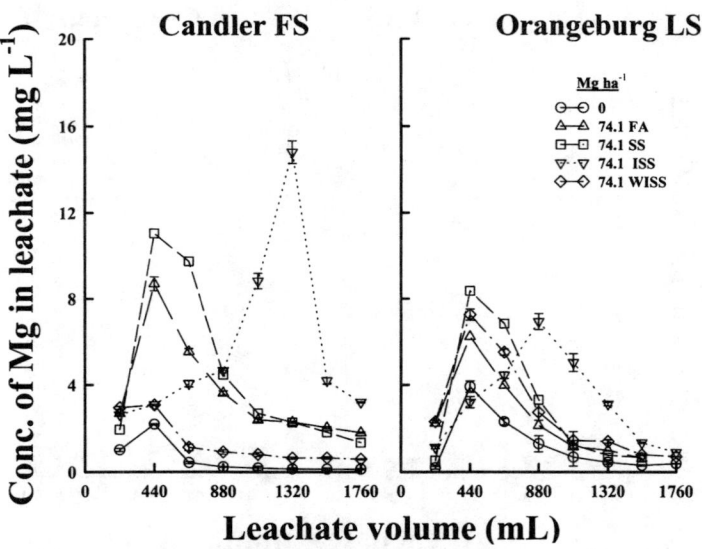

Fig. 3. Comparison of Mg leaching in Candler fine sand and Orangeburg sandy loam amended with single rate of various amendments. The error bars as in Fig. 1.

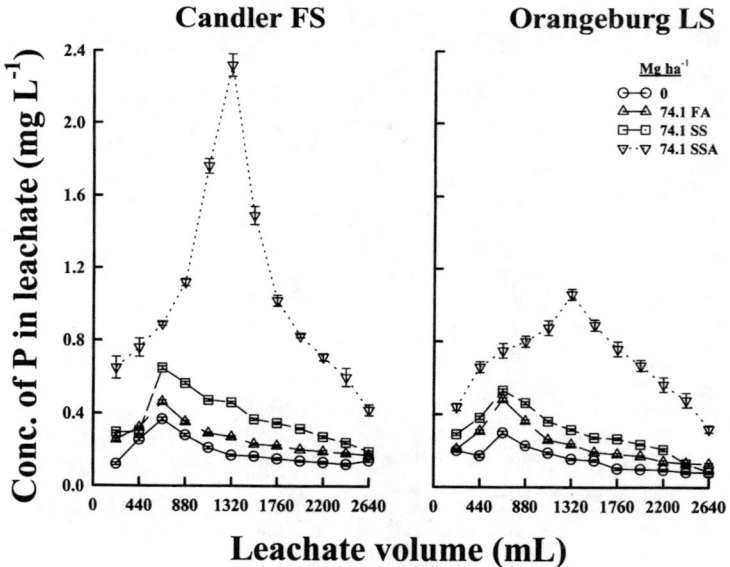

Fig. 4. Comparison of P leaching in Candler fine sand and Orangeburg sandy loam amended with single rate of various amendments. The error bars as in Fig. 1.

five times greater than that in Orangeburg sandy loam. Similarly, P content of SS was almost three times greater than that of P content in FA (Table 1). The observed low concentrations of P could be attributed to the interaction among soil-FA-SS-SSA. These interactions may modify solubility or enhance precipitation of P as Ca $(PO_4)_3$ and Mg $(PO_4)_3$ in both soils and /or other reactions, which may affect the leaching of P.

Concentration of Metals in Leachate

Mean concentrations of metals were calculated from the cumulative amount of metals (Table 2) leached in 2640 mL leachate. The mean concentrations of Cr in the leachate from amended Candler fine sand were in the range of 0.02 to 0.10 μg mL^{-1} where as the concentration of Cd varied in the range of 0.008 to 0.03 μg mL^{-1}. Leachate from Orangeburg sandy loam amended with SSA contained concentrations in the range of 0.009 to 0.08 μg mL^{-1} and 0.02 to 0.04 μg mL^{-1} for Cr and Cd respectively. The critical upper limits set for Cr and Cd for drinking water were 0.1 and 0.005 mg L^{-1} respectively [15]. Accordingly, concentrations of Cr and Cd in the leachate collected from both amended soils used in this study at the single application rate of 74.1 Mg ha^{-1} did not exceed the critical concentrations for drinking water standard.

The concentrations of Zn and Cu in the leachate from the Candler fine sand amended with various amendments varied in the range of 72 to 1022 μg L^{-1} and 186.4 to 1981 μg L^{-1} respectively. The corresponding concentration ranges for Orangeburg sandy loam amended with similar amendments were 34.8 to 262.9 μg L^{-1} and 9.1 to 22.7 μg L^{-1} respectively. The average reported concentration of Zn in fresh water was 15 μg L^{-1} (range 1 to 100) while that of Cu was 3 μg L^{-1} (range 0.2 to 30) respectively [16]. The concentrations of Zn in leachate from unamended soils of Candler fine sand, and Orangeburg sandy loam were 66.7 and 24.6 μg L^{-1} respectively. Similarly, the concentrations of Cu in unamended soils were 168.2 and 2.7 μg L^{-1} respectively. Highly elevated concentrations of Zn and Cu in the leachate from amended Candler fine sand may have originated in part from the soil itself since this soil received periodic fungicide

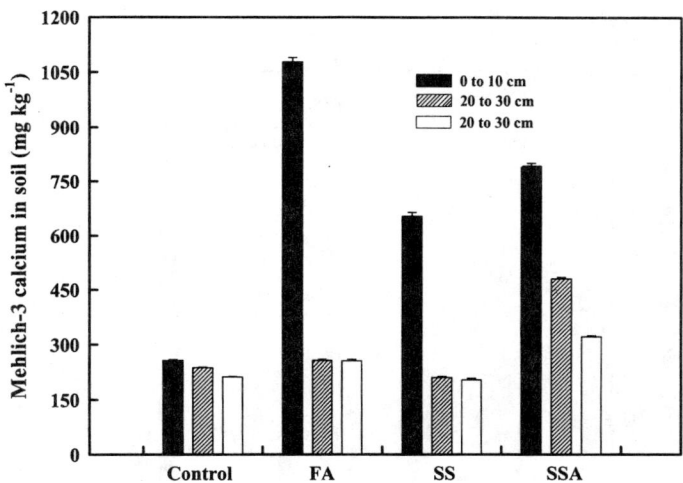

Fig. 5. Concentrations of Mehlich 3 (M-3) extractable soil Calcium (Ca) at various depth sections of Candler fine sand, which received single rate of various amendments. The concentrations were determined after the completion of leaching of soil columns with 2, 640 mL of water. The error bars as in Fig. 1.

application for citrus cultivation. Similarly the elevated concentration of Zn in Orangeburg sandy loam also have originated from the soil itself due to previous treatments made to this soil.

The concentrations of Pb and Ni in the leachate from the Candler fine sand amended with various amendments varied in the range of 165.2 to 344.3 μg L^{-1} and 34.8 to 101.9 μg L^{-1} respectively. The corresponding concentration ranges for Orangeburg sandy loam with same amendments were 374.2 to 798.1 μg L^{-1} and 21.6 to 104.9 μg L^{-1} respectively. The average reported concentration of Pb in fresh water is 3 μg L^{-1} (range 0.06 to 120) while that of Ni is 0.5 μg L^{-1} (range 0.02 to 27) respectively [16]. However, the reported range of Pb concentrations in drinking water are 1 to 20 mg L^{-1} from rural areas and 1 to 40 mg L^{-1} for drinking water from urban areas [17]. The concentrations of Pb in unamended soils of Candler fine sand, and Orangeburg sandy loam were 107.2, and 317.0 mg kg^{-1} respectively and these values lies within the reported range (2 to 300 mg kg^{-1}) for agricultural soils except for Orangeburg sandy loam [13]. Similarly, the concentrations of Ni in unamended soils of Candler fine sand, and Orangeburg sandy loam were 15.9, and 14.4 mg kg^{-1} respectively and these values lies below the reported mean concentration of 20 mg kg^{-1} [13]. In general, concentrations of all the metals (except Zn and Cu) in the leachate varied with various types of amendments but never exceeded the critical value for drinking water standard.

Leaching Potential of Metals

Cumulative leaching of metals varied with both types of amendments and soils used in this study. The leaching of Cr, Cd, and Ni from both soils amended with SSA at 74.1 Mg ha^{-1} was significantly lower than the values obtained with either SS or FA, suggesting that this is a safer amendment in terms of metal leaching. Concentrations of these metals in leachate collected from unamended soils were almost nondetectable with the exception of Ni from Orangeburg sandy loam (Table 2). In addition, there were no substantial differences in cumulative amount of Cr, Cd, and Ni leached between SS and FA amended Candler fine sand and Orangeburg sandy loam (Table 2).

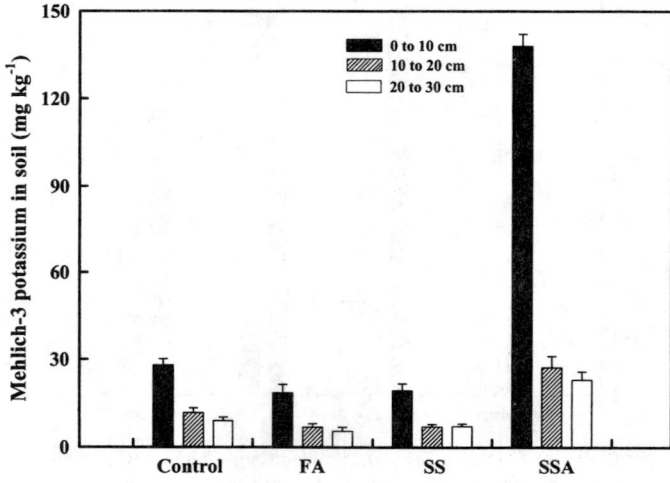

Fig. 6. Concentrations of Mehlich 3 (M-3) extractable soil Potassium (K), the soils and other details as in Fig. 5.

Cumulative leaching of Cu and Zn was significantly greater in amended Candler fine sand compared to that from the amended Orangeburg sandy loam. This indicates that Candler fine sand has received recent fungicides applications prior to the collection of samples for this study, which have added Cu and Zn to the soil. However, there was no substantial increase of Cu and Zn evident in FA amended soils compared to that of their respective unamended soils. Substantial increase of Cu and Zn was evident when these soils (especially Candler fine sand) were amended with SS and SSA. Therefore, it is apparent that the interactions of various amendments were different with different types of soils. These interactions are influence the solubility, complexation, and precipitation of Cu and Zn in both soils and /or other reactions, which may have consequential effect on leaching of Cu and Zn.

Similarly, substantially higher Pb leaching was recorded in Orangeburg sandy loam compared with the Candler fine sand from a citrus grove. This was probably due to higher concentrations of Pb in unamended Orangeburg sandy loam compared to Candler fine sand. In general, about 60 to 70 % of cumulative leaching of most of these metals was accounted in the first 6 to 8 leaching events. Peak leaching of metals occurred at 5th or 6th leaching event irrespective of soils or rates of amendment used (data not presented).

Distribution of Major Nutrient Elements and Metals in the Soil Column

The distribution of elements in soil columns were determined by analyzing soils for Candler fine sand only at 10 cm depth increment sections following the termination of leaching of 2, 640 mL of water. Amending soil with FA, SS, and SSA resulted in increased concentration of soil Ca (M-3 extractable) at a depth of 0-10 cm compared to that of an umamended Candler fine sand. The concentration of Ca at this depth followed the order FA> SSA> SS> control (Fig. 5). The differences in concentration of Ca in the 10-20 cm and 20-30 cm depth sections were not significantly different in FA and SS amended soil columns. However, the concentration of Ca was significantly different at the last two 10 cm depth sections of soil columns amended with SSA (Fig. 5). In contrast, the concentration of Ca was significantly higher in the two lower depth sections of SSA amended Candler fine sand compared to amendment with either FA or SS. This indicates that transport of Ca may be promoted by amending soil with SSA (Fig 5).

The concentrations of M-3 extractable K followed a somewhat similar trend as that of Ca by having higher concentrations in the top 0-10 cm of the soil columns at the end of leaching (Fig 6). The highest concentration of K (140 mg kg^{-1}) was observed in the top 10 cm of the SSA amended soil column. The concentrations of K in the top 10 cm depth section of the unamended control soil columns and those amended with FA, SS were almost 4 to 5 times less than SSA amended soil column. Similarly low K concentrations were observed in the lower profile (10-20 cm and 20-30 cm) of FA or SS amended soil columns.

The concentration of M-3 extractable Mg was the highest in the top 10 cm of the SSA amended soil column. Mg in the top 10 cm depth decreased in the order: SSA > SS > FA > control (Fig. 7). Results also showed that concentrations of M-3 extractable Mg were similar across all the depth sections of unamended control and in FA amended soil columns. This indicates that transport of Mg is not hindered in coarse textured soils by FA amendment. However, amending soil with SS or SSA did not enhance substantial downward transport of Mg and most of the M-3 extractable Mg was present at the top 10 cm depth section of the soil column.

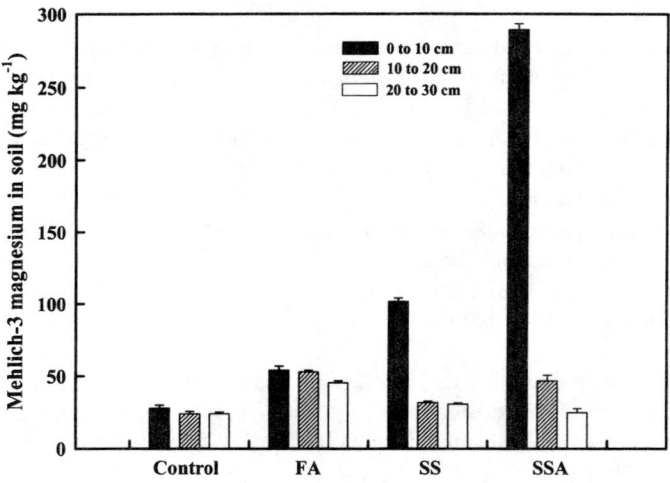

Fig. 7. Concentrations of Mehlich 3 (M-3) extractable soil Magnesium (Mg), the soils and other details as in Fig. 5.

The concentrations of M-3 extractable P in the top 10 cm of the coarse textured soil profile increased in the order: SSA > SS > Control > FA. Mehlic-3 extractable P concentrations were neither affected in the lower depth sections of the unamended control soil columns nor those amended with FA and SS. However, significant downward transport of P was observed in the soil columns amended with SSA after the completion of leaching with 2, 640 mL of water (Fig 8).

Measured concentrations of M-3 extractable various metals indicated that there were no significant differences in the distribution of Cr, Zn, Cd, Pb, Ni, and Cu at various depth sections of the soil columns amended with various types of amendments after the completion of leaching with 2,640 mL of water. Therefore no data is presented.

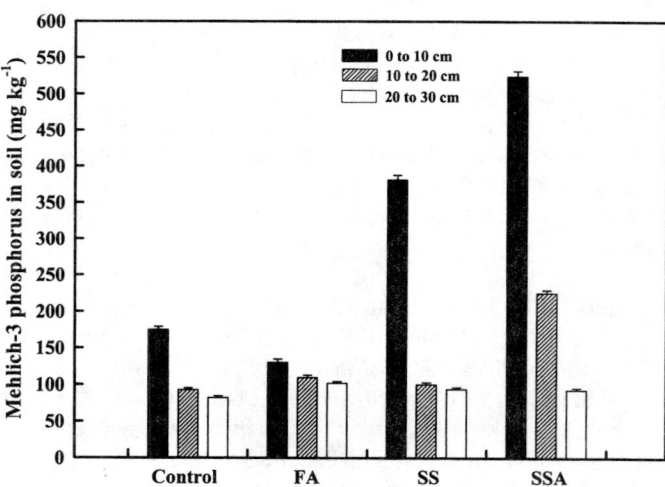

Fig. 8. Concentrations of Mehlich 3 (M-3) extractable soil Phosphorus (P), the soils and other details as in Fig. 5.

CONCLUSIONS

This study strongly suggests that most of the metals leaching potential of sewage sludge ash is lower than either fly ash or sewage sludge and sewage sludge ash could be used at higher rates without deleterious effect. The leaching potential of macronutrient elements were, however, higher with SSA compared FA or SS. Greater cumulative leaching of elements was observed in coarse textured soil than with medium textured soil amended with all amendments. About 60 to 70 % of the cumulative leaching of various elements was over within first 7 to 8 leachate fractions. For all amendments, the concentration of majority of elements monitored did not exceed critical limit or did not exceeded the average concentrations of those respective metals generally found in the natural fresh water environment. With little caution, it may be possible to apply these amendments safely to agricultural land at low to medium rates (100 Mg ha^{-1}) without causing excessive loading of metals into groundwater which are of serious concern. However, field studies with groundwater monitoring are warranted in different soil types prior to making this recommendation.

ACKNOWLEDGEMENT

This study was funded, in part, by the contract number DE-FG09-96SR-18558 U.S.Department of Energy and Environmental Protection Agency. We appreciate the assistance of President street wastewater treatment plant staff, Savannah, Georgia.

REFERENCES

1. Davis, E.C., and Boegly, W.J., Jr., A review of water quality issues associated wit coal storage. *J. Environ. Qual.* 10, 127, 1981.

2. Basta, N.T., Examples and case studies of beneficial reuse of municipal by-products. *in Land Application of Agricultural, Industrial, and Municipal By-Products.* J.F. Power, and Dick W.A. (eds.) SSSA, Madison, WI, 481, 2000.

3. Sajwan, K.S., Paramasivam, S., Alva, A.K., Adriano D.C., and Hooda, P.S., Assessing the feasibility of land application of fly ash, sewage sludge and their mixtures. *Adv. Environ. Res.* (In Press), 2002.

4. Page A.L., Fate and effects of trace elements in sewage sludge when applied to agricultural lands. Environ. Protection Tech. Series, EPA-670 2-74-005 (U.S. Environ. Protection Agency, Cincinnati, OH), 1974.

5. Page, A.L., and Chang, A.C., Overview of the past 25 years: Technical perspective. *in. Sewage sludge: Land utilization and the environment.* C.E.Clapp et al. (eds.), SSSA Misc. Publ. ASA, CSSA, and SSSA, Madison, WI, 3, 1994.

6. Power, J.F., and Dick, W.A., Land Application of Agricultural, Industrial and Municipal By-Products. SSSA Publ. Madison, WI, 2000.

7. Basta, N.T., Land application of biosolids: A review of research concerning benefits, environmental impacts, and regulation of applying treated biosolids. Oklahoma Agric. Exp. Stn. Tech. Bull. B-808, 1995.

8. CAST., Application of sewage sludge to croplands: appraisal of potential hazards of the heavy metals to plants and animals. Council Agr. Sci. and Tech. Rept. 83, 77, 1976.

9. Chaney, R.L., Trace metal movement: Soil-plant systems and bioavailability of biosolids-applied metals. in *Sewage sludge: Land utilization and the environment.* C.E.Clapp et al. (ed.), SSSA Misc. Publ. ASA, CSSA, and SSSA, Madison, WI, 27, 1994.

10. National Research Council., Use of reclaimed water and sludge in food crop production. National Academy of Sciences Press, Washington, D.C., 1996.

11. Water Environment Federation., National outlook-state beneficial use of biosolids activities. Water Environ.Fed., Alexandria, VA., 1997.

12. Adriano, D. C., Page, A.L., Elseewi, A.A., Chang, A.C., and Starnghan, I.R., Utilization and disposal of fly ash and other coal residue in terrestrial ecosystems: A Review. *J. Environ. Qual.* 9, 333, 1980.

13. Adriano, D.C., Trace Elements in Terrestrial Environments. Springer-Verlag, New York, 2001.

14. SAS., User's Guide: Statistics, Release 8.01 edition. Statistical Analysis System Institute Inc., Cary, NC., 2000.

15. Rubenstein, R., and Segal, S.A., Risk assessment of metals in groundwater. in *Risk Assessment of Metals in Groundwater*, H.E. Allen et.al. (eds.), Lewis Publishers, Chelsea, MI, 209, 1993.

16. Bowen, H.J., Environmental Chemistry of the Elements. Academic Press. New York, 1979.

17. Merian, E., Metals and their Compounds in the Environment: Occurrence, Analysis, and Biological Relevance. VCH Pub. Weinheim, Germany, 1991.

ARSENIC SPECIATION IN A FLY ASH SETTLING BASIN SYSTEM

Brian P. Jackson, John C. Seaman, and William Hopkins

Savannah River Ecology Laboratory
University of Georgia
P.O. Drawer E
Aiken, SC 29802

1. ABSTARCT

The sluicing of coal fly ash to settling basins is a major method for disposal of this industrial by-product. Fly ash often contains elevated concentrations of trace elements such as As, Se, and Mo, which can be solubilized upon contact with water and also become elevated in the surficial sediments. Both the soluble and sediment-sorbed trace elements can be bioavailable and potentially toxic to animals inhabiting the ash basins. This study examines the aqueous speciation of As in the surface and interstitial waters and the solid phase As speciation in the sediments of a fly ash basin system. Ion chromatography coupled to inductively coupled plasma mass spectrometry (IC-ICP-MS) was used to determine arsenite As(III), arsenate As(V), dimethylarsenate (DMA), and momomethylarsenate (MMA) in the aqueous samples. Hydoxylamine hydrochloride and oxalic acid extractions were used to assess the proportion of amorphous Fe, amorphous Al and amorphous aluminosilicates in depth sectioned samples of a sediment core taken from the ash basins. The concentration of As solubilized by these extractants was also measured. Surface water As concentrations were low with an average of 13 and 3 $\mu g\ l^{-1}$ determined in the summer and fall 2000. Arsenate was the major As species in the surface waters; DMA and As(III) were detected in the summer sampling but no DMA was detected in the fall sampling. Pore water As concentrations were much higher than the surface waters, reaching a maximum of 110 $\mu g\ l^{-1}$ at a sediment depth of 8-12cm. Arsenate was the major dissolved species at the sediment-water interface but decreased with depth, while the proportion of As(III) increased to a maximum at a depth of 8-12

Chemistry of Trace Elements in Fly Ash, edited by Sajwan *et al.*
Kluwer Academic/Plenum Publishers, 2003

cm. The increase in total dissolved As with depth was mirrored by an increase in soluble Mo and an increase in pH, and the depth of maximum As concentration marked the onset of an increase in soluble Fe. This suggests that the observed increased As solubility may result from the decrease in sorption by amorphous Fe phases due to the onset of reductive dissolution, coupled with the prevalence of As(III), that may be poorly sorbed by the remaining mineral phases in the sediment. This observation was supported by the selective extraction data of the sediment core sections, which indicated that As was mostly bound to amorphous Fe phases in the sediment. The oxalate extraction also showed that a significant proportion of total Al was present as amorphous phases and that < 20% of amorphous Al was present as amorphous aluminosilicates.

2. INTRODUCTION

Large quantities of coal fly ash are generated annually through fossil fuel combustion. In 1998 it was reported that 63 million tons of fly ash were generated in the US.[1] The trace element content of fly ash is highly variable depending on the original coal, combustion conditions, and ash handling practice. However, it is generally the case that trace element concentrations in the ash are higher than average background concentrations in the terrestrial environment. For example average concentrations of As and Se in fly ash have been reported as 2 - 440 and 0.2 – 130 mg kg^{-1}, compared with soil background concentrations of 1 – 50 and 0.1 – 2 mg kg^{-1}, respectively.[2] In addition, volatile trace elements such as Zn, As, Se and Cd are often present as surface condensates on fly ash particles and thus are highly soluble. The two major methods for the disposal of fly ash are landfilling and sluicing to settling basins. Land filling poses potential environmental problems from leaching of contaminants to groundwaters. Aquatic disposal through settling basins presents identical environmental problems to that of landfilling and the additional issue that a wide range of wildlife invariably inhabits these wetland sites and thus become subject to trace element toxicity and act as vectors for the dispersal of increased trace element concentrations within the ecosystem.

In a study of 24 ash samples from the south east, total As concentrations ranged from 12.9 - 321 mg kg^{-1}, while the percent of the total As that was water soluble from these ashes ranged from 0.28-19%, respectively.[3] X-ray absorption fine structure spectroscopy has been used to investigate the speciation of As associated with fly ash and while both As(III) and As(V) oxidation states were detected, the predominant speciation was identified as As(V).[4,5] It has been postulated that As my be present as calcium arsenate in alkaline coal fly ashes as a result of the reaction between CaO and As$_2$O$_3$ in the stacks.[6] Upon contact with water during the sluicing of fly ash a proportion of As will be solubilized, and may undergo changes in speciation depending on the prevailing pH and eH conditions. As(V) has been identified as the major species in coal fly ash leachates in a number of studies.[3,7-9]

A basin settling system removes fly ash from the water column through gravitational settling, hence the sediments of these systems are predominantly fly ash in progressive stages of weathering. The weathering products of fly ash in these environments may thus exert a primary control on As solubility and solid phase speciation. It has been suggested that the secondary precipitates of amorphous Fe oxides may control the solubility of As and Se in acidic fly ashes.[10] Accelerated leaching experiments on coal fly ash identified calcite and gypsum as major solid phases, while Si and Al from the ash may form the amorphous clay minerals allophane and imogolite.[11, 12] The presence of these high surface area, variably charged minerals could affect solubility of both cations and anions in ash basin systems.

This study focuses on the speciation of As in the surface waters of the fly ash basins, the interstitial waters of the sediments, and the possible solid phase speciation of As in the sediments. Aqueous speciation of As is investigated in terms of oxidation state and potential methylated As compounds, while speciation in sediments is investigated in terms of the solid phase speciation/association of As. The study was conducted at the D-area ash basins at the Savannah River Site in Aiken, SC. This ash basin system has been intensively studied over the last thirty years. and increased body burdens of As and Se have been identified in amphibians, reptiles and invertebrates inhabiting this ash basin system.[13,14]

Aqueous speciation analysis at environmentally relevant concentrations requires low detection limits. Liquid chromatography coupled to inductively coupled plasma mass spectrometry (LC-ICP-MS) has been used to investigate As speciation in drinking waters[15] and soil interstitial waters.[16] The main inorganic As species in natural water are As(V) and As(III), with the relative species distribution controlled by pH and Eh conditions. In addition, microbially-mediated methylation products, namely monomethylarsenate (MMA) and dimethylarsenate (DMA), may also be produced in surface and interstitial waters. Because these four As species are anionic, ion exchange chromatography has been used extensively for their separation.[3, 16, 17] While suppressed conductivity has been used as a detection system for As species[18] considerable improvements in detection limits can be achieved with the use of ICP-MS as an elment-specific detector.[3]

3. MATERIALS AND METHODS

3.1 Sample Collection

Samples were collected from the D-Area fly ash basin system and the USDOE Savannah River Site located in Aiken, SC. In this system, fly ash is first slucied to two receiving pits where the majority of the large fly ash particles quickly settle out. Suspended solids then flow through a primary settling basin, a secondary settling basin, then to a swamp area. Discharge from this swamp area flows via Beaver Dam Creek to the Savannah River. Surface water samples were collected from the primary and secondary basins and the swamp in the summer and fall of 2000. Surface water samples were collected in acid washed, 250 ml nalgene bottles that had been thoroughly rinsed with DI water. Pore water samples were initially collected (summer 2000) using a vacuum operated pore water extractor.[19] Upon being returned to the laboratory all aqueous samples were filtered (0.22 μm pore size) and pH and electrical conductivity were determined. An aliquot of the filtered sample was acidified for dissolved elemental analysis, while a sub-sample (not acidified) was taken for speciation analysis. All samples were stored at 4 °C prior to analysis and samples were analyzed for dissolved As species within 24 hrs of collection.

An intact sediment core was also collected during the fall sampling. This core was collected in a 50 cm polyethylene cylinder that was capped after collection in the water column to retain the sediment-interstitial water relationship as best as possible. The core was kept upright, transported back to the laboratory and split into six sections. Saturated sediment samples from each of these sections were loaded in to high speed centrifuge tubes without any further addition of water, and centrifuged at 15,000 rpm to collect interstitial water. Sub samples of each core section were air-dried, ground to pass through a 2mm sieve and subsequently used for determination of total HNO_3 extractable elements, dilute salt extractable elements, NH_2OH/HCl extractable elements, and 0.2 M oxalate/oxalic acid extractable metals.

3.2 Trace element determination

All trace element analysis was conducted by ICP-MS (Elan 6100DRC, Perkin Elmer, Shelton, CT). Quality control for aqueous samples involved initial and continuing calibration verification using a secondary source calibration standard and duplicate analysis of samples. Total HNO_3 extractable trace elements were determined on samples from the sediment core following hot block digestion of 0.25 g of air-dried sediment sample with 10 mls 1:1 $DI:HNO_3$ at 105 °C for 2 hrs followed by two 1 ml additions of

H_2O_2. Duplicate samples of each core section were digested and analyzed by ICP-MS. Standard reference materials were also analyzed as a further quality control measure.

3.3 Chemical extraction of sediment core sections

The proportion of amorphous Al, amorphous Fe, amorphous aluminosilicates and associated trace elements were estimated by chemical extraction techniques. Two sets of triplicate samples of 0.25 g air-dried sediment from each core section were initially extracted with 10 mM NH_4 acetate (pH 6) to remove readily soluble As, Se, Fe, Si, and Al. After shaking for two hours the samples were centrifuged (3000 rpm) and supernatant was decanted for subsequent ICP-MS analysis of As, Fe, Si, and Al. One set of triplicate samples was then extracted with either 0.25M $NH_2OH\cdot$ HCl/0.25M HCl at 50° C for 30 min. (extraction of amorphous Fe (hydr)oxides) or 0.175M $NH_4C_2O_4$/0.1 M oxalic acid (extraction of amorphous Al (hydr)oxides and aluminosilicates) in the dark for 2 hrs. Following extraction, the suspensions were centrifuged and the supernatant was decanted for ICP-OES (Optima 4300DV, Perkin Elmer, Norwalk, CT) analysis of Al, Si and Fe and ICP-MS analysis of trace elements.

3.4 Speciation analysis

The As species arsenite (As III), arsenate (AsV), monomethylarsenate (MMA), and dimethylarsenate (DMA), were separated on an Dionex Ionpac AS7 column using a HNO_3 gradient elution. This separation methodology was originally used to separate these As species from arsenobetaine, an organic form of As commonly found in marine animals,[17] and was further developed to allow for the speciation of the four As species listed above in addition to the As poultry litter feed additives roxarsone and p-arsanilic acid.[20] Eluant was delivered to the As 7 column using a Dionex GP50 pump, and the column was interfaced with the ICP-MS by a length of peak tubing directly to concentric nebulizer of the ICP-MS. Prior to speciation analysis the lens voltage of the ICP-MS was optimized daily for maximum signal intensity at m/z 75 by aspirating a 10 μg l^{-1} standard As solution. A programmable autosampler (AS3500, Thermo Separations) was used to perform automated standard and sample injections and was also programmed to send timed event outputs to trigger data collection by the ICP-MS and to start and reset the gradient conditions of the GP50 pump. With this configuration complete automated analysis by the IC-ICP-MS was possible.

Three point calibration curves were established for each species using serial dilution of a mixed As species standards. The mixed As species calibration standard was

Table 1. IC-ICP-MS gradient elution program for the As7 column (1 % methanol added to mobile phases)

Time	(2.5mM HNO₃) A	(50mM HNO₃) B	flow rate
0.0	100%	0%	1 ml min.$^{-1}$
1.00	100%	0%	1 ml min.$^{-1}$
1.01	0	100%	1 ml min.$^{-1}$
5.99	0	100%	1 ml min.$^{-1}$
6.00	100%	0%	1 ml min.$^{-1}$

prepared daily from stock solutions of the individual As species, which had been prepared from reagent grade salts, or in the case of MMA, had been purchased as a 100 mg l^{-1} stock solution (Crescent Chemical, Haupauge, NY). An injection volume of 100 μl was used for standards and samples. All calibration curves were linear with R^2 values > 0.995. Detection limits for each of the As species was ca. 0.02 μg l^{-1}. The gradient conditions for this separation are given in Table 1 and an example chromatogram showing the separation of a 11 μg l^{-1} mixed As standard is shown in Figure 1. This methodology was found to be both precise and accurate for measurement for the low concentrations of As species as evidenced by the excellent agreement between total As concentration and sum of As species determined by IC-ICP-MS on duplicate field samples taken from the secondary ash basin (Figure 2).

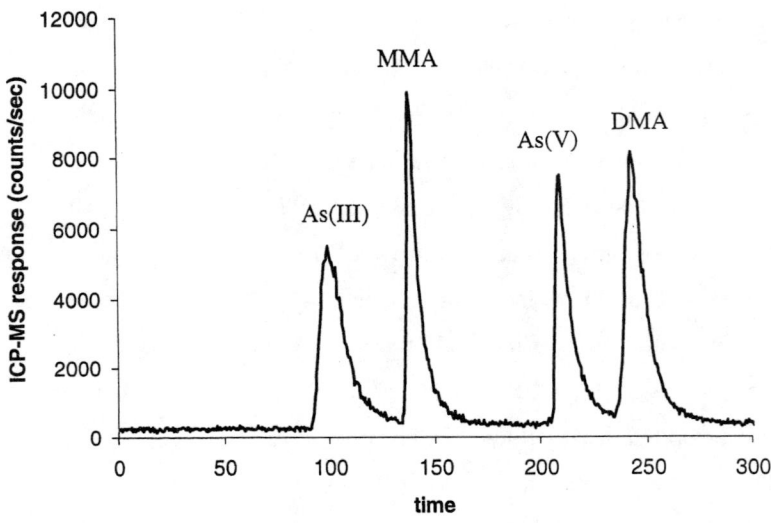

Figure 1. IC-ICP-MS chromatogram of 11 μg l^{-1} mixed As species standard.

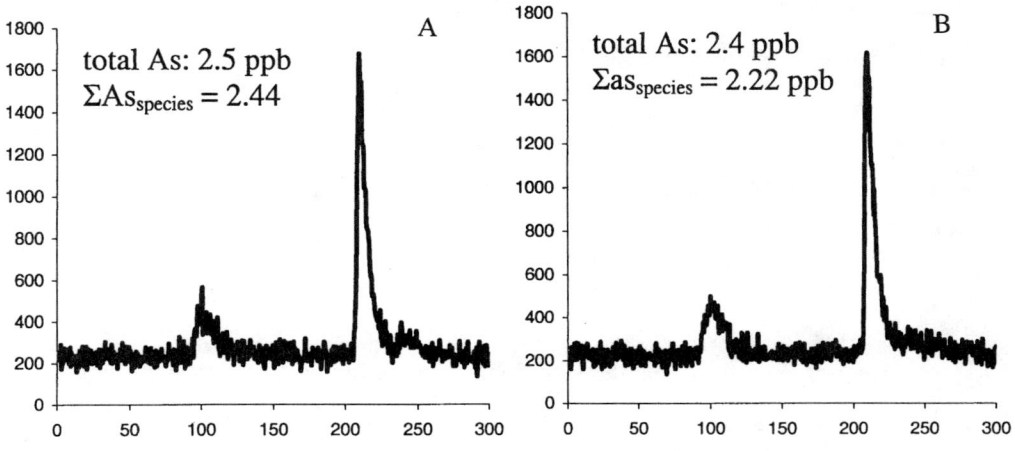

Figure 2. IC-ICP-MS As speciation analysis of duplicate field samples.

4. RESULTS AND DISCUSSION

4.1 Surface water concentration and speciation

The concentration of As throughout the basin system at any one sampling occasion was found to be fairly constant. The results for total As concentration and the corresponding speciation analysis of individual samples collected from the ash basin system during the two sampling dates are given in Table 2.   Even at the low As concentrations found in the ash basin system the results of the speciation analysis were in good agreement with the total As determination with the sum of As species usually within ± 10% of the total As determination. The reproducibility of the speciation analysis on duplicate samples taken from the secondary ash basin in fall 2000 is shown in Figure 2. The concentration of As during the summer sampling was approximately 12 µg l^{-1} and the major As species was As(V) with lower concentrations of As(III) and DMA also detected.   Arsenic solubilized from the fly ash would be inorganic mostly As (V), hence the presence of an organic As compound, DMA, in the surface waters during the summer indicates that biological methylation was occurring during this season.   Total As

Table 2. Total dissolved As and As speciation in ash basin surface waters.

Location	sampling date	total As	As speciation			
			As(III)	As(V)	DMA	ΣAs species
		µg l^{-1}	µg l^{-1}			
Receiving pit	Fall	4.05	0.75	3.6	n.d.	4.35
Primary settling basin	Summer	12	0.33	7.2	1	8.53
Secondary settling basin	Summer	12.6	0.06	9	1	10.06
Secondary settling basin	Fall	2.5	0.76	1.65	n.d.	2.41
Swamp	Summer	13.6	1.1	10.2	0.55	11.85
Swamp	Fall	2.25	0.5	1.8	n.d.	2.3

concentrations were lower in the fall sampling, approximately 3 µg l^{-1} on average. The major As species detected was As(V) but, while smaller concentrations of As(III) were again detected, DMA was not detected in the surface waters during the fall sampling. A seasonal decrease in dissolved As in this ash basin system had previously been reported.[21] However, because the source term (fly ash) of As released into the basin system is variable, depending on the As concentration of the coal, it is not possible to ascribe differences in As concentrations to seasonal biogeochemical cycles within the basin system.

Total dissolved concentrations of As determined in this study are low, and not of particular environmental concern. The maximum determined As concentration of 13 µg l^{-1} was well below the USEPA clean drinking water standard of 50 µg l^{-1} that was applicable at the time of collection and is only slightly above the new proposed limit of 10 µg l^{-1}. The ash basins are not subject to drinking water regulation, and clearly it appears that these settling basins are effective in reducing dissolved As concentrations to acceptably low levels. Previous studies of this ash basin system have reported somewhat higher total dissolved As concentrations. A seasonal study conducted in 1982 reported a mean As concentration for the ash basin system of 46 ± 31µg l^{-1} (N=63).[21] Arsenic concentrations throughout the ash basin system were similar at any one sampling, and the variation in As concentrations arose through seasonal fluctuations. A study conducted in 1990 reported variable As concentrations in the basin system, with values of 7, 46, and 65 µg l^{-1} of dissolved As reported for the primary, secondary, and swamp areas, respectively.[22] The decreased dissolved As concentrations reported in our study may reflect changes in ash handling practices over the last decade but may also be due to differences in the As content of the source coal.

4.2 Pore water As concentrations

Initial pore water sampling using in-situ samplers indicated that dissolved As concentrations in the interstitial waters were much higher than those found in the surface water. Pore water As concentrations of 3 - 91 µg l^{-1} were found in the secondary basin and concentrations of 13 - 122 µg l^{-1} were found in the swamp (data not shown). Some of the variability in As concentrations resulted from the difficulty in using the in-situ samplers at a repeatable and measurable depth. In order to further investigate the As speciation in the sediment as a function of depth an intact sediment core was collected from the ash basin swamp and was sectioned in the laboratory and total trace element analysis and As speciation analysis was determined. The pH and dissolved

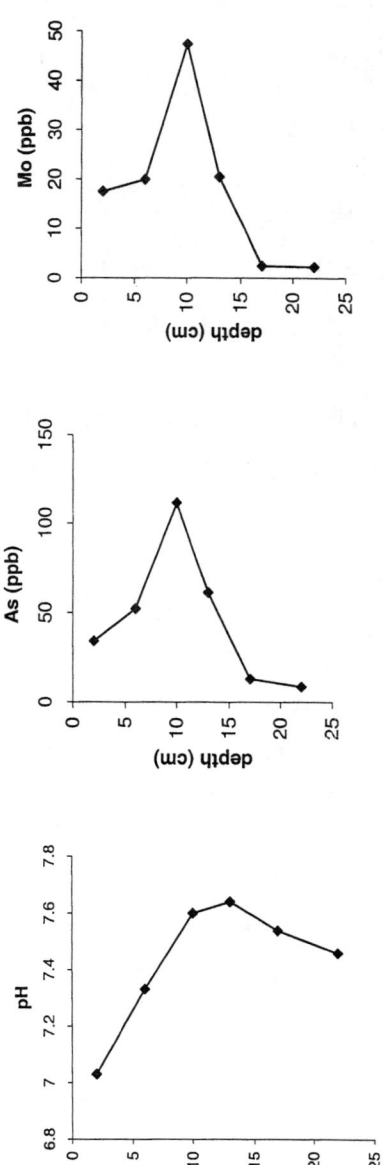
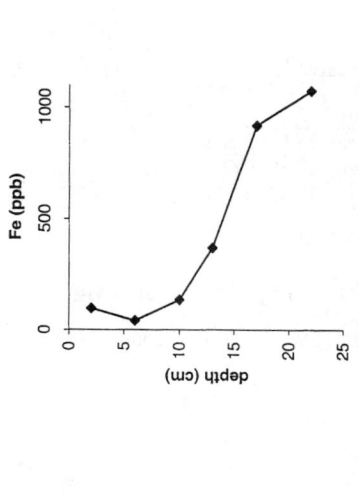
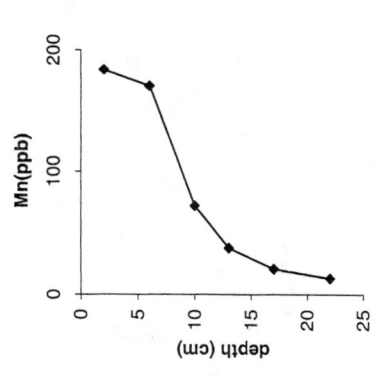

Figure 3. Selected solution chemistry of interstitial water extracted from sectioned sediment core.

Table 3. Total elemental concentrations in the depth-sectioned sediment core determined by HNO$_3$/H$_2$O$_2$ digestion.

Core section	Na	Mg	Al	K	Ca	V	Cr	Mn	Fe	Co	Ni	Cu
						mg kg^{-1}						
0-4 cm	245.9	449.4	15280	600.3	3159	42.94	36.57	102.3	8203	22.51	84.21	192.2
4-8 cm	280.5	494.9	16430	613.5	3194	51.39	39.12	113.0	8952	24.24	87.88	195.8
8-12 cm	264.6	433.5	16650	366.2	2874	33.03	18.60	78.58	5281	7.982	27.58	45.11
12-14cm	239.6	395.9	18450	338.7	2121	17.85	13.61	33.67	2420	2.446	10.15	12.18
14-20 cm	204.9	294.1	15130	265.8	1150	11.83	11.06	16.07	1328	0.923	4.823	4.114
20-23cm	135.8	223.4	9769	192.4	635.5	8.279	8.497	9.938	947.6	0.372	2.204	1.189

Core section	Zn	As	Se	Rb	Sr	Cd	Cs	Ba	Tl	Pb	U
						mg kg^{-1}					
0-4 cm	277.2	64.46	23.20	7.200	98.16	8.821	1.388	405.5	5.712	27.70	5.873
4-8 cm	317.5	79.26	28.02	7.534	114.6	8.930	1.453	455.0	6.379	30.18	6.490
8-12 cm	79.34	20.22	10.10	7.587	95.16	1.990	1.780	276.1	1.542	25.73	3.150
12-14cm	35.64	3.508	3.014	6.298	62.27	0.411	1.607	161.4	0.433	19.25	2.435
14-20 cm	24.58	0.694	1.233	3.986	28.60	0.091	1.338	113.4	0.186	16.24	1.986
20-23cm	7.749	0.194	0.522	2.917	14.98	0.026	1.114	80.62	0.129	12.73	1.156

concentrations of As, Mo, Fe, and Mn are given in Figure 3 and the total (concentrated HNO$_3$) extractable elemental concentrations for the sediment core sections are given in Table 3.

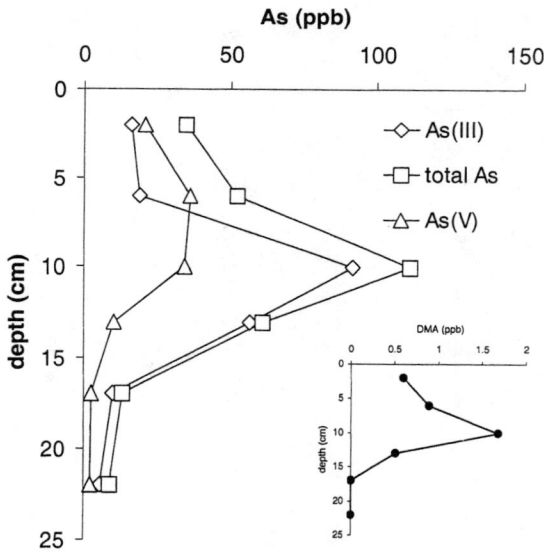

Figure 4. Interstitial water As speciation in sectioned sediment core.

Arsenic concentrations in the interstitial water increase markedly to 100 µg l^{-1} at 10 cm depth in the core, then decrease rapidly to 8 µg l^{-1} at depth 20 cm. The As concentration profile is identical to Mo, and the solubility of both these anions increases concurrently with an increase in pore water pH. An increase in the solubility of oxyanions with a rise in solution pH suggests that the primary sorption mechanism of As and Mo in these sediments is adsorption. Specific adsorption of anions is generally at maximum around their pKa and then decreases with increasing pH as the proportion of negatively charged surface functional groups of variable charged surfaces increases. Both these oxyanions are strongly sorbed by amorphous Fe oxide phases and the observed increase in As solubility under reducing conditions has been ascribed to dissolution of the sorptive Fe phase.[23] However, whether this mechanism is operative here is unclear; soluble Fe concentrations increase with depth suggesting a more reducing environment, but at the peak of As solubility soluble Fe is still fairly low.

Speciation of As in the pore water extracts was examined by IC-ICP-MS (Figure 4). The increase in soluble As occurs concurrently with a change in the relative As species distribution from predominantly As(V) and low solubility at the sediment water-interface to predominantly As(III) and increased solubility at a depth of 10 cm. Interestingly, DMA concentrations, albeit very low in comparison to total soluble As, mirror the As(III) concentration profile and suggest that microbial methylation may also be occurring in this zone. Unlike As(V), which is anionic in the pH range of natural waters, As(III) is a very weak acid that it is undissociated under natural pH conditions (pKa$_1$ = 9). It is for this reason that As(III) has been reported to be less strongly sorbed in soils and sediments than As (V).[24,25] However, other studies have suggested that when amorphous Fe oxides is the sorbent, As(III) is adsorbed as strongly as As(V).[26] Hence, the occurrence of soluble As(III) in sediment systems may also be due to the reductive dissolution of amorphous Fe oxides and reduction of solubilized As(V) rather than an inherent greater solubility of As(III).

The major and trace element composition of the sediment core changes with depth (Table 3) as a result of the input of freshly deposited ash, the weathering of previously deposited ash, and the leaching of soluble elements. The upper 8 cm of the sediment core is approximately 1.5 % Al, 0.8-0.9 % Fe, and has the greatest concentration of trace elements. The concentration of most elements decrease with depth, which indicates a change in mineralogy from Al and Fe oxides to crystalline aluminosilicate clay minerals or quartz, which would not be dissolved by the HNO$_3$ digest procedure used in this study. The decrease in trace element concentrations to low levels at a depth of 23cm indicates that leaching does not appear to be occurring to any great extent. The core section which exhibited the greatest As solubility (i.e. 8 –12 cm) is actually four times lower in total As concentration than the sediment above. Total Fe concentration also decreases at this depth, and soluble Fe concentrations begin to increase at this depth suggesting the onset of reducing conditions. Hence a decease in variable charged oxide solid phases, an increase in solution pH, and the prevalence of dissolved As(III), which may be less highly adsorbed than As(V) by the remaining mineral phases, may explain the soluble As maximum. Soluble As decreases significantly after this depth in accordance with low total As concentrations deeper in the of the sediment core.

4.3 Partial chemical extraction of sediments

After extraction with 10 mM NH$_4$Oac (pH 6) to remove soluble and weakly exchangeable trace elements, samples of the core sections were extracted with either 0.2M oxalic acid/ NH$_4$ oxalate or 0.25M NH$_2$OH· HCl/0.25M HCl. The oxalic acid extraction technique dissolves amorphous Al oxides and hydroxides and the amorphous aluminosilicate minerals allophane and imogolite.[27] The latter two mineral phases have been suggested as secondary weathering products in weathered fly ashes.[11,12] Silicon recovered in this extract is considered quantitative for the amorphous aluminosilicate originally present in the solid sample, while any Al excess in the molar Al:Si > 2 is due to amorphous Al phases.[27] Amorphous Fe content of the sediment cores was assessed with the hydroxylamine extraction.[28] Because the oxalate extraction was conducted in the dark it is also effective for the extraction of amorphous Fe oxides[28] thus allowing a comparison between the two techniques.

The concentration of oxalate-extractable Al is much greater than oxalate-extractable Si throughout the profile (Table 4) and the mole ratio of Al:Si increases greatly at depths > 12 cm. The low extractable Si below 12 cm and the large decrease in acid-extractable trace element concentrations below this depth (Table 3) suggest that the depth of fly ash deposition in swamp area where the core was taken is limited to the upper 12 cm of sediment. The ratio of Al:Si in the upper portion of the core (7-11) is much higher than the 0.5 – 1 range reported for weathered Danish and Indian coal fly

Table 4. Oxalate extractable Al and Si and Al:Si ratio in the depth-sectioned sediment core.

Core section	Al	Si	Al:Si
	% (w/w)		mole ratio
0-4 cm	0.56	0.07	7.8
4-8 cm	0.59	0.08	7.8
8-12 cm	0.55	0.05	11.9
12-14cm	0.44	0.03	15.3
14-20 cm	0.30	0.01	22.6
20-23cm	0.12	0.00	255.6

Table 5. Extractable elemental concentrations expressed as a percentage of the total HNO_3 acid digestion concentrations (given in Table 2).

Core section	Hydroxylamine extract		Oxalate extract	
	Fe	As	Fe	As
0-4 cm	12%	36%	9%	66%
4-8 cm	8%	29%	6%	59%
8-12 cm	6%	51%	5%	94%
12-14cm	10%	64%	9%	100%
14-20 cm	24%	73%	25%	100%
20-23cm	33%	95%	31%	100%

ashes[12] and suggests that amorphous Al oxides phases may be an important solid phase controlling sorption in the upper portion of the sediment.

The % of extractable Fe and As relative to the HNO_3 digestion values for both the hydroxylamine and oxalate extractions are given in Table 5. Both extraction methods gave similar results for extractable Fe, which is in agreement with previous studies that have shown that either extractant is selective for the amorphous Fe phase.[28] The concentration of extractable As differed between the extractions with higher concentrations of As extracted with the oxalate extraction. A similar increase in extraction of As with oxalate has been observed in single and mixed systems of amorphous Fe and goethite.[26] In that case it was shown that oxalate can displace some As sorbed to goethite and prevent readsorption of As to goethite after the dissolution of an amorphous Fe solid. The % extractable Fe is a direct measure of the % amorphous Fe in the sediment profile, and it can be seen that the depth of minimum % amorphous Fe in the solid phase also corresponds to the depth of maximum soluble As. Hence the observed maximum in dissolved As concentrations at this depth appears to be due to a minimum in amorphous Fe solid phase, the prevalence in As(III) as the main As species, and a maximum in solution pH.

5. CONCLUSIONS

Dissolved As concentrations in the basin system were low and the predominant speciation was As(V). Sediment As concentrations were > 90 mg kg^{-1} at the sediment-water interface but decreased to < 10 mg kg^{-1} below 20 cm depth. Dissolved As concentrations in the sediment interstitial waters increased to a maximum of 100 µg l^{-1} at 12 cm and the predominant species at this depth was As(III). Arsenic was associated with amorphous Fe oxides in the sediment core. It appeared that As solubility was controlled by redox conditions in the sediment profile, and the maximum in As solubility was coincident with an increase in soluble Fe and the prevalence of As(III) in solution. Ion chromatography coupled to ICP-MS was demonstrated to be a precise and accurate method for As speciation and provided the low detection limits necessary for environmental analysis.

ACKNOWLEDGEMENTS

This research was supported by financial Assistance Award Number DE-FC09-96SR18546 from the US Department of Energy to the University of Georgia Research Foundation.

REFERENCES

1. ACAA. (1998) 1998 Coal Combustion Product (CCP) Production and Use. American Coal Ash Association, International. Alexandria, VA.

2. Eary, L. E., Rai, D., Mattigod, S. V., and Ainsworth, C.C., Geochemical factors controlling the mobilization of inorganic constituents from combustion residues: review of the minor elements, *J. Environ. Qual.*, 19, 202,1990.

3. Jackson, B. P., and Miller W. P., Arsenic and selenium in coal fly ash extracts by ion chromatography-inductively coupled plasma mass spectrometry, *J. Anal. Atom. Spectrom.*, 13, 1107, 1998.

4. Huggins, F, E., Shah , N., Huffman, G. P., and Robertson, J., D., XAFS spectroscopic characterization of elements in combustion ash and fine particulate matter, *Fuel Processing Technol.* 65-66, 203, 2000.

5. Goodarzi, F., and Huggins, F. E., Monitoring the species of arsenic, chromium, and nickel in milled coal, bottom ash and fly ash from a pulverized coal fired-power plant in western Canada. *J. Environ. Monitoring* 3, 1, 2001.

6. Mahuli, S., Agnihotri, R., Chauk, S., Ghost-Dastidar, A., and Fan, L. S., Mechanism of arsenic sorption by hydrated lime, *Environ. Sci. Technol.*, 31, 3226, 1997.

7. Turner, R. R., Oxidation state of arsenic in coal fly ash in coal fly ash leachate, *Environ. Sci. Technol.*, 15, 1062, 1981.

8. Silberman, D., and Harris, W., R., Determintion of arsenic (III) and arsenic (V) in coal and oil fly ashes, Intern. *J. Environ. Anal. Chem.*, 17, 73, 1984.

9. Wang, J., Tomlinson, M. J., Caruso, J. A., Extraction of trace elements in coal fly ash and subsequent speciation by high-performance liquid chromatography with inductively coupled plasma mass spectrometry, *J. Anal. Atom, Spectrom.* 10, 601, 1995.

10.   Van der Hoek, E. E., and Comans, E., N., J., Modeling arsenic and selenium leaching from acidic fly ash by sorption on iron (hydr)oxide in the fly ash matrix, *Environ. Sci. Technol.* 30, 517, 1996.

11.   Warrren, C. J. and Dudas, M. J., Formation of secondary minerals in artificially weathered fly ash, *J. Environ. Qual.*, 14, 405, 1985.

12.   Zevenbergen, C., Bradley, J. P., Piet Van Reeuwijk, L., Shyam, A. K., Hjelmar, O., and Comans , R. N. J., Clay formation and metal fixation during weathering of coal fly ash. *Environ. Sci. Technol.* 33, 3405, 1999.

13.   Rowe, C. L., Kinney O. M., Flori A. P., Congdon, J. D., Oral deformities in tadpoles (*Rana catesbeiana*) associated with coal ash deposition: effects on grazing ability and growth. *Freshwater Biol.* 36,723, 1996.

14.   Hopkins W A, Mendonca M. T., Rowe C. L, Congdon J. D. 1998. Elevated trace element concentrations in southern toads, *Bufo terrestris,* exposed to coal combustion wastes. *Arch. Environ. Contam. Toxicol.* 35:325-329.

15.   Gallagher, P. A., Schwegel, CA., Wei, X. Y., Creed, J. T., Speciation and preservation of inorganic arsenic in drinking water sources using EDTA with IC separation and ICP-MS detection, *J. Environ. Monitoring*, 3, 731, 2001

16.   Jackson, B. P., and Miller W. P. Soluble arsenic and selenium species in fly ash/organic waste-amended soils using ion chromatography   inductively coupled plasma mass spectrometry. *Environ. Sci. Technol.* 33, 270,1999.

17.   Mattusch, J., Wenrich R., Determination of anionic, neutral and cationic species of arsenic by ion chromatography with ICPMS detection in environmental samples, *Anal. Chem.* 70, 3649, 1998.

18.   McGeehan, S. L., Naylor, D. V., Simultaneous determination of arsenite, arsenate, selenite, and selenate in soil extracts by suppressed ion chromatography. *J. Environ. Qual.* 21, 68, 1992.

19.   Winger, P. V., and Lasier, P., J., A vacuum-operated pore-water extractor for estuarine and freshwater sediments. *Arch. Envrion. Contam. Toxicol.* 21 321, 1991.

20.   Jackson B. P., Bertsch P. M.. Determination of arsenic speciation in poultry wastes by IC-ICP-MS. *Environ. Sci. Technol.* 35, 4868, 2001.

21.   Alberts J. J., Newman M. C., Evans D. W., Seasonal variations of trace elements in dissolved and suspended loads for coal ash ponds and pond effluents. *Water Air Soil Pollut.* 26, 111, 1985.

22.   Sandhu, S. S., and Mills, G. L., Mechanisms of mobilization and attenuation of inorganic contaminants in coal ash basins, in Emerging technologies in hazardous waste management II, Tedder, D. W. and Pohland, F. G. Eds. American Chemical Society, Washington D.C. pp. chap. 17.

23.   McGeehan, S. L., Naylor, D. V., Sorption and redox transformation of arsenite and arsenate in two flooded soils. *J. Environ. Qual.* 58, 337, 1994.

24.   Xu, H. Allard, B., Grimvall, A, Effects of the acidification and natural organic materials on the mobility of arsenic in the environment. *Water Air Soil Pollut.* 57-58, 269, 1991.

25.   Bowell, R. J., Sorption of arsenic by iron oxides and oxyhydroxides in soils. *Appl. Geochem.* 9, 279, 1994.

26.   Jackson, B. P. and Miller, W. P., Effectiveness of phosphate and hydroxide for desorption of arsenic and selenium species from iron oxides, *Soil Sci. Soc. Am. J.* 64, 1616, 2000.

27.   Wada, K. Allophane and imogolite, In Minerals in soil environments, Dixon, J. B. and Weed, S. B. Eds., Soil Science Society of America, Madison, Wi., 1989, 1051.

28.   Chao, T. T., and Zhou, L., Extraction techniques for selective dissolution of amorphous iron oxides from soils and sediments, *Soil Sci. Soc. Am. J.* 47, 225, 1983.

LEAD DESORPTION AND REMOBILIZATION POTENTIAL BY COLLOID PARTICLES IN CONTAMINATED SITES

A.D. Karathanasis

Agronomy Department, University of Kentucky
Lexington, KY 40546

1. ABSTRACT

The utilization of coal combustion byproducts (CCB) alone or in conjunction with other wastes as soil amendments has steadily increased during the last few years. In spite of their beneficial contributions, these amendments, if not monitored, pose a considerable environmental risk because of their high heavy metal concentrations. Lead is one of several toxic metals found in CCB in relatively high quantities. Although Pb has shown substantial attenuation by the soil matrix, it has also exhibited great sorption affinity and transportability in association with colloidal particles. This study investigated the potential of water-dispersible colloids to desorb Pb from contaminated soil particle surfaces and co-transport it to groundwater. The study employed intact soil monoliths contaminated by Pb, which were flushed with colloid suspensions of different mineralogical composition and deionized water (d-H_2O) used as a control. The soil monoliths represented upper solum horizons of an Alfisol and a Mollisol with contrasting macroporosity and organic carbon content. The soil colloids were fractionated from low ionic strength Bt horizons of Alfisols with montmorillonitic, mixed, and illitic mineralogy and variable physicochemical and surface charge properties. The results indicated a sharp decrease, to near zero, of Pb desorbed by deionized water-flushing solutions after 3 pore volumes of leaching, but a continuous desorption and transport of Pb in the presence of colloids. The colloid-induced desorption and remobilization of Pb was in the range of 10-60% of the initial eluent Pb concentration. Colloids with high surface charge (montmorillonitic) and small size diameter showed a greater Pb desorption and transport potential, but the amount of remobilized Pb was the result of contributions by both ion exchange and physical exclusion processes. These findings have important ramifications on assessing and predicting contamination risks and developing remediation strategies.

2. INTRODUCTION

In recent years, improper disposal of various waste materials has posed serious threats to surface and groundwater supplies and developed into a global scale soil and water pollution problem.[1,2] A number of studies have shown considerable advantages in the utilization of waste amendments, including CCB, first as a way of disposal, but also as sources of nutrients and for improvements of soil physical and chemical properties.[3] However, most of these wastes contain elevated concentrations of heavy metals, which under certain conditions may cause increased contamination and toxicity risks. Although most heavy metals are generally considered to be relatively immobile in most soils for short periods, their mobility under certain solution- and soil chemical conditions may exceed ordinary rates and pose a significant threat to groundwater quality.[4,5,6] This threat has been substantiated by recent research evidence showing that water-dispersed colloidal particles migrating through soil macropores and fractures can significantly enhance metal mobility, causing dramatic increases in transported metal load and migration distances.[7,8,9,10,11]

Lead is considered one of the most potentially hazardous metals on a global scale because of its prominent showing in the enrichment factors, transfer rates and its toxic effects to ecology and human health. In spite of the low solubility and high attenuation by organic and mineral soil matrices, which inhibit the apparent mobility of Pb in soil environments, it has been found that a significant fraction of Pb can be transported in association with particulate material or colloid lead carbonate, lead oxide and lead hydroxide minerals.[12,13] Due to the larger overall surface charge of the colloid particles, metals show a greater affinity to be adsorbed and carried by them rather than be attenuated by the stationary soil matrix. This may result in gross underestimation of groundwater contamination levels and metal migration distances.

220

Dorr and Munnich found that the downward migration of Pb in European forest soils receiving waste amendments was rather independent of pH and soil type, but strongly associated with the movement of particulate organic material with which they formed stable organic complexes.[14] Similar studies in forest soils of the Northeastern United States have shown that the maximum Pb transfer rate predicted from lysimeter measurements (considering only soluble Pb) accounted for only 22% of the annual actual transport.[15] The remaining unaccounted Pb fraction was hypothesized to be transported by a mobile organically-enriched solid phase mostly of colloidal dimensions. Alva et al. found a substantial increase in Pb elution in a loamy sand and a fine sand soil amended with fly ash-sewage sludge mixtures.[16]

Grolimund et al. demonstrated that colloid-facilitated transport can be the dominant transport pathway for a strongly sorbing contaminant such as Pb.[17] Between 3-7% of the total amount of sorbed Pb was transported by colloidal particles in their experiments, underestimating the solute model-predicted amount by up to 7 orders of magnitude, depending on the eluent ionic composition. Similar trends were reported by Kretzschmer and Sticher for Pb-facilitated transport by humic-coated hematite colloidal particles.[18] Karathanasis also found drastically higher colloid-bound than soluble Pb levels transported through undisturbed soil columns subjected to various loads of soluble or colloid Pb, simulating waste amendment applications.[13]

Therefore, substantial evidence exists to support preferential sorption of soluble Pb present in the macropore space of soils receiving waste applications by suspended colloids and facilitated transport to greater soil depths and distances. This mobilization may be enhanced even further by the high pH of some of the wastes, including some CCB, which may reduce somewhat the soluble Pb pools, but may increase dramatically colloid dispersion and mobilization of particulate Pb. Under alkaline conditions, in weakly buffered soils, suspended colloids may even preferentially desorb Pb already attenuated by the soil matrix and remobilize it to groundwater, thus making the conditions even more dangerous.

The objectives of this study were: (a) to assess the potential of colloid particles suspended in macropores of undisturbed soil monoliths to desorb Pb from contaminated soil matrix surfaces and co-transport it to groundwater, and (b) evaluate the effect of colloid and soil matrix properties on the extent of Pb desorption and remobilization.

TABLE 1 Physicochemical and mineralogical properties of soils and colloids

Properties	Soils		Colloids		
	Loradale	Maury	Beasley	Loradale	Shrouts
Clay (%)	21	35	--	--	--
Hydraulic Conductivity (cm min⁻¹)	0.3 ± 0.2	2.6 ± 1.7	--	--	--
Bulk Density (g cm^{-3})	1.4	1.6	--	--	--
Mean Colloid Diameter (nm)	--	--	220	300	270
Organic Carbon (%)	2.1	0.5	0.8	3.4	0.8
pH	6.3	5.8	6.2	6.7	5.8
CEC (cmol kg^{-1})	25.2	21.9	63.4	81.8	46.4
Extractable Bases (cmol kg^{-1})	15.0	10.1	26.5	29.2	17.3
Extractable Fe (mg g^{-1})	6.5	8.3	15.9	15.9	16.4
Extractable Al (mg g^{-1})	4.4	2.8	6.1	5.2	9.2
Surface Area (m^2 g^{-1})	--	--	386	186	123
Electrophoretic Mobility (μm cm v^{-1} s^{-1})	--	--	-1.8	-1.9	-1.6
Smectite + Vermiculite (%)	--	--	60	--	17
HISM + HIV (%)	--	--	--	44	--
Illite (%)	--	--	20	15	60
Kaolinite (%)	--	--	16	35	20
Quartz (%)	--	--	4	6	3

*HISM = hydroxyinterlayered smectite; HIV = hydroxyintrlayered vermiculite

3. MATERIALS AND METHODS

3.1 Intact Soil Monoliths

Intact soil monoliths were taken from upper Bt horizons of Maury (fine, mixed, mesic Typic Paleudalfs) and Loradale sil (fine-silty, mixed, mesic Typic Argiudolls) soils. These two soils were selected because they have considerably different hydraulic conductivities and OC contents. The upper Bt-horizon depth was sampled to represent a rooting-depth subsurface soil layer. Each monolith was prepared by carving the soil into a cylindrically-shaped pedestal of 13 cm diameter and 20 cm length and encasing with an equal length of polyvinylchloride (PVC) pipe of 16 cm diameter. The size of the monoliths was selected to compensate for spatial variability, especially in soil hydraulic conductivity. The annulus between the soil monolith and the PVC pipe was sealed with expansible polyurethane foam. The monoliths were left in the field overnight to allow the foam to dry before they were separated from their base and transported to the laboratory. Physicochemical and mineralogical properties of the soils used in the monolith experiment are reported in Table 1.

3.2 Colloid Fractions

Water-dispersible colloids (WDC) were fractionated from upper Bt horizons of three soils representing the series: Beasley sil (fine, montmorillonitic, mesic Typic Hapludalfs), Loradale sil (fine-silty, mixed, mesic Typic Argiudolls), and Shrouts sicl (fine, illitic, mesic Typic Hapludalfs). These soils were selected for the diverse mineralogical and physicochemical composition of their colloidal fraction. The extraction of the WDC fractions was accomplished by mixing about 10 g of moist soil with 200 mL of d-H_2O (without addition of dispersing agent) in plastic bottles, shaking overnight, centrifuging at 750 rpm for 3.5 min, then decanting. The concentration of the colloid fraction was determined gravimetrically, and before it was stored as a stock suspension, 0.002% (by weight) of NaN_3 was added to suppress microbial activity. Subsamples of stock colloid suspensions were air-dried, gently crushed, and passed through a 0.23 mm opening diameter sieve for characterization. Physicochemical and mineralogical properties of the colloid fractions are shown in Table 1.

3.3 Metal Solutions

An aqueous solution of 100 mg/L was prepared from a $PbCl_2$ reagent (> 99% purity, Aldrich Chemicals, Milwaukee, WI). This solution was used in the contamination phase of the leaching experiments. The same $PbCl_2$ reagent was used to prepare the equilibrium solutions for the adsorption isotherm experiments.

3.4 Leaching Experiments

3.4.1 Preparation Phase

Prior to setting up the leaching experiment, the soil monoliths were saturated from the bottom upward with d-H_2O to remove air pockets. Then, about five pore-volumes of d-H_2O containing 0.002% of NaN_3 were introduced into each monolith (downward vertical flow) using a peristaltic pump at a constant flux (2.2 cm/h) to remove loose material from the pores of the soil monoliths and suppress biological activity.

3.4.2 Contamination Phase

Four duplicate monoliths from each soil were leached with a 100 mg/L Pb flushing solution at a rate of 2.2 cm/h for 350 to 400 pore volumes to achieve a certain level of Pb contamination. This level was considered reached when the eluted Pb attained a concentration of about 5 mg/L, which corresponded to about 33-40% saturation of the soil matrix.

3.4.3 Flushing Phase

A flushing solution/suspension was applied to each contaminated monolith at a constant flux of 2.2 cm/h for the next 25-28 pore volumes. The flushing solution (control) consisted of d-H_2O, while three flushing suspensions consisted of 300 mg/L colloid suspension (one for each soil and colloid type) in d-H_2O. Eluents were monitored periodically with respect to volume, colloid, and Pb concentration. Breakthrough curves (BTC) were constructed based on reduced Pb and colloid concentrations (ratio of effluent concentration to influent concentration = C/C_o) and pore-volume (flux averaged volume of solution pumped per monolith pore volume). A value of C_o 5 mg/L was used for Pb, and C_o = 300 mg/L was used for colloids.

Colloid concentrations in the eluent were determined by placing 200 mL of the sample into a Bio-Tek multichannel (optical densitometer with fiber-optic technology; Bio-Tek

Instruments, Inc., Winooski, VT) microplate reader and scanning at 540 nm. Total Pb concentration in the eluents was allocated to solution phase and colloidal phase (colloid-bound contamination). The eluent samples were centrifuged for 30 min at 3,500 rpm to separate the soluble contaminant fraction from the colloid-bound contaminant fraction. The soluble Pb fraction was analyzed by ICP-spectrometry.

3.5 Metal Adsorption Isotherms

Most eluted colloid samples contained moderate to high colloid concentration (> 50 mg/L), and their eluted colloid-bound Pb was extracted with 1 N HCl-HNO$_3$ solution and analyzed by ICP spectroscopy.[19] For the few eluted samples with colloid concentration < 50 mg/L, the experimental and analytical uncertainty was too high to rely on direct extraction determinations. Therefore, for these samples the colloid-bound Pb fraction was calculated from adsorption isotherms generated from batch experiments. The amount of the colloid-bound metal eluted in the leaching experiments was calculated by extrapolating the metal equilibrium concentration in the eluent to the adsorption capacity of the colloid and multiplying by the colloid concentration in the eluent. Agreement between extracted and isotherm-estimated Pb on low colloid concentration eluted samples was verified with extractions of selected samples following a concentration pretreatment.

The batch equilibrium experiments were carried out using 50 mL test tubes. A 250 mg of air-dried colloid sample was added to each test tube along with 35 mL of adsorbate metal solution containing 0-20 mg/L of Pb. After 24 h or shaking, the samples were centrifuged at 3,500 rpm for 30 min and the supernatants were analyzed with the respective analytical methods described earlier. Similar experiments were used for development of Pb whole soil isotherms in order to compare metal affinity differences between colloids and soil matrices.

4. RESULTS AND DISCUSSION

4.1 Adsorption Isotherms

Isotherms of Pb sorption by soils and colloid fractions were prepared by plotting equilibrium concentrations of Pb against the calculated amount adsorbed in the solid phase. The data conformed well to the Freundlich equation (Table 2). The Maury soil exhibited about three

TABLE 2. Freundlich equation parameters for Pb absorbed by soils and colloids

	Kd	1/n	R
Soils			
Loradale	0.60	0.49	0.97
Maury	1.75	1.09	0.98
Colloids			
	11.43	0.90	1.00
Beasley			
Loradale	15.29	0.97	1.00
Shrouts	4.15	0.30	1.00

times greater sorption affinity for Pb than the Loradale soil, even though the latter contained four times higher OC content. This is attributed to the higher amounts of Fe and especially Mn-oxide concentrations of the Maury soil, which are known to be excellent scavengers for Pb. The lower pH of the Maury soil (5.8) compared to that of Loradale (6.3) may have also accentuated the very high sorbate-sorbent affinity.

The sorption affinity of the colloids followed the sequence Loradale > Beasley > Shrouts. In all cases, the affinity of the colloids for Pb was greater than that shown by the soil matrix, suggesting preferential sorption by colloids as a plausible mechanism for enhanced Pb transport through soil macropores to the groundwater. The Kd of the Loradale colloid was about 25 times higher than that of the Loradale soil matrix, probably due to its higher organic carbon content. Lead sorption affinity for the Beasley and Shrouts colloids was 19 and 7 times greater than for the Loradale soil matrix, and 6.5 to 2.4, higher, respectively, for the Maury soil matrix. The < 1 values of $1/n$ in the Freundlich equation for soil colloids suggest a decreasing energy of sorption with increasing surface coverage, especially in the illitic colloid (Shrouts).

4.2 Lead Sorption During the Saturation Phase

Both monoliths exhibited a strong affinity for Pb showing complete attenuation until breakthrough traces occurred at about 200 pore volumes for Maury and 240 pore volumes for Loradale. The sorption phase continued until breakthrough levels of Pb just exceeded 5 mg/L. At that stage, the corresponding Pb saturation level was estimated to be approximately 33% and 40% for the Loradale and Maury monoliths, respectively. However, under the pH conditions of the soil monoliths (6.0), retention of Pb through metal hydrolysis is also possible.[20] This hydrolysis may be accompanied by precipitation of metal hydroxides in a manner not easily distinguishable from adsorption reactions. Geochemical modeling of aqueous phase chemical equilibria using the MINTEQA2 program indicated that conditions were favorable for the formation of Pb oxides and hydroxides.[21]

Following the Pb saturation phase, the monoliths were flushed with distilled water at 360 (Maury) and 400 (Loradale) pore volumes to displace excess soluble Pb remaining in the soil pores. Leaching with water continued for approximately three more pore volumes, and ended when the eluent Pb concentration was stabilized around 5 mg/L. The total quantity of Pb removed during the H_2O flushing stage accounted for less than 1% of that applied to the monoliths. Eluted Pb concentrations in the last collected samples were 4.9 and 4.6 mg/L for the Maury and Loradale monoliths, respectively. These final values were used as the influent concentrations (Co) for construction of the BTC at the flushing phase of the experiment.

4.3 Colloid Elution

No measurable colloid elution in any of the columns was detected during the Pb sorption phase or during flushing with d-H_2O solutions (controls). Colloid breakthrough in colloid

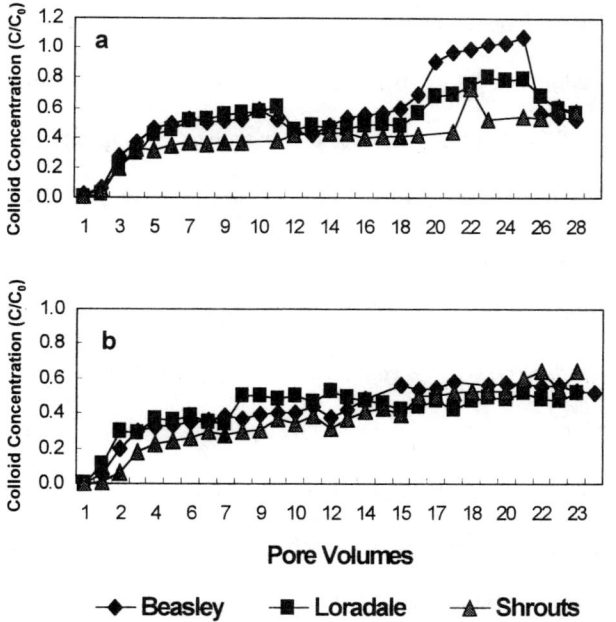

Fig. 1. Colloid elution through Maury (a) and Loradale (b) soil monoliths during the colloid-flushing phase.

flushing suspensions was greater than anticipated and substantially higher in the Maury than in the Loradale monolith (Fig. 1). The reduced colloid elution through the Loradale monolith is likely the consequence of its lower hydraulic conductivity and macroporosity, which resulted into more effective straining of the suspended colloids. Greater colloid elution through the Loradale monolith may have also been inhibited by higher eluent EC levels (155µS/cm) compared to the 50-60 µS/cm EC values observed in eluents of the Maury monoliths.[22] According to van Olphen decreases in the repulsive potential between colloid surfaces occur with increases in solution ionic strength, which may result in some colloid coagulation and increased filtration by the soil matrix.[23]

Colloid elution in colloid flushing suspensions increased sharply during the first 5 pore volumes to about 0.35 and 0.50 C/C_o, in the Loradale and Maury monoliths, respectively. Thereafter, the Loradale monoliths experienced a gradual increase in colloid breakthrough, which reached maxima at the end of the experiment ($C/C_o = 0.55$). There were no significant differences in the breakthrough of different colloids, except for an initial lower elution of the illitic colloids. In contrast, the Maury monoliths experienced a small drop in colloid breakthrough from 0.60 to 0.45 C/C_o between 11 and 15 pore volumes, which was followed by another surge up to the 25th pore volume. During this surge, there were clear differences between colloid breakthrough C/C_o maxima, following the sequence montmorillonitic (1.0) > mixed (0.75) > illitic (0.55). This is attributed to mineralogical composition and surface charge differences, as well as the smaller mean colloid size diameter of the montmorillonitic colloid (Table 1).[24] A decrease in colloid mobility with increasing diameter through porous media has been documented by several researchers.[25,26] Karathanasis noted that the mobility of mixed mineralogy and kaolinitic Pb-saturated soil colloids was about 50% and 10%, respectively, of that observed for montmorillonitic colloids, mainly due to double layer suppression and coagulation effects.[13] Therefore, the larger particle size of the floccules may be responsible for the reduced elution of the mixed and the illitic colloid. The colloid breakthrough thresholds observed after the 5th or 25th (Maury) pore volume is attributed to a steady state porosity reached by the monoliths as a function of colloid flux and colloid filtration rates that compromised a

Maury Soil Columns

Loradale Soil Columns

Beasley Colloids

Beasley Colloids

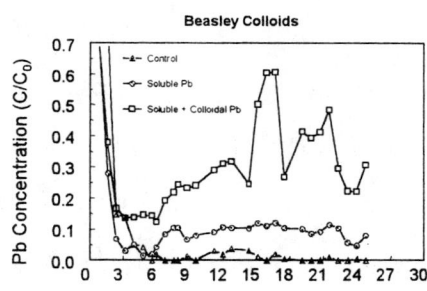

Loradale Colloids

Loradale Colloids

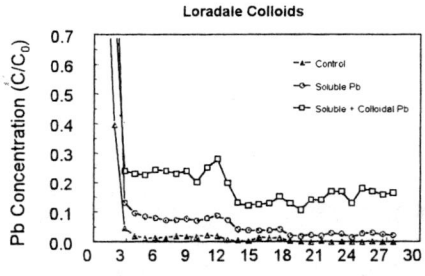

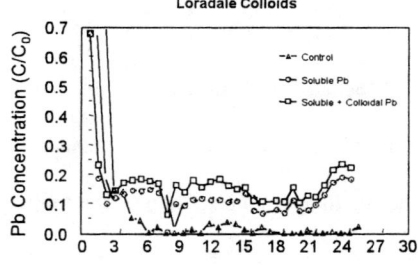

Shrouts Colloids

Shrouts Colloids

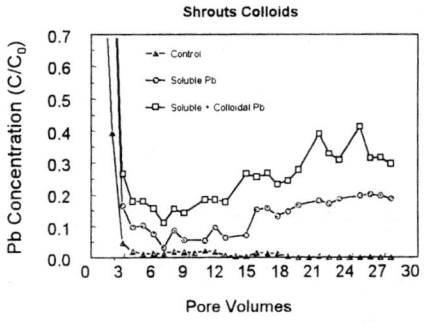

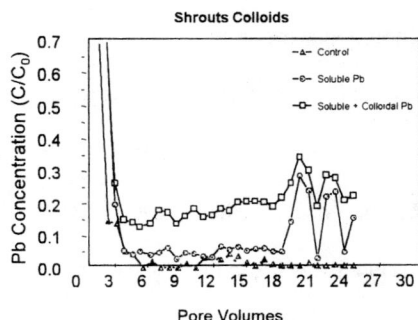

Fig. 2. Lead elution in d-H2O (control) and colloid suspensions flushed through Maury and Loradale soil monoliths.

portion of the originally available colloid flow paths. Finally, the colloid elution resurgence between 15 and 25 pore volumes in the Maury monolith could be the result of flow path rearrangements, due to flux-related detachment of pore wall particles or some biological activity within the columns.

4.4 Lead Elution

Lead elution by d-H_2O flushing solutions (controls) decreased drastically to near 0 after six pore volumes, suggesting absence of soluble Pb in the macropore space and total inability of d-H_2O to desorb Pb previously attenuated in the soil matrix (Fig. 2). In contrast, Pb elution as soluble or total (soluble + colloid bound) by colloid flushing suspensions continued throughout the leaching cycle in all soil monoliths. While the soluble Pb fraction in the eluents was relatively stable between soils and colloids in the range of 0.05-0.15 C/C_o, the total Pb and therefore, colloid-bound fraction, varied significantly averaging between 0.2 and 0.5 C/C_o (Fig. 2).

Most BTC showed considerable asymmetry, which is attributed not only to preferential flow, but to extensive chemical interaction with the soil matrix. This interaction is anticipated considering the variable affinity of soil matrices and colloids for Pb (Tab. 2). In addition, even though the colloid input flow velocity was maintained at a constant rate throughout the experiment, the variable filtration rate of the colloids by the soil matrix and the resulting reduction of macroporosity due to partial clogging may have altered the solution flow path and soil hydraulic conductivity.[27] These irregular decreases in flow velocity within the matrix may have also enhanced the formation of soluble metal-organic complexes, due to increased interaction time and reduced mass transfer resistance for Pb dissolution.

In all cases, the total Pb fraction was considerably higher than the soluble fraction during the colloid application cycles, showing good correlation with colloid breakthrough trends. Since there was no soluble source of Pb in the macropore space, this is the strongest evidence yet that the higher affinity of the colloids for Pb over that of the soils resulted in competitive sorption between the two solid phases, which allowed Pb to be stripped from the soil matrix and adsorbed onto the migrating colloids (Fig. 3). This mechanism was supported by the identical

230

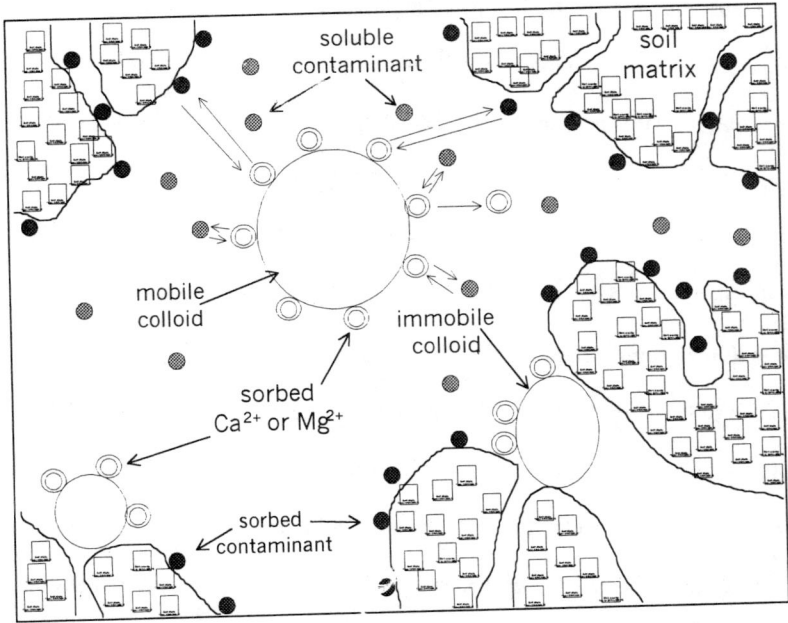

Fig. 3. Diagram showing potential competitive sorption interactions between the Pb-saturated soil matrix and the flushing colloids causing desorption and remobilization of Pb in the macropore space.

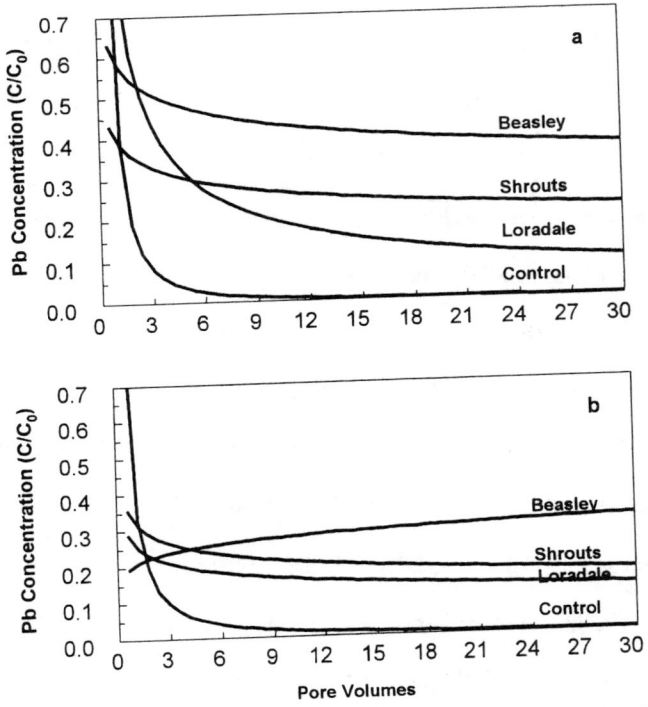

Fig. 4. Power function-fitted BTC for total (soluble+colloid-bound) Pb desorbed and remobilized from Maury and Loradale soil monoliths by d-H2O (control) and colloid flushing suspensions.

mineralogical composition of the eluted compared to the input colloids, suggesting that *in*-situ colloid generation and detachment within the soil monoliths, and therefore, contribution to the eluted Pb was negligible. Mills et al. suggested that competitive sorption exchanges of a metal between two solid phases may continue until a state of equilibrium is established.[28] Exchange equilibrium rates are relatively fast, and likely within the range of residence times spent by the colloids within the monoliths. Since the number of interactive exchange sites available on the eluting colloids is limited compared to the sites available within the matrix of the entire soil monolith, the extent of Pb desorption is controlled by the concentration of the colloid eluted through the soil monolith, and the accessibility of interactive sites within the monolith and soil matrix.[29] The association of the eluted Pb with the eluted colloids was assessed with HCl-HNO_3 extractions.[19] The results indicated that exchange sites on the eluted colloids were between 35 and 60% saturated with Pb.

The range in Pb desorption and remobilization by the colloid flushing suspensions was colloid- and soil monolith-specific. Total Pb elution was greater overall through the Maury soil monoliths, especially for the montmorillonitic (Beasley) and illitic (Shrouts) colloids (Fig. 4), even though the lower Pb sorption affinity of the Loradale soil would have made it easier to desorb Pb from its matrix. Apparently, the lower colloid breakthrough caused by the lower hydraulic conductivity and macroporosity of the Loradale soil limited further Pb mobilization, in spite of the greater potential suggested by the Kd coefficients. The similar Pb elution by the mixed mineralogy (Loradale) colloids in both soil monoliths may suggest additional Pb contributions in the form of organic complexes to the lead load transported in the high O.C. content Loradale soil.[30] Indeed, nearly 85% of the total Pb transported in the presence of the mixed mineralogy colloids through the Loradale soil monolith was in the soluble fraction (Fig. 2). The greater potential of the illitic (Shrouts) colloids to desorb Pb from the soil matrix compared to the mixed mineralogy (Loradale) colloids, in spite of their 4-fold lower Kd, and their smaller surface charge and surface area is surprising. This suggests that other physicochemical parameters and possibly physical exclusion mechanisms may also exert considerable influence on the overall colloid behavior.

232

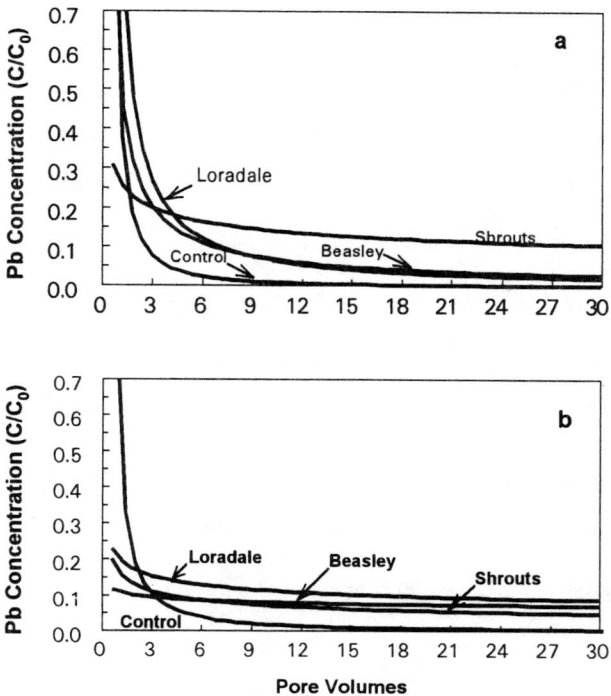

Fig. 5. Power function-fitted BTC for soluble Pb desorbed and remobilized from Maury and Loradale soil monoliths by d-H2O (control) and colloid flushing solutions.

The presence of colloids in the flushing suspensions enhanced the transport of both soluble and colloid-bound Pb fractions. Since the soluble source of Pb in the macropore space was negligible, as indicated by the control solutions (d-H_2O), the additional soluble Pb eluted in the presence of colloids must have been caused by colloid-induced desorption from the soil matrix. Weakly held outer-sphere Pb complexes sorbed on the colloids or soil matrix may be easily converted to soluble forms through ionic strength changes or organometallic interactions induced by continuous flow rate and flow path changes within the soil matrix.[31] Furthermore, direct ion exchange reactions between soluble cations present in the colloid suspensions and Pb sorbed in the soil matrix may also contribute a portion to the eluted soluble Pb fraction. Even though the differences in the eluted soluble Pb fraction between colloids were small, elution was highest in the presence of illitic colloids passing through the Maury monoliths, and the mixed colloids passing through the Loradale monoliths (Fig. 5).

5. CONCLUSIONS

The potential for desorption and remobilization of Pb from contaminated soils by *ex-sit* mineral colloids migrating through the macropore space was clearly demonstrated in this experiment. The magnitude of desorption and remobilization was dependent upon the physicochemical characteristics of the colloids and the soil matrices. The high sorptive affinity for Pb and the small particle size diameter of the montmorillonitic colloids, especially in soils with greater macroporosity, contributed to significant enhancement in desorption and transport of soluble and colloid-bound Pb compared to that generated by d-H_2O flushing solutions or other colloid types. The mechanism involves competitive sorption exchange of Pb between the migrating colloids and the soil matrix and direct ion exchange between soluble cations of the colloid suspensions and outer-sphere Pb complexes of the soil matrix.

Considering the extent of metal impacted sites worldwide, involving different types of soil clays, and the common practice of covering some contaminated areas with clay caps or liners, the findings of this experiment have significant ramifications. Mobilized clay colloids generated through natural leaching processes have the potential to strip metals retained by the soil matrix and enhance their migration to deeper depths. Therefore, these processes should be given serious consideration in assessing groundwater contamination risks and in developing effective remediation strategies.

REFERENCES

1. Adriano, D.C. *Trace elements in the terrestrial environment.* Springer-Verlag, New York, 1989.

2. Alloway, B.J. (ed.) *Heavy metals in soils.* Blackie Academic & Professional, London, 1995.

3. Sajwan, K.S., A.K. Alva, and R.F. Keejer. *Biogeochemistry of trace elements in coal and coal combustion byproducts.* Kluwer Academic/Plenum Publishers, New York, 359, 1999.

4. Jorgensen, S.S. Mobility of metals in soils. *Folia Geogr. Danica* 19, 104, 1991.

5. Kaplan, D.I., P.M. Bertsch, and D.C. Adriano. Facilitated transport of contaminant metals through an acidified aquifer. *Ground Water* 33, 708, 1995.

6. Maskall, J., K. Whitehead, and I. Thornton. Heavy metal migration in soils and rocks at historical smelting sites. *Environ. Geochem. Health* 17, 127, 1995.

7. Puls, R.W., R.M. Powell. Transport of inorganic colloids through natural aquifer material: Implications for contaminant transport. *Environ. Sci. Technol.* 26, 614, 1992.

8. Liang, L. and J.F. McCarthy. 1995. Colloidal transport of metal contaminants in groundwater. *In* H.E. Allen et al. (ed), *Metal speciation and contamination of soils.* Lewis Publ., Boca Raton, FL, 87, 1995.

9. Ouyang, Y., D. Shinde, R.S. Mansell and W. Harris. Colloid-enhanced transport of chemicals in subsurface environments: a review. *Critical Reviews in Environ. Sci. Technol.* 26, 189, 1996.

10. Ryan, J.N. and M. Elimelech. Colloid mobilization and transport in groundwater. Colloids and Surfaces, A: *Physicochemical and Engineering Aspects* 107, 1, 1996.

11. Karathanasis, A.D. Subsurface migration of Cu and Zn mediated by soil colloids. *Soil Sci. Soc. Am. Journal* 63, 830, 1999.

12. EPA., Air quality criteria for lead. *EPA 600/8-83-018F,* Research Triangle Park, NC, 1996.

13. Karathanasis, A.D. Colloid-mediated transport of Pb through soil porous media. *Intern. J. Environ. Studies* 57,579, 2000.

14. Door, H. and Munnich, K.O. Lead and Cs transport in European forest soils. *Water, Air, Soil Pollut.* 57-58, 809, 1991.

15. Wang, E.C. and G. Benoit. Mechanisms controlling the mobility of lead in the Spodosols of a northern hardwood forest ecosystem. *Environ. Sci. Technol.* 30, 2211, 1996.

16. Alva, A.K., S. Paramosivan, O. Prakash, and K.S. Sajwan. Effects of flyash and sewage sludge amendments on transport of metals in different soils. *In* Sajwan et al. (ed.) *Biogeochemistry of trace elements in coal and coal combustion byproducts.* Kluwer Academic/Plenum Publishers, New York, 207, 1999.

17. Grolimund, D., M. Borkovec, K. Barmettler and H. Sticher. Colloid-facilitated transport of strongly sorbing contaminants in natural porous media: A laboratory column study. *Environ. Sci. Technol.* 30, 3118, 1996.

18. Kretzschmar, R., and H. Sticher. Transport of humic-coated iron oxide colloids in a sandy soil: Influence of Ca^{2+} and trace metals. *Environ. Sci. Technol.* 31, 3497, 1997.

19. EPA. Methods for the determination of metals in environmental samples. Method 200.2 EPA/600/R-94/111. Washington, D.C., USA, 1994.

20. Lindsay, W.L. *Chemical equilibria in soils.* John Wiley & Sons, New York, USA, 1979.

21. Allison, J.D., D.S. Brown, and K.J. Novo-Gradac. MINEQA2/PRODEFA2, a geochemical assessment model for environmental systems: Version 3.0 user's manual. USEPA, Environmental Research Laboratory, Athens, GA, USA, 1990.

22. Roy, S.B., and D.A. Dzombak. Chemical factors influencing colloid-facilitated transport of contaminants in porous media. *Environmental Science Technology* 37, 656, 1997.

23. van Olphen, H. *An introduction to Clay Colloid Chemistry.* 2nd ed. John Wiley & Sons, New York, NY, USA, 1977.

24. Seta, A.K., and A.D. Karathanasis. 1997. Stability and transportability of water-dispersible soil colloids. *Soil Science Society of America Journal* 61, 604, 1997.

25. Kaplan, D.I., P.M. Bertsch, D.C. Adriano. Mineralogical and physicochemical differences between mobile and nonmobile colloidal phases in reconstructed pedons. *Soil Sci. Soc. Am. Journal* 61, 641, 1997.

26. Ronen, D., M. Magaritz, U. Weber, A.J. Amiel, and E. Klein. Characterization of suspended particles collected in groundwater under natural gradient flow conditions. *Water Resources Research* 28, 1279, 1992.

27. Davis, A., and I. Singh. Washing of Zinc (II) from contaminated soil column. *Journal Environmental Engineering*, 121, 174, 1995.

28. Mills, W.B., S. Liu, and F.K. Fong. Literature review and model (COMET) for colloid/metals transport in porous media. *Ground Water* 29, 199, 1991.

29. Elliott, H.A., M.R. Liberati, and C.P. Huang. Competitive adsorption of heavy metals by soils. *Journal Environmental Quality* 15, 214, 1986.

30. Verloo, M. and A. Cottenie. Stability and behaviour of complexes of Cu, Zn, Fe, Mn, and Pb with humic substances of soils. *Pedobiologie* 22, 174, 1972.

31. Ji, G.L., and H.Y. Li. ˙Electrostatic adsorption of cations. *In* Yu T.R. (ed.), *Chemistry of Variable Charged Soils*. Oxford University Press, Oxford, UK, 64, 1997.

SELENIUM AND MOLYBDENUM ADSORPTION ON KAOLINITE CLAY MINERAL COATED WITH HYDROUS OXIDES OF IRON AND ALUMINUM

D.K. Bhumbla[1], S.S. Dhaliwal[1], K.S. Sajwan[2], and B.S. Sekhon

[1]Division of Plant and Soil Sciences, West Virginia University, Morgantown WV 26505-6108, USA

[2]Department of Natural Sciences and Mathematics, Savannah State University Savannah, GA 31404, USA

ABSTRACT

Reclamation of abandoned acid mine soils is often limited by lack of suitable soil materials for establishing self-sustaining vegetation. Fly ash has been used to create technogenic soils for reclamation of abandoned mine lands. Fly ash contains toxic trace elements such as selenium and molybdenum that can leach into the groundwater from ash treated soils. In abandoned acid mine lands, oxidation of pyrite generates acidity and dissolved Fe and Al. Neutralization of pyrite oxidation generated acidity with alkalinity from fly ash produces coatings of iron and aluminum hydrous-oxides on the surface of soil clay minerals. Kaolinite is the dominant clay mineral in these soils and Fe and Al coated kaolinite can retain Se and Mo. Laboratory investigations were conducted to determine the effect of three levels of oxide coatings on kaolinite, on Se and Mo retention capacity of coated kaolinite. Adsorption of selenite and molybdate occurred on hydrous oxide coated kaolinite above pH_o. Significant adsorption of selenate only occurred at pH values lower than pH_o of the concerned solid phase. Coating of kaolinite with Fe as well as Al hydrous oxide increased the anion retention capacity of the mineral. The effect was, however, more for Fe than Al hydrous oxide. Surface sites for the adsorption of anions were limited. Results of this investigation show that adsorption will play a very important role in controlling the mobility of oxyanions in fly ash treated mine soils containing relatively high amounts of Fe and Al hydrous oxide coatings.

INTRODUCTION

Fly ash, a power plant product, has been used as an alkaline amendment for the reclamation of surface-mined acidic mine soils[1, 2]. Fly ashes have several characteristics which are beneficial for the establishment of vegetation on the disturbed mine lands[3-6]. The desirable role of fly ash in establishment and long-term sustenance of vegetation on acidic mine soils has been demonstrated in a number of investigations[7-9]. However, fly ashes contain elevated levels of trace elements[10] and the mobilization of oxyanion-forming trace elements (Mo and Se) from fly ash undergoing weathering in acidic mine soil environment is one of the major concerns in the use of fly ash in reclamation.

Forms of Mo and Se present in fly ash are easily accessible to solution and are expected to be readily released into solution in a weathering environment[10]. The anionic species of Mo and Se in a near surface aqueous environment like fly ash treated acidic mine soil pose an environmental risk. These elements being bio-toxic may pollute the food chain via plant uptake and the ground water through leaching.   Therefore, it is very important to understand the processes involved in retention/mobility of these elements in environments used for the disposal of fly ash.

Adsorption by soil solid phases is the major process, which may attenuate aqueous concentrations of Mo and Se below levels considered environmentally safe[10]. The solid phases responsible for the adsorption of Mo and Se are those with ligand-exhangeable surface OH⁻ groups[11, 12]. Such solid phases occur in relatively high contents in tropical region soils and are responsible for their high anion retention capacities[11, 13]. The temperate regions soils with their dominantly permanent charge clay mineralogy usually lack large anion retention capacities. Consequently, a bulk of the literature on anion adsorption has involved the investigation of solid phases representative of the tropical region soils[13, 14]

Acidic mine soils formed as a result of pyrite oxidation contain unusually high concentrations of aqueous Fe^{3+} and aluminum[15]. Reclamation of these soils with alkaline fly ash results in the neutralization of soil acidity forcing aqueous Fe^{3+} and Al^{3+} into the solid phase. In a soil system with predominance of permanent charge clay minerals, these newly formed solid phases containing Fe and Al will be deposited on the surface of clay minerals. These two dimensional deposits of Fe and Al hydroxy material formed on the surface of clay minerals are called sesquioxide coatings[16]. Because of unusually high concentration of aqueous Fe and Al in acidic mine soils, unusually high amounts of sesquioxide coatings may be generated in acidic mine soils reclaimed with fly ash. Iron and Al hydrous oxide coatings are known to modify the surface charge characteristics of the host solid phase making it more like those of the pure Fe and Al hydrous oxide solid phases[16]. Therefore, these new solid phases formed in fly ash treated mine soils may play an important role in controlling the aqueous concentrations of Mo and Se in these soils.

To understand the mechanism of a natural phenomenon, it is necessary to obtain definitive information by working with simplified systems allowing unambiguous interpretations of results. Once the fundamental information is available, this information can be used to interpret the results of investigations involving more complex systems representative of the natural conditions. Therefore, to understand the anion retention behavior of fly ash treated soils, one may select pure clay minerals representing solid phases in these soils and artificially coat them with Fe and Al hydroxy material in the laboratory.

Molybdenum and Se are adsorbed by hydroxylated surface through a ligand exchange mechanism involving A type surface hydroxyl groups coordinated to a single cationic atom in the structure[17]. This adsorption process is dependent upon the nature of the adsorbent surface and the system pH[11]. System pH affects adsorption of Mo and Se by regulating the equilibrium among various anionic aqueous species and by changing the surface charge on the adsorbent. In systems where solid phases consist of oxide minerals like goethite and gibbsite carrying a positive variable charge over a large pH range, adsorption continues to occur at high pH values because anionic species with various degrees of protonation can be adsorbed on to a positively charged surface[18]. However, the solid phases consisting dominantly of permanent charge clay minerals and/or sesquioxide coated permanent charge clay minerals will develop a net negative variable charge at relatively low pH values. Although anion adsorption could occur on negatively charged variable charge surfaces, protonated aqueous anionic species are required for adsorption to occur[19]. Therefore, pH changes are expected to have more drastic effect on the adsorption of various anions by solid phases that develop net negative variable charge at low pH. The pure solid phases consisting of crystalline forms of both Fe and Al hydrous oxides have high anion retention capacities. Although the content of Fe and Al hydroxy material occurring as coatings may be small, they may impart high anion retention capacities to the host solid phases. Occurrence of the hydroxy material as a two-dimensional layer on the surface of high surface area clay minerals should result in a very high specific surface area of this material. The present investigation was conducted to evaluate the effect of various amounts of Fe and Al hydrous oxide coatings on the anion retention characteristics of common clay minerals. Molybdenum and Se were selected for this investigation because of their importance in fly ash treated mine soils. Kaolinite is the dominant clay minerals in soils developing

Table 1. Important physico-chemical characteristics of kaolinite.

Characteristic	Mineral Kaolinite
Identification	KGa-1,
N_2 Surface area	23.50 ± 0.06 (m^2g^{-1})
Cation Exchange capacity	3.3 (c $mol_c kg^{-1}$)
SiO_2	43.9 %
Al_2O_3	38.5 %
Fe_2O_3	0.98 %
MgO	0.03 %

on recently surface mined sites, so kaolinite was used as the model clay mineral in these investigations.

MATERIALS AND METHODS

Well-crystallized kaolinite used in this investigation was obtained from Source Clay Repository of Clay Mineral Society. Some of the important physico-chemical properties of this mineral are given in Table 1. The < 2 μm fraction of the mineral was separated by centrifugation method and this fraction was used in further studies.

1. Preparation of Coated Samples

The procedure used for the preparation of Fe and Al hydrous oxide coated samples was similar to that used by Hendershot and Lavkulich (1983)[16]. Samples were prepared with two levels of each of Fe and Al hydrous oxide coatings. Two coating levels were 0.5 and 4.0% $Fe(OH)_3/Al(OH)_3$ by weight. The coatings were generated by titrating appropriate concentrations of $FeCl_3$ and $AlCl_3$ solutions with NaOH in the presence of the clay mineral using 250-mL centrifuge tubes as reaction vessels. The samples were titrated to pH 7.0, shaken overnight on a reciprocating shaker, and if necessary the pH was readjusted to 7.0 on the next day. This process was continued until the pH was virtually constant. The samples were centrifuged, the supernatant was acidified to pH 2.0 and analyzed for Fe/Al to make sure that the entire amount of Fe/Al was precipitated. After centrifugation, the solid was transferred to dialysis tubing and dialyzed against deionized distilled water until Cl⁻-free. The samples were transferred to polyethylene beakers, dried at 40°C, and aged by saturating with water and re-drying at 40°C a total of 10 times.

1.1. Adsorption Studies

The adsorption isotherms for various samples were obtained from a batch investigation. For this purpose 45 mg solid samples was weighed into 50-mL centrifuge tubes and to each tube 0.05 M NaCl solution was added along with predetermined amount of HCl/NaOH to give a definite pH and a volume of 29.7 mL. The head-space in the centrifuge tubes was filled with argon and the tubes were tightly capped and shaken on a reciprocating shaker for 30 minutes. Different concentrations of solutions of Na salts of selenite, selenate, and molybdate were added to the centrifuge tubes as 300 μL volumes so that the final volume in each centrifuge tube was exactly 30 mL. There were 3 different initial pH values and 11 different initial anion concentrations for each sample. Initial anion concentrations varied from 0.4 to 40 mmol m^{-3}. The contents of centrifuge tubes were shaken

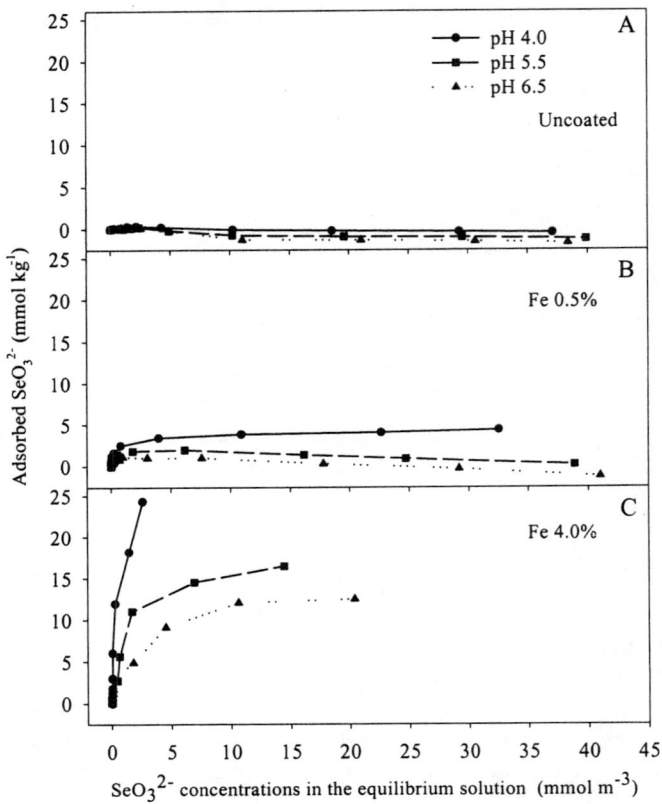

Fig. 1. Selenite adsorption on the kaolinite clay mineral coated with hydrous oxides of iron at the rates of 0%, 0.5%, and 4.0% Fe; and at three levels of equilibrium pH 4.5, 5.5, and 6.5

for exactly 12 hours followed by centrifugation. The supernatant was divided into two portions one each for reading pH and analyzing for the anion concentration. The portion saved for pH was maintained under argon environment at all times. Anion concentrations in the supernatant samples with < 100 ugL^{-1} of an anion were determined by using Graphite Furnace Atomic Absorption Spectrophotometer. The Inductively Coupled Plasma Emission Spectrometer was used for analyzing the supernatant samples with > 100 µgL^{-1} of an anion. All the samples were analyzed as such without any further dilutions. The amount of an anion adsorbed was calculated by the difference between the initial and the final concentrations of the equilibrium solution.

RESULTS AND DISCUSSION

The adsorption isotherms for selenite at different pH values and different levels of Fe and Al hydrous oxide coatings are given in Fig. 1 & 2. The amount of selenite adsorbed by uncoated kaolinite was very low. Also the adsorption occurred only at low equilibrium concentration of selenite. At higher equilibrium concentrations, there was an apparent decrease in the total amount of selenite adsorbed and the amount adsorbed became negative with further increase in equilibrium concentration. The total amount of selenite adsorbed also decreased with increase in pH. Also there was a decrease in equilibrium concentration above which the apparent adsorption began to decrease. These results indicate that selenite reacted specifically with the surface sites. The surface sites, which could be involved in this reaction, were very limited and became saturated at relatively low equilibrium concentration. With increase in pH, there was a decrease in the number of sites which was reflected in decrease in total amount adsorbed, decrease in equilibrium concentration up to which positive adsorption occurred. The increase in negative adsorption with increase in pH

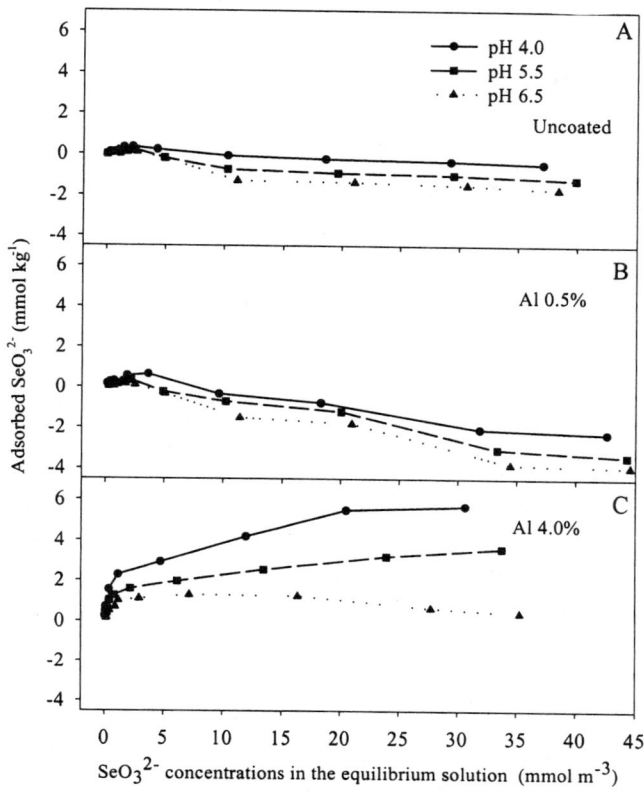

Fig. 2. Selenite adsorption on the kaolinite clay mineral coated with hydrous oxides of aluminum at the rates of 0%, 0.5%, and 4.0% Al; and at three levels of equilibrium pH 4.5, 5.5, and 6.5

indicates that there was an increase in the number of negatively charged sites. Coating of kaolinite surface with Al as well as Fe hydrous oxides increased the total amount adsorbed as well as the equilibrium concentration up to which the positive adsorption occurred (Figs. 1 & 2). Amounts of selenite adsorbed by both Al and Fe coated kaolinite samples were much higher than one would expect if the two phases existed independently of each other[23, 24]. Therefore, coating of kaolinite with small amounts of Fe and Al hydrous oxides produced new solid phases with adsorption characteristics markedly different from the pure forms of the either phase.

Although both Fe and Al hydrous oxides increased selenite adsorption by kaolinite, their effect was not the same (Fig. 3). In terms of molar concentrations the levels of Al hydrous oxide coating were higher, but Fe hydrous oxide coating increased the adsorption to a larger extent. This indicates that Al hydrous oxide coatings created lower number of ligand-exchageable sites than those created by equivalent amounts of Fe hydrous oxide coatings. Aluminum is known to polymerize in solution before complete precipitation[20]. Apparently, precipitation of polymeric species produces some sort of three-dimensional order on the surface of the clay mineral reducing the number of Al atoms exposed on the surface. This would in turn reduce the number of singly coordinated OH groups involved in ligand exchange mechanism of anion adsorption[17]. Gibbsite has been observed to adsorb lower amount of selenite compared to goethite[21]. Iron hydrous oxide coated clay samples have been shown to have higher surface area compared to Al hydrous oxide coated clay samples[16]. Another interesting observation was the negative adsorption of selenite on Al hydrous oxide coated samples (Fig. 2). At 0.5% coating level (Fig 2), the negative adsorption produced by Al hydrous oxide coated sample was higher than that produced by uncoated sample (Fig. 2). This suggests that surface of Al coated samples contains a higher amount of negative charge in the region of negative

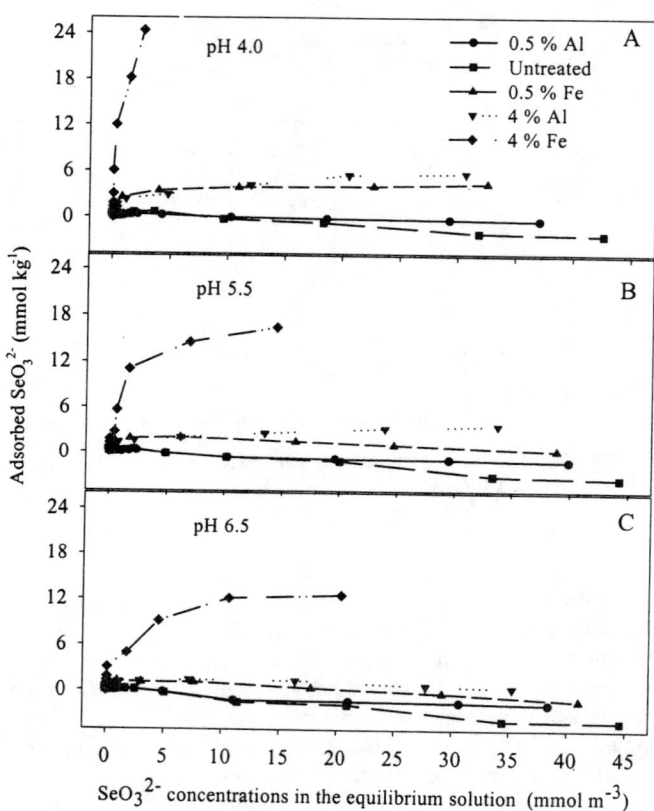

Fig 3. Adsorption of selenite on kaolinite coated with three levels of hydroxyl-oxides of
Fe and Al (0%, 0.5%, and 4.0%; and at three levels of equilibrium pH (4.0, 5.5, and 6.5)

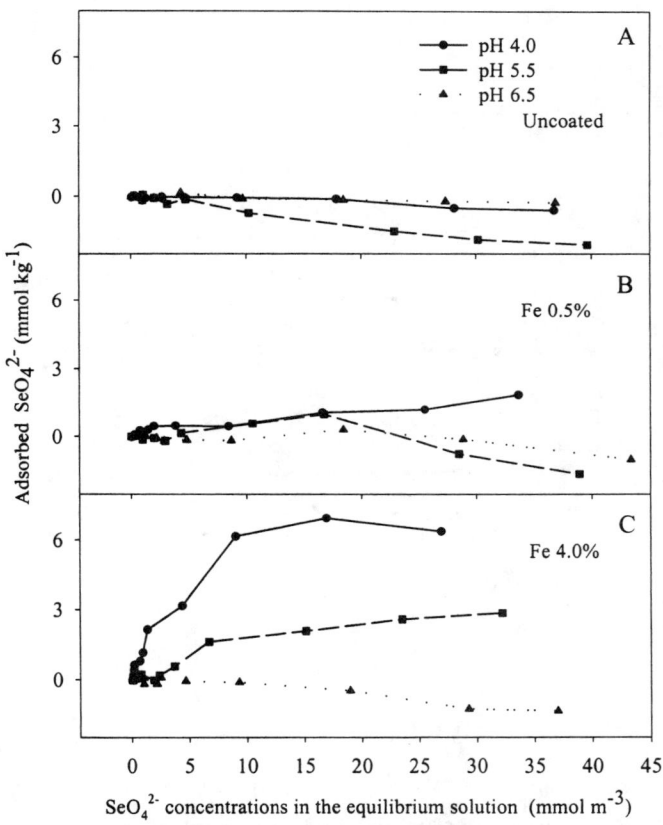

Fig. 4. Selenate adsorption on the kaolinite clay mineral coated with hydrous oxides of iron at the rates of 0%, 0.5%, and 4.0% Fe; and at three levels of equilibrium pH 4.5, 5.5, and 6.5

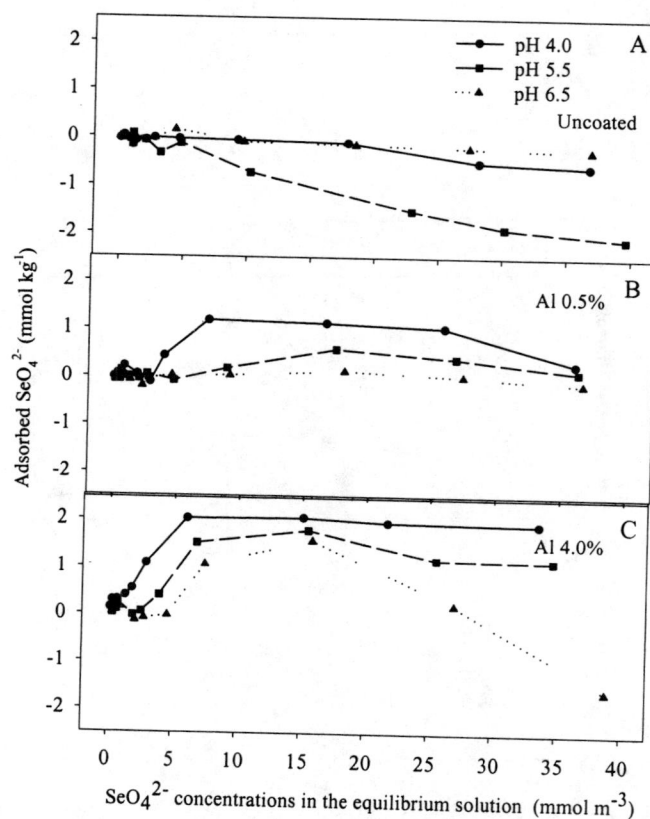

Fig. 5. Selenate adsorption on the kaolinite clay mineral coated with hydrous oxides of aluminum at the rates of 0%, 0.5%, and 4.0% Al; and at three levels of equilibrium pH 4.5, 5.5, and 6.5

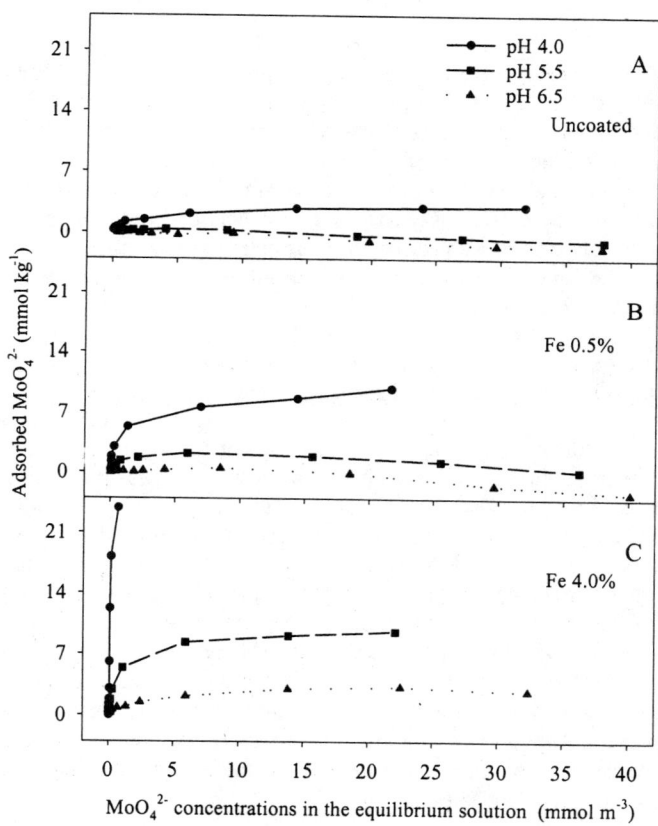

Fig. 6. Molybdate adsorption on the kaolinite clay mineral coated with hydrous oxides of iron at the rates of 0%, 0.5%, and 4.0% Fe; and at three levels of equilibrium pH 4.5, 5.5, and 6.5

adsorption compared with the uncoated sample. The only plausible explanation for this to happen seems to be that adsorption by Al coated sample imparted negative charge to the sites of adsorption. Since the Al hydrous oxide coated sample had more number of adsorption sites compared to those on uncoated sample, it could have higher amount of negative charge as well. If the adsorption should always produce negative sites, Fe coated sample would also show higher negative adsorption. However, this did not happen because negative adsorption produced by Fe coated sample was not higher than that produced by uncoated sample (Fig. 1). Specific adsorption of anions by hydroxylated surfaces can occur by three mechanisms described as schemes 1, 2, and 3 by Hingston et al. (1972)[11]. Schemes 2 & 3 convey a negative charge to the surface and scheme 3 can actually produce a negatively charged surface. Results of this investigation seem to suggest that in adsorption of selenite by Fe hydrous oxide coated samples schemes 1 and 2 predominate whereas in case of Al hydrous oxide coated samples scheme 3 plays an important part in addition to schemes 1 and 2 (Fig. 2).

The isotherms for the adsorption of selenate by kaolinite at different pH values and different amounts of Fe as well as Al hydrous oxide coatings are given in Figs. 4 & 5. The positive adsorption of selenate only occurred on surfaces with a net positive charge. There was virtually no adsorption of selenate on the uncoated mineral (Fig. 4) whose pH_o was 4.26. Actually, the surface charge characteristic data indicate that even the uncoated mineral should have a net positive charge at pH 4.0. Virtually zero adsorption of selenate suggests that the number of positively charged sites is very limited and Cl⁻ competes with selenate for adsorption at low concentrations of selenate. As the concentration of selenate increases, there is probably some adsorption of selenate, which is masked by the effect of negative adsorption.

Coating of surface with Al as well as Fe increased the net positive charge. This is reflected in adsorption of relatively higher amounts of selenate by the coated samples (Figs. 4 and 5). There was an initial decrease in the adsorption of selenate even by coated samples indicating that at very low selenate concentrations Cl⁻ competes with selenate for the adsorption sites. The increase in adsorption of selenate at higher concentrations, however, indicates that selenate has a greater affinity for positively charged hydroxylated surfaces than chloride. The concentration of Cl⁻ in the equilibrium solution was 0.05 M whereas that of selenate varied from 0.38 to 40 µM. Therefore, selenate probably adsorbs by ligand exchange but only with aqua groups occurring at positively charged sites on the hydroxylated surfaces. Infrared studies have shown SO_4^{2-} to be adsorbed by a ligand exhange mechanism at protonated surfaces[22]. Sulfate has also been shown to form binuclear complex with surfaces, which do not produce a negative charge and do not lower the ZPC. Results of this investigation indicate that selenate probably adsorbs by a mechanism similar to that for sulfate. This seems to be so because although the adsorption for selenate was less than that for selenite, the negative adsorption was also less for selenate (Figs. 1-5). This pattern also indicates that adsorption of selenite does impart a negative charge to the surface which has been observed in a number of studies with the pure hydrous oxides.

Adsorption data for molybdate is given in Figs. 6 & 7. Adsorption of molybdate exhibited a general trend similar to that for selenite (Figs. 1 & 2). However, adsorption of molybdate at pH 4.0 was invariably higher than that for selenite. The explanation for this may lie in the fact that pK_2 value for conjugate acid of molybdate is about 4.5. Therefore, aqueous species of molybdate at pH 4.0 will be a combination of $HMoO_4^-$ and MoO_4^{2-} representing a very favorable condition for specific adsorption on to a surface carrying a combination of protonated and neutral OH sites. Selenite will dominantly occur as $HSeO_3^-$ aqueous species, pK_1 and pK_2 of its conjugate acid being 2.75 and 8.5, respectively. Therefore, although the molar concentration of molybdate and selenate may be the same, at pH 4.0 concentration of adsorbing species will be higher for molybdate. This effect of the concentration of various aqueous species and the type of surface sites present is further demonstrated by a more drastic decrease in the adsorption of molybdate than selenite as the pH increases. With increase in pH concentration of the aqueous molybdate species will shift more towards MoO_4^{2-} with concurrent decrease in surface sites favoring the adsorption of this species. This dual effect will result in a drastic decrease in adsorption of molybdate with increase in pH.

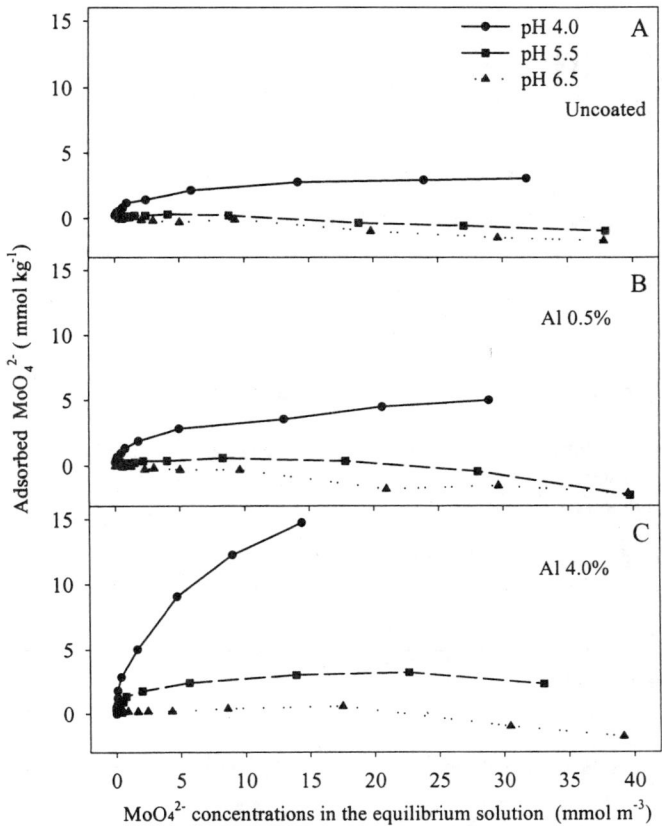

Fig. 7. Molybdate adsorption on the kaolinite clay mineral coated with hydrous oxides of aluminum at the rates of 0%, 0.5%, and 4.0% Al; and at three levels of equilibrium pH 4.5, 5.5, and 6.5

Increase in pH should increase the concentration of $HSeO_3^-$ aqueous species of selenite adsorption of which does not require the hydroxylated surface to be positive. Although adsorption of selenite will decrease with increase in pH simply because of decrease in the number of adsorption sites, the effect of pH should not be as drastic as for molybdate.

SUMMARY AND CONCLUSIONS

Investigation of the adsorption of Se (IV), Se (VI), and Mo by Fe and Al hydrous oxide coated kaolinite show that the general principles governing the adsorption process are same as those for pure forms of hydrous oxides. The effect of pH is more drastic than that for pure hydrous oxides because the charge characteristics of hydrous oxide coated clays and pure hydrous oxides markedly differ. Adsorption of selenite and molybdate occurred on hydrous oxide coated kaolinite above pH_o. Significant adsorption of selenate only occurred at pH values lower than pH_o of the concerned solid phase. Coating of kaolinite with Fe as well as Al hydrous oxide increased the anion retention capacity of the mineral. The effect was, however, more for Fe than Al hydrous oxide. Surface sites for the adsorption of anions were limited. High affinity between these sites and the anions was indicated by adsorption at extremely low equilibrium concentrations and very steep slope of the adsorption isotherms until the adsorption sites were apparently saturated. Adsorption of Mo and Se (IV) imparted negative charge to the surface whereas adsorption of Se (VI) didn't. Beyond the equilibrium concentrations where the surface adsorption sites were saturated, negative adsorption of all anions occurred. Results of this investigation show that adsorption will play a very important role in controlling the mobility of oxyanions in fly ash treated mine soils containing relatively high amounts of Fe and Al hydrous oxide coatings.

REFERENCES

1. Phung, H. T., Lund, L. T., and Page A. L., Potential use of fly ash as a liming material, in *D.C. Adriano and L. Brisbin (eds.) Environmental chemistry and cycling processes,* U.S. Dept. Energy, Augusta, GA, 504, 1978.

2. Taylor, E. M., and Shuman, G. E., Fly ash and lime amendment of acidic coal soils to aid revegetation, *J. Environ. Qual.,* 17, 120, 1988.

3. Adams, L. M., Capp, J. P., and Gillmore, D. W., Coal mine spoil and refuse bank reclamation with power plant fly ash, in *Proc. Third Mineral Waste Utilization Symp.,* U.S. Bur. Mines and I.I.T. Res. Inst., Chicago, IL., 105, 1972.

4. Martens, D. C., and Plank, C. O., Basic soil benefits from ash utilization, in *Proc. internat. ash utilization symp.,* U.S. Bur. Mines, U.S. Dept. Interior, Pittsburgh, PA., 269, 1973.

5. Capp, J. P., and Gillmore, N. W., Fly ash from coal burning power plants: An aid in revegetating coal mine refuse and spoil banks, in *First symp. on preparation plant and mine refuse disposal,* National Coal Assoc., Louisville, KY., Washington, DC., 200, 1974.

6. Capp, J. P., Power plant fly ash utilization for land reclamation in eastern United States, in *Reclamation of Drastically Disturbed Soils,* Am. Soc. Agron., Madison, WI., 339, 1978.

7. Fail, J. L., and Wochok, Z. S., Soybean growth on fly ash-amended strip mine spoils, in *Plant Soil,* 48, 473, 1977.

8. Keefer, R. F., Singh, R. N., Doonan, F., Khawaja, A. R., and Horvath, D. J., Application of fly ash and other wastes to mine soil as an aid in revegetation, in *Proc. fifth internat. ash utilization symp.,* Atlanta, GA., 840, 1979.

9. Singh, R. N., Bhumbla, D. K., Keefer, R. F., and Horvath, D. J., Improving crop production by altering chemical properties of mine land with industrial waste, in *Proc. internat. symp. on Nutrient management for sustained productivity,* Vol. 1, Ludhiana, India., Dept. Soils, Punjab Agric. Univ., 366, 1992.

10. Eary, L. E., Rai, D., Mattigod, S. V., and Ainsworth, C. C., Geochemical factors controlling the mobilization of inorganic constituents from fossil fuel combustion residues: 2, Review of the minor elements, *J. Environ. Qual.,* 19, 202, 1990.

11. Hingston, F. J., Posner, A. M., and Quirk, J. P., Anion adsorption by goethite and gibbsite, I, The role of the proton in determining adsorption envelopes, *J. Soil Sci.,* 23, 177, 1972.

12. McKenzie, R. M., The adsorption of molybdenum on oxide surfaces, *Aust. J. Soil Res.,* 21, 505, 1983.

13. Barrow, N. J., Reaction of anions and cations with variable-charge soils, *Adv. Agron.,* 38, 183, 1985.

14. Parfitt, R. L. Anion adsorption by soils and soil materials, *Adv. Agron.,* 30, 1, 1978.

15. Brown, A. D., and Jurinak, J. J., Mechanism of pyrite oxidation in aqueous mixtures, *J. Environ. Qual.,* 18, 545, 1989.

16. Hendershot, W. H., and Lavkulich, L. M., Effect of sesquioxide coatings on the surface charge of standard mineral and soil samples, *Soil Sci. Soc. Am. J.,* 47, 1252, 1983.

17. Parfitt, R. L., and Russell, J. D., Adsorption on hydrous oxides, IV, Mechanisms of adsorption of various ions on goethite, *J. Soil Sci.*, 28, 297, 1977.

18. Hingston, F. J., Posner, A. M., and Quirk, J. P., Anion binding at oxide surfaces-the adsorption envelopes, *Search,* 1, 324, 1970.

19. Hingston, F. J., A review of anion adsorption, in *M.A. Anderson, and A.J. Rubin (eds.) Adsorption of inorganics at solid-liquid interfaces*, Ann Arbor Science, Ann Arbor, MI., 51, 1981.

20. Hsu, P. H., Aluminum hydroxides and oxyhydroxides, in *J.B. Dixon and S.B. Weed (eds.) Minerals in soil environment*, Soil Sci. Soc. Am., Madison, WI., 99, 1977.

21. Hingston, F. J., Posner, A. M., and Quirk, J. P., Anion adsorption by goethite and gibbsite, II, Desorption of anions from hydrous oxide surfaces, *J. Soil Sci.*, 25, 16, 1974.

22. Parfitt, R. L., and Smart, R. Sc. T., The mechanism of sulfate adsorption on iron oxides, *Soil Sci. Soc. Am. J.*, 42, 48, 1978.

INFLUENCE OF COAL ASH/ORGANIC WASTE APPLICATION ON DISTRIBUTION OF TRACE METALS IN SOIL, PLANT, AND WATER

Yuncong Li[1], Min Zhang[1], Peter Stoffella[2], Zhenli He[2], Herbert Bryan[1]

[1]Tropical Research and Education Center, IFAS
University of Florida
Homestead, FL 33031
U.S.A. and
[2]Indian River Research and Education Center, IFAS
University of Florida
Ft. Pierce, FL 34954
U.S.A.

ABSTRACT

This study was conducted to evaluate effects of coal ash mixture (coal ash, biosolids and yard waste compost ratio of 1:1:1 v/v) on accumulation of trace metals in soil and their distribution in crop leaves and fruits, and its leaching potential into groundwater. Coal ash mixture was applied at rates of 0, 25, 75 Mg ha^{-1}. Samples of soil and tomato (*Lycopersicon esculentum* Mill. cv. 'Sanibel') tissue (leaves and fruits) were collected and analyzed for trace metals (Zn, Cu, Mn, Fe, Cd, Pb, Ni, and Mo). Zero-tension pan lysimeters were installed to monitor water quality. Application of the coal ash mixture significantly increased AB-DTPA extractable Fe, Ni and Mo in treated soils and concentrations of Mn and Mo in tomato leaves. Only concentrations of Fe and Mo in tomato fruits from plots treated with this coal ash mixture were greater than those from control plots. Application of 75 Mg ha^{-1} coal ash mixture significantly increased total amounts of Zn, Pb and Mo collected in lysimeter water during 12 months following application. Concentrations of trace metals analyzed in this study were very low. The maximum concentrations of Mn, Pb and Cd in lysimeter water samples from both treated and control plots were only occasionally greater than MCL (Manxmen Contaminant Level for drinking water). Therefore, appropriate application of coal ash mixture should not lead to any significant detriment to soil, food, and groundwater.

INTRODUCTION

Power plants in the United States in 2000 used over 857 million Mg (tons) of coal, and three of the 70 major electric utility coal consumers are located in Florida, namely Tampa Electric Co., Florida Power Corporation, and Gulf Power Company.[1] The data released by the American Coal Ash Association showed that over 108 million tons of coal combustion products including about 80 million tons of fly and bottom ash were produced in 2000. On the national level, less than 30 percent of coal ash produced was utilized. Agricultural use of coal ash was only 0.02 million tons per year, or about 0.02% of coal ash produced.[2]

Coal ash use as a soil amendment for agriculture has a large potential. There are a number of benefits that result from the application of coal ash to agricultural soils. The benefits include improvment of soil texture, modification of soil pH, and provision of essential plant nutrients for crop production.[3, 4] Townsand and Hodgson [5] reported that the particle fractions of coal ash samples ranged from 45-70% silt and 1-4% clay. The fine-sized ash particles should increase the total porosity of the soil. Ghodrati et al.[6] reported that moisture holding capacity in a soil amended with 30% fly ash was increased from 12% to 25%, and that the rate of water flow through the soil were reduced three-fold. Coal ash consists of over 40 elements, and most of the trace elements are essential nutrients for plant growth. Application of coal ash increases bioavailability to plants of nutrients in soils. Adriano et al.[7] studied the effects of coal ash on soil chemical properties, growth and mineral nutrition of corn (*Zea mays* L.) and bush bean (*Phaseolus vulgaris* L.), and reported that the application of fly ash increased concentrations of extractable K, Ca, Mg, Cu, Fe, Mn, and Zn in the soil, and concentrations of K and Fe in corn shoots, and K, Ca, Mg, S, and Cu in bush bean shoots. Similar results had been reported for barley (*Hordeum vulgare* L. var. 'Leduce'),[8] rice (*Oryza sativa* cv. 'PR 106'), wheat (*Triticum aestivum* L.),[9] tomato (*Lycopersicon esculentum* Mill. cv. 'Pusa Ruby')[10] and other vegetable crops.[11] Application of coal ash also increased content of boron in alfalfa (*Medicago sativa* L.)[12] in corn.[13] Therefore coal ash can be a source of nutrients for crops and an amendment for Florida's sandy or coarse soils.

Coal ash contains no, or very little, nitrogen and phosphorus, which are essential to crop growth. Mixing coal ash with an organic waste such as biosolids, animal manure, and compost has been found to supply N and P, and to improve the bioavailability of other nutrients.[14] Amendment of a soil with a mixture of fly ash and organic waste significantly increases nutrient uptake and yield of tall wheat grass (*Agropyron elongatum*).[15] Menon et al.[16] reported that application of fly-ash amended compost improves yields of collard greens and mustard greens grown on sandy loam soil, and increased the concentrations of K, Ca, Mg, S, Zn and B in amended soils. Soil microbial activity was increased by mixing fly ash and sewage sludge into soil.[17, 18] Probably the addition of organic materials to coal ash binds trace metals, and reduces their leaching potential.[19,20] The disposal of urban yard debris is a well-known and serious problem in Florida. Florida alone generates over 3 million tons of yard waste annually. The past methods of urban yard debris disposal were incineration or landfilling. Both of these methods are very expensive and environmentally unfavorable. Recently Florida enacted a state law that prohibits landfilling of yard waste. Available disposal alternatives for urban yard debris are mulch and compost production. Many pubic and private landfill operators in Florida, as well as elsewhere in the U.S., have expanded in yard debris mulching or composting facilities. Research has demonstrated that yard waste compost can serve as a soil amendment to increase organic matter, improve microbial activities in soil, provide nutrients, and ultimately improve plant growth and yield.[21]

Even the utilization of coal ash or the co-utilization with organic waste (biosolids and yard waste composts) can significantly improve soil fertility, however, environmental concern is always an issue in the land application of coal ash products. Of particular concern is the possibility of trace metal accumulation in treated soils [22] and the possible uptake and concentration in the edible parts of plants. Metals released from fly ash may be leached into groundwater.[19,20] Sometimes an environmental agency may require relevant assessments before the coal ash use can be permitted in agriculture. In U.S., coal combustion by-products are generally exempt from hazardous waste regulations and some states have elected to regulate these materials as solid, specific or industrial wastes. Under Florida regulations, coal ash and other wastes generated from the combustion of coal and other fossil fuels are exempt from regulation as hazardous waste. However, the state environmental agency is opposed strongly to the unregulated use of coal waste products. To meet this stricture it is necessary to conduct environmental monitoring, and to obtain a permit for experiment plots that will be treated with pure coal ash. To meet requirements in this project, we used commercially available coal ash products. Our research results may be of value to the regulatory agency in evaluate their practices.

The objective of this study was to evaluate the effects of co-utilization of coal ash and organic waste (biosolids and yard waste compost on the accumulation of selected metals (Zn, Cu,

Table 1. Chemical composition of coal ash/biosolids, yard waste, mixture and soil used in this study.

Materials or soil	pH	Organic C	N	Zn	Cu	Mn	Fe	Cd	Pb	Ni	Mo
		----- g kg⁻¹-----		----------------------- mg kg⁻¹ ------------------------							
Yard waste (YW)	7.4	154.9	9.2	184	33	67	1990	0.1	44	4	3
Coal ash/biosolids (CA/BS)	7.7	828	7.7	120	395	67	21533	1.3	15	39	9
Coal ash mixture CA/BS/YW	7.4	103.7	9.3	121	301	98	17400	1.3	23	31	8
Soil	7.6	28.9	0.9	79	537	623	20533	1.3	19	18	1

Mn, Fe, Cd, Pb, Ni, and Mo) in soil, their distribution in tomato leaves and fruits, and their potential to leach into groundwater.

MATERIALS AND METHODS

The field experiment was initiated in 1999 at the University of Florida, Tropical Research Education Center, Homestead, FL. The soil is a Krome very gravelly loam (loamy-skeletal, carbonatic, hyperthermic Lithic Udorthents) containing 58 %, 19 %, 15 % and 8 % gravel (>2mm), sand, silt and clay, respectively. The coal ash/biosolid product (composted mixture of coal ash and biosolids, 1:1) was obtained from the Ft. Meade processing facility, Florida-N-Viro, Sarasota, FL. Yard waste compost was collected from the City of St. Petersburg, FL. Coal ash product and yard waste compost were then mixed to produce the final material (coal ash mixture) with 1:1:1 ratio of coal ash, biosolids and yard wastes. Soil samples from the experimental site, coal ash/biosolids, yard waste compost, and final coal ash mixture were analyzed for pH (H_2O), organic carbon, total nitrogen and trace metals. Total carbon and nitrogen was determined using a Vario Max CNS analyzer (Elmentar Americas, Inc., Mt. Laurel, NJ). Inorganic carbon was analyzed with a titration method [23] and organic carbon was calculated as the differences between total carbon and inorganic carbon in the material. Total concentrations of metals were determined using the USEPA 3050A method. [24] Chemical composition of yard waste, coal ash product, the mixtures and the soil used in this study are shown in Table 1. Coal ash/biosolids, yard waste compost, final mixture have pH values similar to that of the soil used in this study. The final mixture had levels of organic C and N about 3.6 and 10 times higher, respectively, than those in the soil. Concentrations of Zn, Pb, Ni and Mo in the final mixture are higher than those in the soil, while concentrations of Fe, Cu, and Mn were lower in the coal ash mixture. Concentrations of trace metals analyzed in this study were far below the maximum concentrations specified in the USEPA 503 regulation for land application of biosolids. [25]

The experimental plots were laid out on a split plot design with 2 irrigation treatments as main plots (data not presented), and different rates of coal ash mixture (coal ash/biosolids/yard waste) as subplots with 4 replications. Subplots consisted of three rates of coal ash mixture (0, 25, and 75 Mg ha⁻¹ as dry weight basis). Traditional raised-beds with drip irrigation for vegetable production were established in December 1999. Beds were 180 cm from center to center with a width of 95 cm wide and a height of 15 cm. Coal ash composts were applied using a compost spreader. Materials were rototilled 10 cm deep and beds were then refinished and covered with plastic mulch. Tomatoes (cv. 'Sanibel') seedlings were transplanted on February 2000 in a single row in the center of each bed with 50 cm between plants. Tomatoes were managed with standard practices [26] and harvested in May 2000.

Soil samples were taken at 0-20 cm depth prior to tomato planting and after harvesting, air-dried and extracted with AB-DTPA (ammonium bicarbonate - diethylenetriaminepentacetic acid), which is a method to determine plant nutrients available in calcareous soils. [27] The extracts

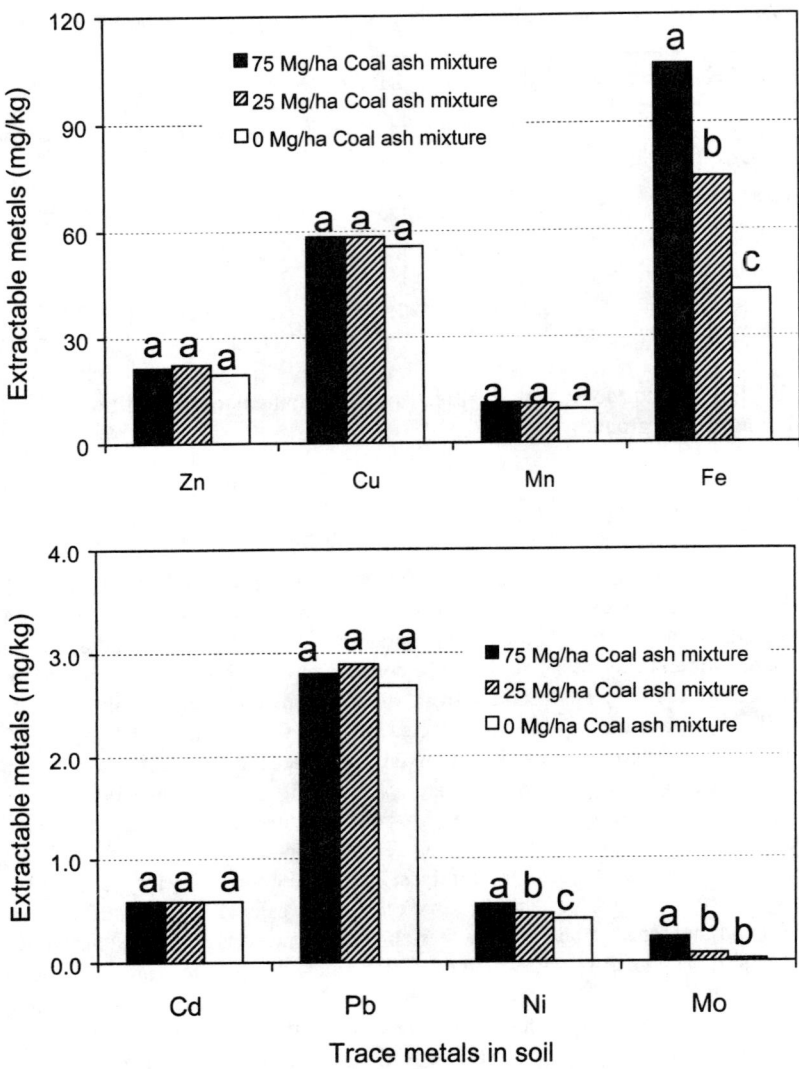

Figure 1. Ammonium bicarbonate-DTPA extractable trace metals in soil amended with coal ash mixture. Letters above the bars indicate significant differences between mean values of the same metal of the different rates of coal ash mixture at $P <0.05$.

were analyzed for selected metals (Zn, Cu, Mn, Fe, Cd, Pb, Ni, and Mo) using inductively coupled argon plasma spectroscopy (ICPAEC; Model 61E Thermo Jarrell Ash Inc., Franklin, MA).

Newly mature leaves were collected during period when the fruit was developing. Three tomato fruits from each plot were also collected on May 11, 2000. The tissue samples were washed, oven dried (70C for 72 hr), and analyzed for selected trace metals.

Thirty zero-tension pan lysimeters were constructed and installed in the field for water quality monitoring. A plastic cylinder with 34-cm dia. and 11 cm deep was used to collect rain or irrigation water, which leached down through the crop root zone. Water collected in the pan was drained through a plastic tube (I.D. 9.5mm) into a water collector. The latter was constructed from 17-cm PVC (I.D. 15 cm) and capped on both ends. Water samples were collected from lysimeters monthly. Water sample volume was measured immediately following each sampling and the solution was analyzed for selected trace metals.

All data were analyzed with SAS. [28] The analysis of variance (ANOVA) was conducted to check for differences in the concentrations of metals in soil, plant tissue and water samples. Means were separated with Duncan's multiple range tests.

Other elements (nitrogen, phosphorus, calcium, magnesium, potassium, aluminum, boron and silicon) in soil, plant tissue and water samples were also analyzed. Tomato numbers, total weight and color of fruit from each plot were also recorded. The latter data will be presented elsewhere.

RESULTS AND DISCUSSION

Metals in soils

Total amounts of AB-DTPA extractable Fe, Ni and Mo in the soils treated with 75 Mg ha^{-1} coal ash mixture were significantly greater than those for the treatment of 25 Mg ha^{-1} coal ash mixture and control (Fig 1). These results are affected by the concentrations of these metals in original soils and coal ash mixture. The concentrations of Ni and Mo in coal ash mixture were about 1.7 and 8 times higher than these in soil (Table 1).

Total Fe contents were very high in both untreated soil (20.5 g kg^{-1}) and coal ash mixture (17.4 g kg^{-1}). Soil Fe is mainly in the form of iron oxides and not readily extractable by AB-DTPA.[29] The application of a large amount (75 Mg ha^{-1}) of coal ash mixture directly increased AB-DTPA extractable iron. It is also possible that increasing soil organic matter through application of coal ash mixture increased iron availability. Iron deficiency is major problem for crops grown on calcareous soils. Using coal ash mixture could help to increase iron bioavailability. Sikka and Kansal [9] reported significantly increase of DTPA-extractable Fe in coal ash treated soils.

Coal ash mixture application did not increase AB-DTPA extractable Zn and Pb even though the total Zn and Pb in coal ash mixture are about 1.5 and 1.2 times, respectively, higher than those in soils. Total input of Zn and Pb were only 9 and 1.7 kg ha^{-1}, respectively, at the application rate of 75 Mg ha^{-1} of the coal ash mixture, which was not enough to increase bioavailable Zn and Pb in the soil. Likewise this rate of application of the coal ash mixture did not significantly increase the concentrations of extractable Cu, Mn, and Cd.

Metals in tomato leaves

The concentration of Mo in tomato leaves was 3 times higher in soils treated with coal ash compost than in the control (Fig. 2). The maximum concentrations of Mo in leaf samples were also much higher from amended soils than from control plots (Table 3). Plants accumulated more Mo in leaves because the AB-DTPA extractable Mo concentration had been elevated in soils amended with 75 Mg ha^{-1} coal ash mixture. Molybdenum is required for human and animals. Relatively high Mo levels in leaves of crops consumed by humans or animals may be nutritionally beneficial. The application of coal ash alone or mixed with organic waste has been reported to increase Mo concentrations in leaves of maize (Zea mays L.) [30] and of barley. [8]

The reduction of Mn concentrations in tomato leaves following the application of the coal ash mixture can not be explained by the levels of AB-DTPA-extractable Mn found in treated soils. Indeed there was no significant difference between the levels of AB-DTPA extractable Mn in treated and untreated soils. Nevertheless, the background level of Mn in untreated soil is more than six times higher than in the coal ash mixture (Table 1). This indicates that plants growing in treated soils take up more Mn than predicted based on levels extracted by AB-DTPA.

The treatments with the coal ash mixture did not significantly affect the concentrations of Zn, Cu, Fe, Cd, Pb, and Ni in tomato leaves, while others have reported increased concentrations of iron and decreased concentrations of Zn in tissue samples of rice and wheat. [9]

Table 2. Concentrations of selected trace metals extracted by AB-DTPA from soils collected after tomato harvest.

	Application rate of coal ash mixture (Mg ha[-1])		
	0	25	75
	------ mg kg[-1] ------		
Zn	21.3	22.3	19.5
Cu	58.4	58.3	55.6
Mn	11.3	11.1	9.46
Fe	106.3a[1]	75.0b	43.0c
Cd	0.58	0.58	0.58
Pb	2.80	2.89	2.68
Ni	0.55a	0.46bc	0.41c
Mo	0.23a	0.08b	0.02b

[1]Means followed by similar letter within each row are not significantly different at p=0.05 level of probability.

Table 3. Maximum concentrations of selected trace metals in tomato leaves collected during tomato growing season.

	Application rate of coal ash mixture (Mg ha[-1])		
	0	25	75
	------ mg kg[-1] ------		
Zn	17.6	30.4	22.4
Cu	25.6	22.0	23.6
Mn	171.6	114.8	81.2
Fe	80.4	84.0	88.0
Cd	0.80	0.80	0.80
Pb	2.00	0.40	0.80
Ni	0	2.80	1.60
Mo	0.80	4.40	2.40

Table 4. Maximum concentrations of selected trace metals in tomato fruits.

	Application rate of coal ash mixture (Mg ha[-1])		
	0	25	75
	------ mg kg[-1] ------		
Zn	23.2	28.0	27.2
Cu	16.8	13.6	12.8
Mn	23.2	16.4	15.6
Fe	32.4	41.6	37.2
Cd	0	0	0
Pb	0	0	0
Ni	0	0	0
Mo	0.40	0.80	0.80

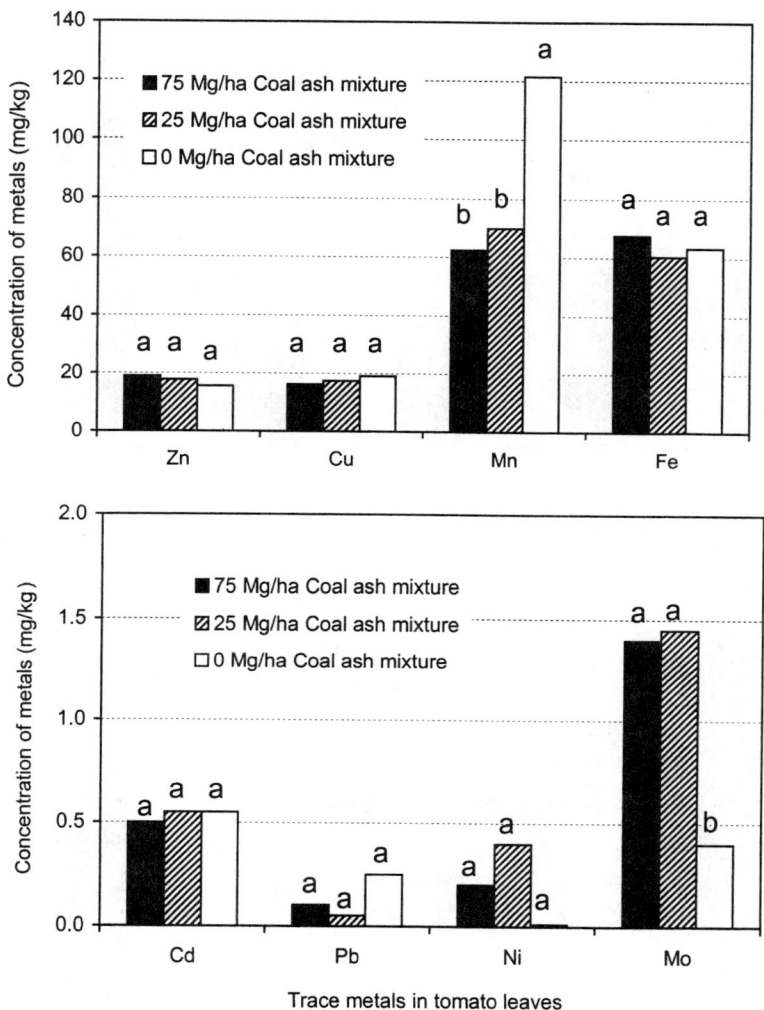

Trace metals in tomato leaves

Figure 2. Concentrations of trace metals in tomato leaves collected during the growing season. Letters above the bars indicate significant differences between mean values of the same metal of the different rates of coal ash mixture at $P < 0.05$.

Metals in tomato fruits

Fruits, a consumable part of plants, normally accumulate lower amounts of trace metals than other plant organs. However, the levels of trace elements in fruits pose the greatest concern for regulatory and consumers when crops are grown in soils amended with biosolids or coal ash. However we did not find significant changes of mean concentrations of metals, except those of Fe and Mo in tomato fruits following the application of the coal ash mixture (Fig. 3). Concentrations of Cd, Pb, and Ni in tomato fruits were below detection limits. Molybdenum in fruits from plots without coal ash mixture amendment ranged from 0-0.4 mg kg^{-1} with a mean of 0.2 mg kg^{-1}, while fruits from plants grown in soils amended with 75 Mg ha^{-1} coal ash mixture had minimum, maximum and mean Mo concentrations of 0, 0.8 and 0.67 mg kg^{-1}, respectively (Table 4 and Fig 3). There is currently no recommended daily allowance (RDA) for Mo in food. We also have no knowledge of the health effects of high Mo in fruits. Molybdenum is known to be essential for the proper functioning of certain enzyme-dependent processes, including the metabolism of iron. Molybdenum also forms part of several enzymes. Including those needed to

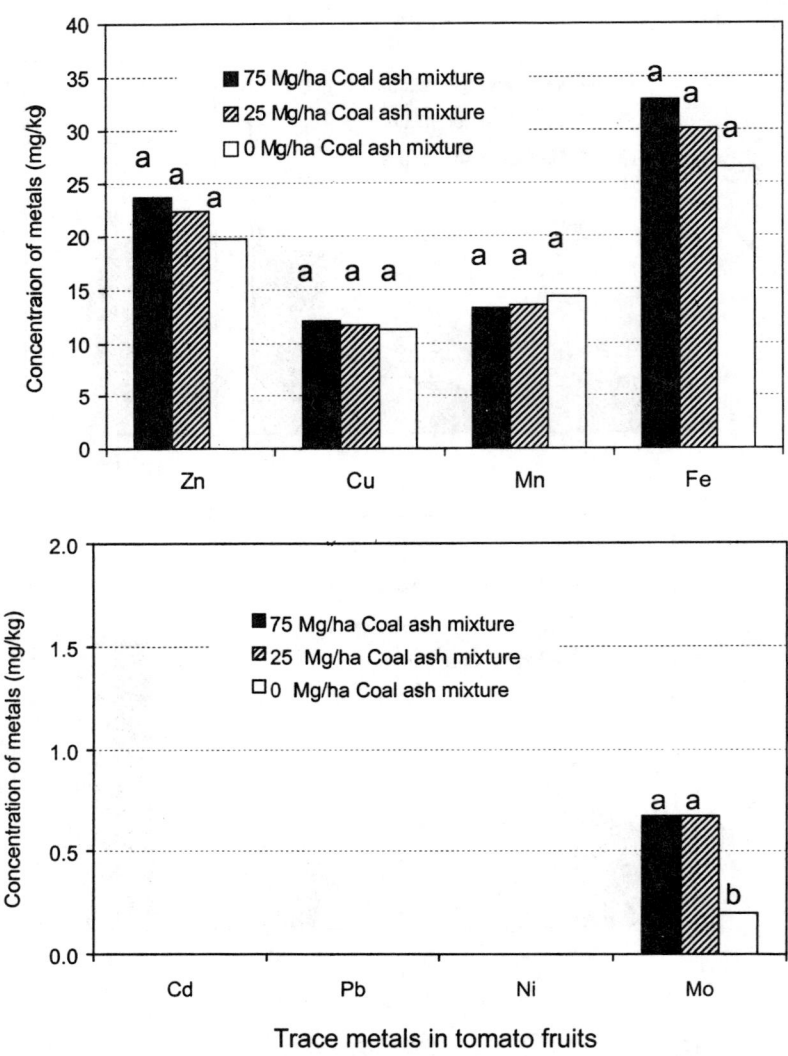

Trace metals in tomato fruits

Figure 3. Concentrations of trace metals in tomato fruits grown on a calcareous soil amended with or without coal ash mixture. Letters above the bars indicate significant differences between mean values of the same metal of the different rates of coal ash mixture at $P < 0.05$.

convert purine into uric acid. [31] The National Academy of Sciences [32] has suggested that between 75-250 µg/day of Mo are safe and adequate in the diet for a normal adult person, and this is equivalent to 0.11-0.37 kg of tomato, based on the mean concentrations of Mo in fruits from the 75 Mg ha[-1] treatment. Molybdenum toxicity is extremely rare in the United States. Neverthelss, Mo is one of the trace metals in edible part of crops that should be monitored carefully when coal ash products are used to amend agricultural soils.

Concentrations of another trace metal, Fe, were significantly increased from 26.5 mg kg[-1] (control) to 32.9 mg kg[-1] in tomato fruit following the application of 75 Mg ha[-1] of the coal ash mixture. Such enrichments of Fe in tomato fruit should be considered as beneficial in increasing of nutritional value of the produce.

Metals in lysimeter water

Cumulative amounts of metals in lysimeter water samples during the 12 months of the study ranged from 0.002 kg ha[-1] for Cu to 0.25 Mg ha[-1] for Zn (Fig. 4). The amounts of Zinc, Pb

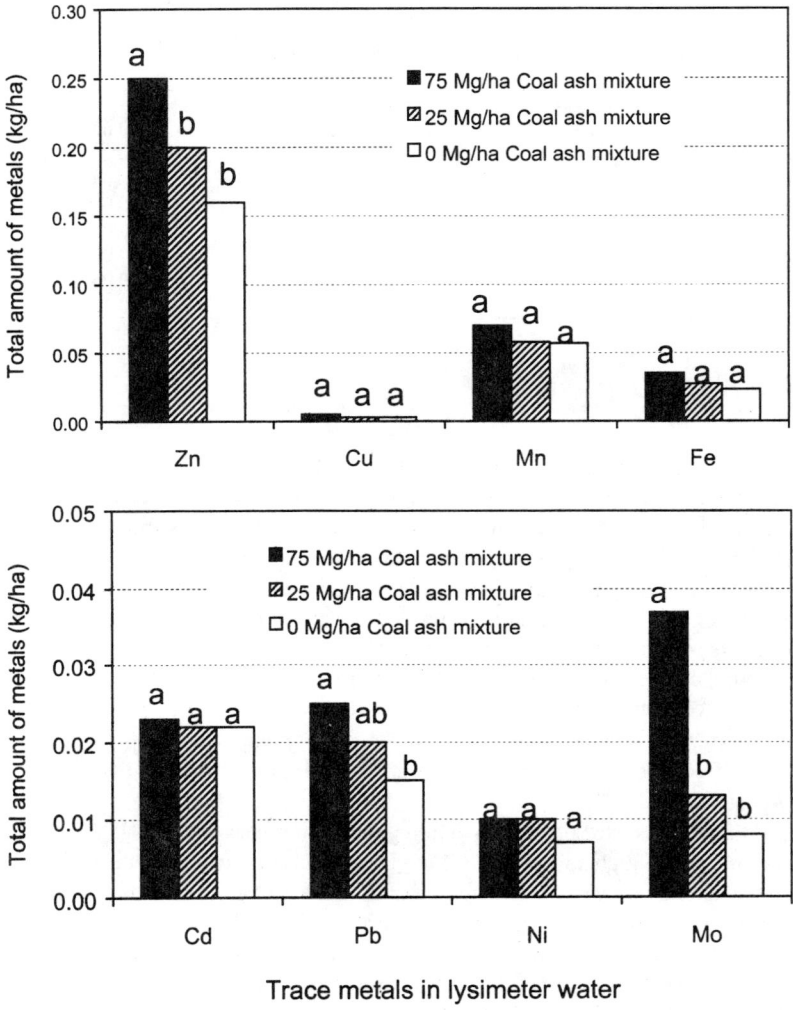

Trace metals in lysimeter water

Figure 4. Total amounts of trace metals in lysimeter water samples during 12 months. Letters above the bars indicate significant differences between mean values of the same metal of the different rates of coal ash mixture at $P < 0.05$.

and Mo leached from soils treated with 75 Mg ha^{-1} coal ash compost were significantly higher than those from untreated soils. However, concentrations of all metals in leachates were very low and most values were below the detection limits (data not presented). The maximum (peak) concentrations of trace metals from lysimeter samples are presented to show the possibility of leaching of these metals in to groundwater (Table 5). The maximum concentrations of Zn, Fe and Cu were below MCL (Maximum Contamination Level), which is the highest level of a contaminant that is allowed in drinking water in the US (Table 6). The experimental site is a Krome very gravelly loam soil, which is a typical calcareous soil with soil pH 7.6 and consists of 69% calcium carbonate. These metals probably were precipitated or adsorbed by the soil as they moved through soil profile down to groundwater. On the other hand, the maximum concentration of Mn in lysimeter water from control plots was 0.20 mg L^{-1}, while MCL for Mn is only 0.05 mg L^{-1}. Similar Mn concentrations in groundwater from area near the experimental site were reported previously. [33] The high levels of Mn in the soil and the bedrock probably are the cause of the high concentrations of Mn in groundwater. Concentrations of Mn in unfarmed natural calcareous soils in the area are about 11 times higher than the mean concentration of Mn in 448 Florida surface soils. [34] However, Mn is in secondary drinking water standards, which are not enforceable.

Table 5. Maximum concentrations of trace metals in lysimeter water.

	Application rate of coal ash mixture (Mg ha^{-1})		
	0	25	75
	mg L^{-1}		
Zn	0.30	0.15	0.02
Fe	0.19	0.20	0.20
Cu	0.02	0.31	0.04
Mn	0.20	0.02	0.01
Pb	0.02	0.01	0.01
Cd	0.01	0.01	0.03
Ni	0.01	0.53	0.06
Mo	0.73	0.10	0.16

Similar results were observed for Pb and Cd (Table 5 and 6). Coal ash mixture treatments neither affected the cumulative amounts, nor the maximum concentrations of these two trace metals. However, the concentrations of both metals in water samples from control plots and Cd from coal ash mixture treated plots exceeded the MCLs. Howie [33] reported that lead concentrations in the Biscayne aquifer beneath of the south Miami-Dade area ranged from <0.001 to 0.016 mg L^{-1}. Just as Mn in local water is a cause for concern, soil Pb and Cd are significant factors in the water quality in the area. The concentrations of Pb and Cd in untreated soils were about 4 and 325 times, respectively, greater than those in Florida surface soils. [35]

A peak Ni concentration of 0.53 mg L^{-1} was detected in water samples from the 25 Mg ha^{-1} coal ash mixture treatment (Table 5). However, the cumulative amounts of Ni in lysimeter water were not affected by the treatments (Fig 4). Nickel is not included in the current National Drinking Water Standards. [36]

Molybdenum was expected to move into groundwater because of its high concentration in coal ash mixture and the high soil pH. [30,37] Molybdenum is anionic and more soluble at high pH. The cumulative amount of Mo in lysimeter water from the 75 Mg ha^{-1} treatment was about 4.6 times greater than in lysimeter water from the control (0.037 vs. 0.008 kg ha^{-1}) (Fig. 4). However, the highest Mo peak was detected in water samples from control plots (Table 5). Jackson et al. [30] reported that the concentration of Mo in lysimeter water under soil treated with fly ash and poultry litter was as high as 0.15 mg kg^{-1}. Mo is also not regulated under the current drinking water standards.

Table 6. Current drinking water standard for trace metals in US. [36]

	MCL/MCLG[1]	Secondary standard[2]
	mg L^{-1}	
Zn		5
Fe		0.3
Cu	1.3/1.3	1.3
Mn		0.05
Pb	0.015/zero	
Cd	0.005/0.005	

[1]MCL (Maximum Contaminant Level) is the highest level of a contaminant that is allowed in drinking water and is an enforceable standard; MCLG (Maximum Contaminant Level Goal) is the level of a contaminant in drinking water below which there is no know or expected risk to health and is a non-enforceable public health goal.
[2]EPA recommends secondary standards for drinking water.

SUMMARY

Under the conditions of this study, the application of a coal ash mixture (25 or 75 Mg ha^{-1}) to a calcareous soil increased AB-DTPA extractable Fe, Ni and Mo in soil. The high concentrations of Ni and Mo in the coal ash mixture increased their bioavailability in the treated soil. With respect to Fe, organic components in the coal ash mixture could be a reason for substantial increases in the amount extractable with AB-DTPA. Iron availability is very important to crops grown on calcareous soils. Leaf analyses showed increased concentrations of Mo and Mn as a result of soil amendment with the coal ash mixture. Concentrations of Mo and Fe in tomato fruits were elevated through application of the coal ash mixture. But these two elements are not considered to cause heath problems. A high application rate of coal ash mixture leads to increased cumulative amounts of Zn, Pb, and Mo in lysimeter water compared to the control. However, concentrations of these trace metals usually are relatively low, occasional peaks exceeded the MCL. These peak concentrations occurred both in treated and control plots.

ACKNOWLEDGMENTS

This research was supported by the Florida Agricultural Experiment Station and a grant from the U.S. Department of Energy, Federal Energy Technology Center, through it's Cooperative Agreement No. DE-FC26-98FT40028 with West Virginia University Research Corporation. However, any opinions, findings, conclusions, or recommendations expressed herein are those of the authors and do not necessarily reflect the views of WVU or DOE. University of Florida Agricultural Experiment Station Journal Series No. R-08676.

REFERENCES

1. USDOE (U.S. Dept. of Energy), *Annual Electric Generator Report, Electric Power Mor..hly*, DOE/EIA-0226 (01/03), Energy Information Administration, Washington, DC. March 2001.
2. ACAA (American Coal Ash Association), *CCP Production and Use Survey – 2000*, American Coal Ash Association, Syracuse, NY., 2001.
3. Bilski, J.J., Alva, A.K., and Sajwan, K.S., Fly ash, in *Soil Amendments and Environmental Quality*, Reckcigl, J.E., Ed., CRC Press, Inc., Boca Raton, Fla., 1995, 327.
4. Korcak, R. F., Utilization of coal combustion by-products in agriculture and horticulture, in *Agricultural Utilization of Urban and industrial By-Products*, Karlen, D.L., Wright, R.J., Kemper, W.D., Eds., American Society of America (Special Publication No. 58), Madison, WI, 1995, 107.
5. Townsand, W.N. and Hodgson, D. R., Edaphological problems associated with deposits of pulverized fly ash, in *Ecology and Reclamation of Devastated Land, Vol. I.*, R.J. Hutuik and Davis, G., Eds., Gorden and Breach, London, 1973.
6. Ghodrati, M., Sims, J. T., and Vasilas, B. L., Evaluation of fly ash as a soil amendment for the Atlantic coastal plain: II Soil hydraulic properties and elemental leaching. *Water Air Soil Pollu.* 81, 349, 1995.
7. Adriano, D. C., Woodford, T. A., and Ciravolo, T.G., Growth and elemental composition of corn and bean seedlings as influenced by soil application of coal ash. *J. Environ. Qual.* 7, 416, 1978.
8. Sale, L. Y., Naeth, M. A., and Chanasyk, D. S., Growth response of barley on unweathered fly ash-amended soil, *J. Environ. Qual.*, 25, 684, 1996.
9. Sikka, R. and Kansal, B.D., Effect of fly-ash application on yield and nutrient composition of rice, wheat and on pH and available nutrient status of soils. *Bioresource Technology*, 51, 199, 1995.
10. Khan, M. R., Khan, M. W., The effect of fly ash on plant growth and yield of tomato. *Environ. Pollution.*, 92,105, 1996.

11. Cary, E.E., Gilbert, M., Bache, C.A., Gutenmann, W.H., and Lisk, D.J., Elemental composition of potted vegetables and millet grown on hard coal bottom ash-amended soil, *Bull. Environ. Contam. Toxicol.*, 31, 418, 1983.

12. Mulford, F.R. and Martens, D.C., Response of alfalfa to boron in fly ash. *Soil Sci. Soc. Am. Proc.* 35, 296, 1991.

13. Kukier, U. and Sumner, M. E., Boron availability to plants from coal combustion by-products. *Water Air Soil Pollu.* 87, 93, 1995.

14. Schumann, A. W., and Sumner, M. E., Chemical evaluation of nutrient supply from fly ash-biosolids mixtures, *Soil Sci. Soc. Am. J.*, 64, 419, 2000.

15. Wong, J.W. C. and Su, D. C., The growth of *Agropyron elongatum* in an artificial soil mix from coal fly ash and sewage sludge. *Bioresource Tech.* 59, 57, 1997.

16. Menon, M.P. Sajwan, K. S., Ghuman, G. S., James, J., and Chandra, K., Elements in coal and coal combustion residues and their potential for agricultural crops, in *Trace elements in coal and coal combustion residues,* R.F. Keefer and Sajwan, K. S., Eds., Lewis Publishers, Chelsea, MI. 1993, 259.

17. Pichtel, J.R. Microbial respiration in fly ash-sewage sludge amended soils. *Environ. Pollu.* 63, 225, 1990.

18. Wong, J.W. C. and Lai, K. M., Effect of an artificial soil mix from coal fly ash and sewage sludge on soil microbial activity. *Bio. Fert. Soils* , 23, 420, 1996.

19. Alva, A.K., Paramasivam, S., Prakash, O., Sajwan, K.S., Ornes, W.H., and Van Clief, D., Effects of fly ash and swage sludge amendments on transport of metals in different soils, in *Biogeochemistry of Trace Elements in Coal and Coal Combustion Byproducts*, Sajwan K.S., Alva, A.K., and Keefer, R.F., Eds., Kluwer Academic/Plenum Publishers, New York, 1999a, 193.

20. Alva, A. K., Bilski, J. J., Sajwan, K. S., and Van Clief, D., Leaching of metals from soils amended with fly ash and organic byproducts, in *Biogeochemistry of Trace Elements in Coal and Coal Combustion Byproducts*, Sajwan K.S., Alva, A.K., and Keefer, R.F., Eds., Kluwer Academic/Plenum Publishers, New York, 1999b, 193.

21. Li, Y.C., Stoffella, P., and Bryan, H.H., Management of organic amendments in vegetable system production in Florida. *Soil Crop Sci. Soc. Florida Proc.* 59, 17, 2000.

22. Jackson, B.P., Miller, W.P., Soil solution chemistry of a fly ash-, poultry litter-, and sewage sludge-amended soil, *J. Environ. Qual.*, 29, 430, 2000.

23. NSSC (National Soil Survey Center). Soil Survey Laboratory Methods Manual. *Soil Survey Investigation Report*, No. 42. USDA-NRCS, Washington, D.C. 1996.

24. USEPA (U.S. Environ. Protection Agency), Acid digestion of sediments, sludges, and soils, *USEPA SW-S846*; Ch 3.2 method 3050A; U.S. Environmental Protection Agency, Gov, Print Office, Washington, D.C., 1990.

25. USEPA, Part 503 standards for the use or disposal of sewage sludge. *Fed. Regis.*, 58, 9387, 1993.

26. Li, Y.C., Bryan, H.H, Klassen, W., Lamberts, M., and Olczyk, T., Tomato production in Miami-Dade County, Florida, *UF Coop Ext. Ser. Fact Sheet*, 2002.

27. Hanlon, E. A., Schaffer, B., Ozores-Hampton, M., and Bryan, H.H., Ammonium bicarbonate-DTPA extraction of elements from waste-amended calcareous soil, *Commun. Soil Sci. Plant Anal.* 27, 2321, 1996.

28. SAS Institute. *SAS User's Guide*, SAS Institute, Cary, NC., 1996.

29. Zhou, M. and Li, Y. C., Phosphorus-sorption characteristics of calcareous soils and limestone from the southern Everglades and adjacent farmlands. *Soil Sci. Soc. Am. J.*, 65, 1404, 2001.

30. Jackson, B. P., Miller, W. P., Schumann, A. W., and Sumner, M. E., Trace element solubility from land application of fly ash/organic waste mixture, *J. Environ. Qual.*, 28, 639, 1999.

31. Sardesai, V. M., Molybdenum: an essential trace element. *Nutr. Clin. Pract.*, 8, 277, 1993.

32. National Academy of Sciences, *Recommended Dietary Allowances*, 10th ed. Washington, D.C., National Academy Press, 1989.

33. Howie, B., Effects of dried wastewater-treatment sludge application on ground-water quality in south Dade County, Florida, *Water-Resources Investigations Report* 91-4135, U.S. Geological Survey, Denver, CO., 1992.
34. Chen, M., Ma, L. Q., and Li; Y.C., Concentrations of P, K, Al, Fe, Mn, Cu, Zn, and As in Soils from South Everglades. *Soil Crop Sci. Soc. Florida Proc.* 59, 124, 2000.
35. Chen, M., Ma, L. Q., and Harris, W. G., Baseline concentrations of 15 trace elements in Florida surface soils, *J. Environ. Qual.* 28, 1173, 1999.
36. USEPA, *Current drinking water standard*, Office of Water, U.S. Environmental Protection Agency, Gov, Print Office, Washington, D.C., 2002.
37. Fleming, L. N., Abinteh, H. N., and Inyang, H. I., Leaching pH effects on the leachability of metals from fly ash, *J. Soil Contamination*, 5, 1, 1996.

THE EFFECT OF FLUE GAS DESULFURIZATION RESIDUE ON CORN (*Zea mays* L.) GROWTH AND LEACHATE SALINITY: MULTIPLE SEASON DATA FROM AMENDED MESOCOSMS

T. Punshon, J.C. Seaman, and D.C. Adriano

Savannah River Ecology Laboratory, University of Georgia, Drawer E, Aiken, SC 29802, USA.

1. ABSTRACT

The environmental effects of applying a weathered flue gas desulfurization residue (FGD) to soil was monitored in a mesocosm experiment conducted over several cropping periods. Dry biomass and elemental composition of crop plants were measured, as well as the quality and chemical composition of soil and leachate collected from treated mesocosms. Data collected in the first year following FGD amendment showed no effect on the germination of corn (*Zea mays* L. var Dekalb DK-683), soybean (*Glycine max* L. Merr. Var. Haskell Pupa 94), radish (*Raphanus sativus* L. var. Sparkler), and cotton (*Gossypius hirsutus* L. var. Deltapine 51) and a significant stimulation in biomass. Metal and metalloid enrichment of plant tissues, specifically As, B, Se and Mo was also significant. Application of FGD residue drastically altered the pH of the soil and the salinity of the leachate. Studies were continued into a second season to monitor the duration of beneficial and deleterious effects arising from FGD amendment, as it is expected that the majority of environmental effects will occur in the initial season following application. Second season data using a monoculture of corn showed no significant stimulation or inhibition of biomass, in contrast to findings of the initial year. Concentrations of metals and metalloids within plant tissues in the second season were lower, although still elevated above control concentrations. Electrical conductivity of mesocosm leachate; elevated from a control level of 0.05 dS m^{-1} to 3.4 dS m^{-1} with addition of 100 tons FGD acre^{-1}, was still elevated 550 days after application. Repeated monitoring of leachate salinity showed evidence of only a slight decline 928 days after application. Soil data collected at the end of the second growth season showed that Se concentration had fallen below detection limits, and levels of As had also fallen by approximately 29%. The duration of environmental effects from FGD residue application can be summarized in terms of their half-lives, i.e. the length of time required for a 50% reduction in altered environmental parameters toward control levels. When half-lives for plants, soil and leachate are compared, the increase in soil pH and leachate salinity have the longest half-lives, and stimulation of plant biomass the shortest.

2. INTRODUCTION

There is a need to investigate the re-use of coal combustion by-products on a more realistic scale, both in terms of the size of experimental systems and the duration over which their effects are monitored. Previous studies have attempted to identify

Chemistry of Trace Elements in Fly Ash, edited by Sajwan *et al.*
Kluwer Academic/Plenum Publishers, 2003

potential environmental effects of coal combustion by-products (CCBs) using small scale growth chamber or greenhouse experiments[1-10], however, comparative studies using the same FGD residue material, applied to the same soil but using different experimental systems (i.e. greenhouse compared to large mesocosm units) have given conflicting results.[11] It is likely that the elucidation of many environmental side effects such as leaching of soluble salts and potentially toxic elements through the soil profile may require larger systems, realistic weather conditions and longer periods of time to allow more accurate predictions to be made.

The re-use of excess CCBs, specifically application to land, has been the focus of a considerable body of research[2, 12-16]. Of the diverse products arising from coal processing and combustion[9], FGD residue is currently of increasing significance because of more stringent air quality standards[17, 18]. Due to its relative novelty amongst the family of CCBs, potential re-use opportunities of FGD residues have not been fully realized; only about 10% of FGD materials are beneficially re-used[17]. In the past, material has been treated as waste and landfilled, although exhaustion of landfill space the elevation of and associated elevated costs have made the search for beneficial uses of high-volume waste materials is a priority.

FGD residue (the solid material arising from the removal of SO_x from the flue gas of coal fired power plants) generally contains Ca-S containing compounds[19]. There are two main types of FGD by-product, based on the type of desulfurization process used; namely wet or dry FGD. Wet FGD residue, produced in the highest quantity, is usually dewatered via centrifugation prior to use and stabilized with fly ash. Dry FGD residues are produced from technologies developed for retrofitting existing coal burning plants, the most common of which is fluidized bed combustion desulfurization. The common element to both procedures is the introduction of a Ca based sorbent into the flue gases, which reacts with the SO_x, removing it from the gas stream. In wet extraction slaked lime or limestone slurry is used, whereas dry extraction generally uses hydrated lime [$Ca(OH)_2$], limestone ($CaCO_3$) or dolomite [$CaMg(CO_3)_2$]. In the latter case, a variable product results[17].

FGD has the potential to provide essential plant nutrients, in that it contains varying – often appreciable – concentrations of calcium sulfate ($CaSO_4 \cdot 2H_2O$)[20] and can therefore have similar effects as agricultural gypsum when applied to land[17]. It is a particularly useful amendment for alleviating problems with excessively low soil pH[21], as well as reducing P solubility in situations where excessive P in the soil leaches out into the surrounding water bodies causing eutrophication[22]. Application of FGD can increase water infiltration in compacted soils and improve the aggregate stability of sodic soils[22]. Other additional benefits include pozzolanic and anhydrate reactions, buffering capacity of fly ash mixtures and as a possible source of Mg[17].

The resolution of longer-term environmental side effects resulting from FGD application can at present only come from long-term field experiments. Work is underway to develop a rapid screening method that will preclude the need for time-consuming experimental trials, and simulate soil-specific results under conditions of long-term weathering[23]. Until a satisfactory extraction technique is developed, however, the variability of FGD materials and the soil to which it is applied require that studies evaluating its safety be conducted on a case-by-case basis. FGD residue composition is influenced by factors such as the source of the coal, the stabilization, treatment and storage procedures, as well as the type of flue gas scrubbing system employed. Furthermore, innovations in the flue-gas scrubbing systems are likely to change the quality of FGD residues for some time to come[24]. It is now well known that FGD materials can bring about effects ranging from the inhibition of germination and growth in crop plants, to the enrichment of plant tissue, soil and groundwater with potentially toxic elements, usually As, B, Se and Mo. As the material weathers, however, both hazards and benefits are depleted[17], and studies have not shown conclusively the comparative rate at which these important changes once again approach levels seen in unamended soil in a naturally weathered (i.e. field mesocosm or field study) system. This work describes second season data from a long-term mesocosm experiment in terms of observed changes in beneficial and deleterious environmental effects in comparison with first season data[11, 19].

3. METHODS AND MATERIALS

3.1. Mesocosm establishment

Fresh FGD residue was collected from a coal-fired power plant in Cope, South Carolina. The material was air-dried and mixed with locally obtained Orangeburg soil (fine-loamy, siliceous, thermic, Typic Paleudult) in a large volume soil homogenizer to obtain treatment mixes which were equivalent to 0, 25, 50, 75 and 100 tons FGD acre^{-1}, (0, 55.5, 111.1, 166.6 and 222.2 Mg FGD ha^{-1}). The Orangeburg soil had a pH$_{water}$ (1:1) of 5.4 (±0.7), 0.5% organic C with 84.8% sand, 10.9% silt, and 4.2% clay. Mesocosm units were constructed from galvanized iron cattle tanks measuring 2.4 m in diameter and 60 cm deep, equivalent to an area of 4.67 m^2 or 0.0012 acres[19]. FGD material used in this study was a low-grade gypsum product, containing 2% S and 9% ash, produced from a dry scrubber process (*see* Punshon et al., 2001 for elemental composition). The material was particularly enriched with B: 6.068 mg kg^{-1} (water soluble) with 42.15 mg kg^{-1} in total (HF + aqua regia extraction).

Each FGD-soil mix was represented by four randomly arranged replicates. After they were set up, the units were allowed to equilibrate for 4 mo, during which times samples of mesocosm leachate were periodically collected from ports fitted at the base of each unit for pH and electrical conductivity (EC) determination. This monitoring was carried out following rainfall events 9, 27, 43, 84, 86, 96, 121, 550 and 928 days after initial incorporation. Details of first season planting activities and subsequent growth and elemental composition are given in Punshon et al.[19].

Mesocosms were established in March 1998, with second season planting commencing in June 1999 involving planting units with a monoculture of corn (*Zea mays* L. var. Dekalb DK-683) at a rate of 30 individual plants per mesocosm. The mesocosms

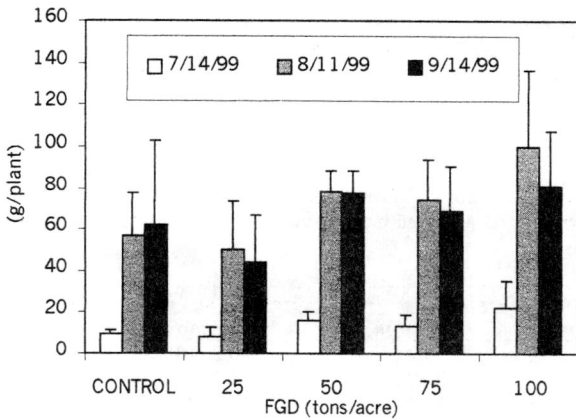

Figure 1. Biomass (g DW plant^{-1}) of corn (*Zea mays* L.) grown on FGD-amended soil in the second growth season. (Data are means ± SD where $n = 4$).

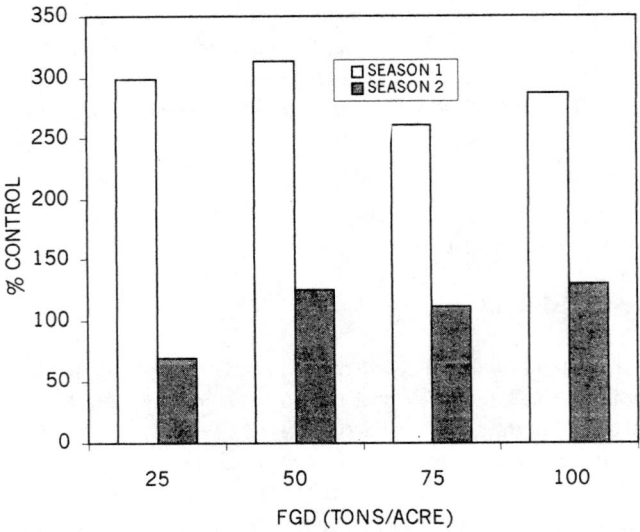

Figure 2. Comparative growth of corn (*Zea mays* L.), expressed as % of control growth in season 1 and 2.

were destructively harvested to obtain elemental composition of plant tissues on three occasions; July 14, August 11 and September 14, 1999.

3.2. Monitoring Methodology

Soil and leachate pH were determined using an Orion 250+ glass electrode, using milli-Q water and soil at a ratio of 1:1. Elemental composition of all environmental samples was determined using a Perkin Elmer Sciex Elan Inductively Coupled Plasma-Mass Spectrometer. Soil and FGD residue extractions determined total [HF (40% v/v) + H_3BO_3] and extractable [0.1 M HNO_3][25] elements. B concentration was determined using a hot-water extraction technique[2].

At each of the three destructive harvesting dates, corn plants were severed at the root-shoot junction and separated into leaf (+ petioles) and kernels, weighed to determine dry biomass before oven-drying (48h @ 60°C). Material was gound to a fine powder (1mm stainless steel screen) and digested in 10 ml HNO_3 (+ 30% H_2O_2) by microwave (CEM Corporation, MDS-2000) in Teflon™ PFA vessels.

Table 1. Concentration (mg kg^{-1} DW) of boron, selenium and arsenic in kernels of (*Zea mays* L.) grown in FGD-amended mesocosm soils for a period of 3 mo. (Data are means ± SD, where n=4).

FGD (t acre^{-1})	------------------------------mg kg^{-1} DW ------------------------		
	Boron	**Selenium**	**Arsenic**
0	0.62 (0.5)	0.26 (0.3)	0.10 (0.08)
25	2.39 (0.6)	1.79 (1.6)	0.25 (0.1)
50	2.07 (1.0)	1.87 (1.0)	0.13 (0.01)
75	4.10 (2.7)	2.27 (0.9)	0.16 (0.05)0
100	2.10 (0.2)	3.42 (0.6)	0.16 (0.02)

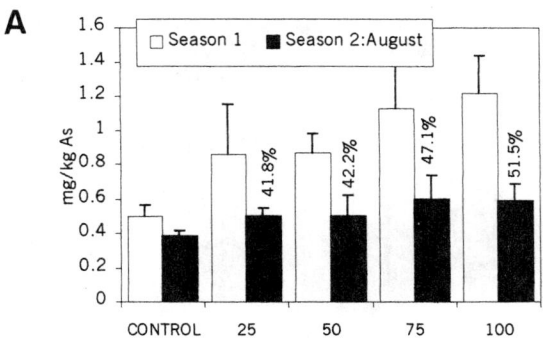

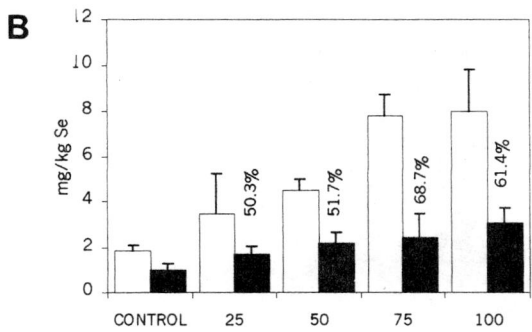

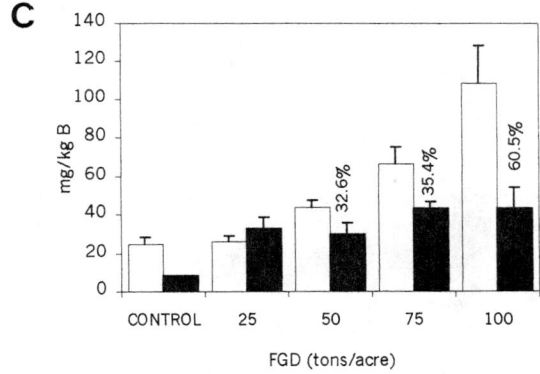

Figure 3. Concentration of As (3A), Se (3B) and B (3C) within corn tissues grown in FGD amended soil in season 1 and season 2 (mg kg^{-1}). Data are means ± SD, where n=4.

4. RESULTS

4.1. Growth of *Zea mays* L.

Growth of corn plants harvested on the three successive harvesting dates is shown in Figure 1, and indicates a slight, but not statistically significant stimulation in biomass as compared to control plants. In all FGD treatments with the exception of the control biomass in the third successive harvest are lower than the second. Growth of corn in season one was approximately 300% of control plants, compared to a maximum of 130% in the second season (Figure 2).

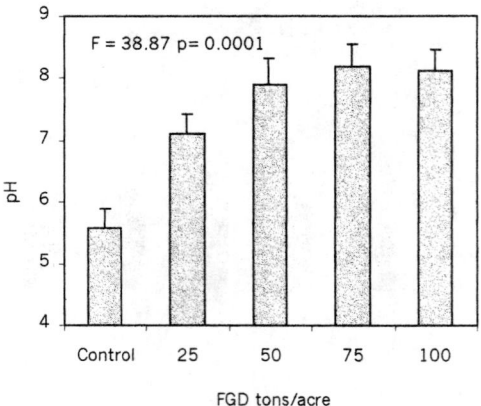

Figure 4. pH of FGD-amended mesocosm soils taken at the beginning of season 2. Data are means ± SD, where n=4, with one-way ANOVA notation.

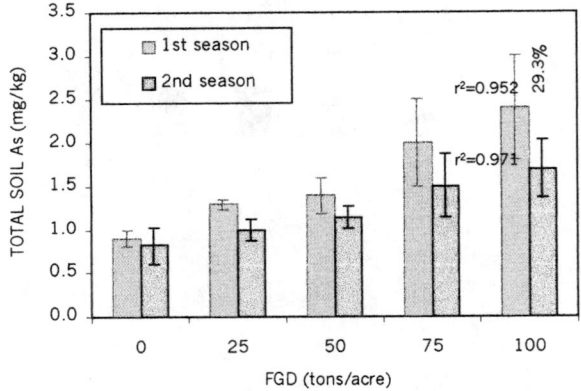

Figure 5. Total concentration of arsenic (mg kg^{-1}) in mesocosm soils in season 1 and 2 following amendment with FGD residue. (Data are means ± SD where n=4).

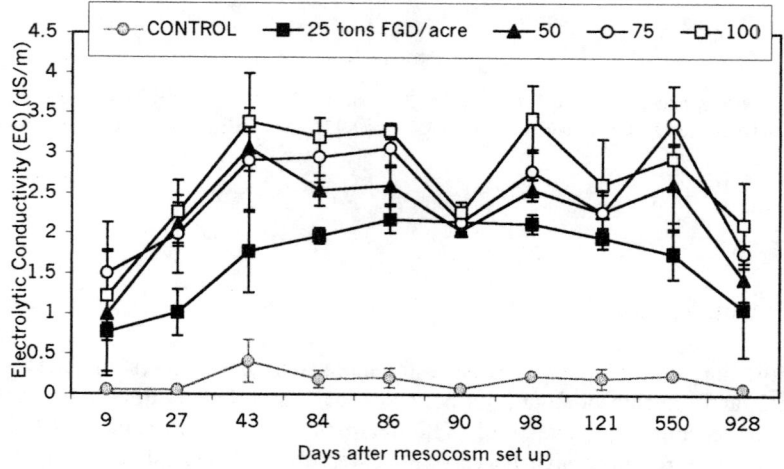

Figure 6. Electrical conductivity of leachate collected from FGD amended mesocosms (Data are means ± SD where n=4).

4.2. Elemental Composition of Plant Tissue

Corn grown in FGD-amended soil in season two contained higher concentrations of As, Se, Mo, B and Ca in the leaf tissue than control plants, with higher concentrations of Se, Ca, and B also detected within the kernels of plants grown in FGD amended soil (Table 1). When first and second season leaf tissue concentrations are compared, the concentration of As, Se and B fall considerably; on average the concentration of enriched metals fell by between approximately 30 – 60%. In general, levels of trace elements showed the greatest reduction between seasons 1 and 2 in the 100 tons FGD acre^{-1} application rate, with a ≈ 50% drop in As (Figure 3A) and ≈ 60% for both Se and B at this level (Figure 3B and 3C).

4.3. Soil Quality

Analysis of mesocosm soils for total elemental composition at the end of the second growing season showed a reduction in soil Se concentration to below detection limits (0.6 µg kg^{-1}), and a reduction in As. Total As (mg kg^{-1}) content of amended soils the preceding year are compared (Figure 4). Reduction in soil As concentrations were between 18 – 29%, with the highest reduction observed at the highest rate of FGD amendment. Other developments in soil chemistry as a result of weathering were a substantial decrease in the concentration of Na in the soil, and a slight increase in Mg. All other elements were not significantly different from control soil. Soil pH remained highly elevated at 8.1 (Figure 4) in the second growth season.

4.4. Leachate Quality

Initial monitoring of mesocosm leachate indicated a substantial rise in EC, although no concurrent changes in leachate pH. Elevated leachate salinity was maintained throughout the second growth season, and began to show indication of a downward trend 928 days after treatment. Figure 6 shows the progress of leachate salinity throughout the entire test period.

Leachate quality immediately after application of FGD material showed enrichment with B and Se, and second season data indicate a decline in concentrations of both elements This decline appears more pronounced for Se, where leachate collected from 25 and 50 tons FGD acre^{-1} contain no more Se than control mesocosms after 927 days (Figure 7A). FGD material appeared to raise the B concentration of leachate water to a maximum 121 days after application, and although this level had dropped by 893 days, it was still greatly elevated above control concentrations. After 928 days of exposure, B content for the different FGD residue application rates were not significantly different from one another (Figure 7B), although the 75 and 100 tons acre^{-1} contained the highest levels of B, with approximately 0.62 and 0.72 mg L^{-1} respectively. The concentration of soluble salts such as Ca and K in the leachate was also elevated by FGD addition, and these levels fell steadily from approximately 94 days onward.

5. DISCUSSION

The application of FGD materials to land intended for crop production carries with it obvious benefits and drawbacks. Research presented here supports the majority of studies performed on FGD residues in that the application of these products greatly elevates soil pH, soluble salt content of leachate and the concentration of potentially toxic

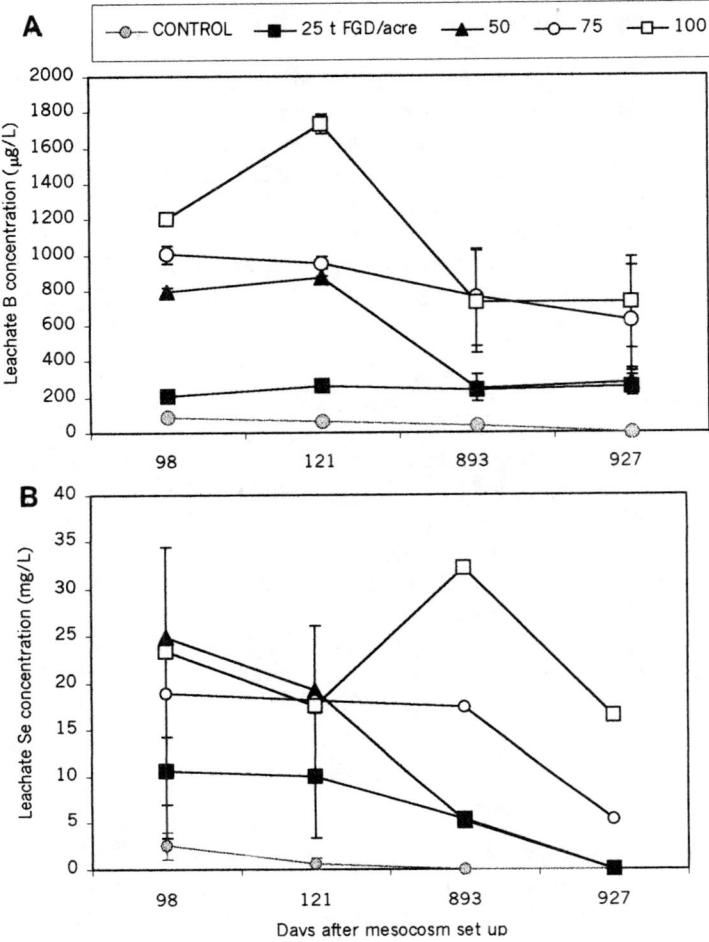

Figure 7. Se (A) and B (B) concentration in sequential mesocosm leachate samples (μg L^{-1})
(Data are means ± SD where n=4).

trace elements within the tissues of crop plants grown on the amended soil. Evaluation of these effects in order to guide the decision making process with respect to safety and recommended application rates must take into account the relative length of time over which changes in environmental quality are likely to remain. The use of FGD by products in place of commercial gypsum is entirely feasible when the purpose of the application is to ameliorate problems associated with the compaction of sodic soils (i.e. soil particle dispersion)[22]. The material used in this case provided a stable elevation of soil pH that was not observed to fall within the duration of the study. However, effects which would further support beneficial use e.g. an observed fertilizer effect - rapidly disappeared, leaving behind considerable enrichment of soil, leachate and plant tissue with As, Se and B, with varying rates of depletion. Short term growth chamber or greenhouse studies may misinterpret the use of the material in this case as being suitable for use as a fertilizer, because of the immediate boost to crop growth, probably associated with the addition of Ca and S, and other micronutrients such as Cu and Zn, which could lead to misuse. In terms of the length of time taken to observe a 50% reduction in the environmental effect of FGD application on plant tissues, soil and leachate, far the most persistent changes in environmental quality occur in the soil and leachate – taken as an indication of groundwater effects.

This suggests that FGD residues used for the correction of mineral deficiencies in plant tissues may be effective only over a relatively short period of time, and that repeated application of FGD residues may present problems in the increase of the metal load of the soil and salinity of the groundwater. Application rates used in this mesocosm study are far higher than those generally reported by other workers[10], and agronomically feasible application rates are far lower, and these high application rates may explain the extent of the increase in soil pH (Figure 4). Some concern must be attached to the enrichment of the soil with As, and the relatively slow rate at which it is depleted – at about 30% per year following application. Arsenic was not detected within the leachate water, and therefore the assumption may be that it remains bound to the soil, where a fraction is still available for plant uptake. Selenium – present in the soil in the first year also appears to have moved through the soil profile to a degree where it is no longer above ICP-MS detection limits, although it remains present in the leachate.

The beneficial re-use of FGD products by application to land remains a viable option, although with the current grade of by-products, care should be taken to weigh positive and negative effects, and the different periods of time over which they persist. Risks associated with long term enrichment with potentially toxic trace elements, and enhancement of groundwater salinity remain a pertinent issue in the safety of FGD residue application.

AKNOWLEDGEMENTS

The authors would like to acknowledge the laboratory assistance of Brad Reinhard. This research was supported by Financial Assistance Award Number DE-FC09-96SR18546 from the DOE to the University of Georgia Research Foundation.

REFERENCES

1. Sale, L.Y., Naeth, M.A., and Chanasyk, D.S., Growth response of Barley on Unweathered Fly Ash Amended soil. *J. Environ. Qual.*, 25, 684-691. 1996.
2. Kukier, U. and Sumner, M.E., Boron availability to plants from coal combustion by-products. *Water, Air & Soil Pollution*, 87, 93, 1996.
3. Kukier, U., Sumner, M.E., and Miller, W.P., Boron release from fly ash and its uptake by corn. *J. Environ. Qual.*, 23, 596-603. 1994.
4. Walker, W.J. and Dowdy, R.H., Elemental Composition of Barley and Ryegrass Grown on Acid Soils Amended with Scrubber sludge. *J. Env. Qual.*, 9, 27, 1980.
5. Stehouwer, R.C., Sutton, P., Fowler, R.K., and Dick, W.A., Minespoil amendment with dry flue gas desulpfuization by-products: Elemental solubility and mobility. *J. Environ. Qual.*, 24, 165, 1995.
6. Stehouwer, R.C., Sutton, P., and Dick, W.A., Transport and plant uptake of soil-applied dry flue gas desulfurization by products. *Soil Science*, 161, 562, 1996.
7. Clark, R.B., Zeto, S.K., Ritchey, K.D., and Baligar, V.C., Boron accumulation by maize grown in acidic soil amended with coal combustion by-products. *USDA Publication*, 1995.
8. Clark, R.B., Zeto, S.K., Ritchey, K.D., and Baligar, V.C., Boron accumulation by maize grown in acidic soil amended with coal combustion products. *Fuel*, 78, 179, 1999.
9. Carlson, C.L. and Adriano, D.C., Environmental impacts of coal combustion residues. *J. Environ. Qual.*, 22, 227, 1993.
10. Sloan, J.J., Dowdy, R.H., Dolan, M.S., and Rehm, G.W., Plant and soil responses to field-applied flue gas desulfurization residue. *Fuel*, 78, 169, 1999.

11.	Punshon, T., Knox, A.S., Adriano, D.C., Seaman, J.C., and Weber, T.J., *Flue Gas Desulfurization residue (FGD): Potential Applications and Environmental Issues.*, in *Biochemistry of Trace Elements in Coal and Coal Combustion Byproducts.*, K.S. Sajwan and R.F. Keefer, Eds., Lewis Publishers: Boca Raton, FL., 1999, 7.

12.	Adriano, D.C., Page, A.L., Elseewi, A.A., Chang, A.C., and Straughan, I., Utilization and disposal of fly ash and other coal residues in terrestrial ecosystems. A review. *J. Environ. Qual.,* 9, 333, 1980.

13.	Brown, J., Ray, N.J., and Ball, M., The disposal of pulverized fuel ash in water supply catchment areas. *Water Res.*, 10, 1115, 1976.

14.	Cervelli, S., Petruzzelli, and Perna, A., Fly ashes as an amendment in cultivated soils. I. Effect on mineralisation and nitrification. *Water, Air & Soil Pollution*, 33, 331, 1987.

15.	Wright, R.J., Kemper, W.D., Millner, P.D., Power, J.F., and Korcak, R.F., *Agricultural Uses of Municipal, Animal, and Industrial Byproducts,* . 1998, US Department of Agriculture. p. 127.

16.	Punshon, T., Adriano, D.C., and Weber, J.T., *Restoration of Eroded Land Using Coal Fly Ash and Biosolids.,* . 1999, Electrical Power Research Institute: Pal Alto, CA.

17.	Dick, W.A., Hao, Y.-L., Stehouwer, R.C., Bigham, J.M., Wolfe, W.E., Adriano, D.C., Beeghly, J., and Haefner, R.J., *Beneficial Uses of Flue Gas Desulfurization By-Products: Examples and Case Studies of Land Application.*, in *Beneficial Uses of Land Applied Agricultural, Industrial and Municipal By-Products,* 1999, .

18.	Santhanam, C.J., Lunt, R.R., Johnson, S.L., Cooper, C.B., Thayer, P.S., and Jones, J.W., Health and Environmental impacts of increased generation of coal fly ash and FGD sludges. *Environmental Health Perspectives*, 33, 131, 1979.

19.	Punshon, T., Adriano, D.C., and Weber, J.T., Effect of flue gas desulfurization residue on plant establishment and soil and leachate quality. *J. Environ. Qual.*, 30, 1071, 2001.

20.	Miller, W.P., *Environmental Considerations in Land Application of By-Product Gypsum.* in *Agricultural Utlization of Urban and Industrial By-Products.*, American Society of Agronomy: Madison, WI., 1995, 183.

21.	Chen, L., Dick, W.A., and Nelson, S., Flue gas desulfurization addition to acid soil: alfalfa productivity and environmental quality. *Environ. Pollut.*, 114, 161, 2001.

22.	Clark, R.B., Ritchey, K.D., and Baligar, V.C., Benefits and constraints for use of FGD products on agricultural land. *Fuel*, 80, 821, 2001.

23.	Knox, A.S. and Ziemkiewicz, P.F. Accelerated procedure for assessing the risks associated with amending agricultural soils with flua gas desulphurization solids. in *Sixth International Conference on the Biogeochemistry of Trace Elements*. University of Guelph, ICOBTE, 2001, 241.

24.	Baege, R. and Sauer, H., Recent developments in CFB-FGD technology. *VGB Powertech*, 80, 57, 2000.

25.	Alloway, B.J., *Heavy metals in soils.* 2nd ed, London: Blackie Academic & Professional. 1995.

USE OF COAL COMBUSTION BY-PRODUCTS (CCBP) IN HORTICULTURAL AND TURFGRASS INDUSTRIES

Maxim J. Schlossberg, William P. Miller, and Stanislaw Dudka[1]

[1]University of Georgia
Department of Crop and Soil Sciences
Athens, GA 30602-7272 U.S.A.

ABSTRACT

Coal combustion by-products (CCBP) include fly ash, bottom ash (cinders) and various desulfurization by-products. They contain plant nutrients, have variable capacities to neutralize soil acidity, and may improve physical properties of mineral soils. They have been usefully applied in agricultural, horticultural, turfgrass and land reclamation settings. However, inherent traits of CCBP such as bulkiness, excess trace metal concentrations, inconsistent availability of P, and low content of N and K make CCBP an untenable fertilizer supplement. Likewise, utilization of municipal biosolids (sewage sludge) can be problematic due to trace metal levels and undesirable nutrient ratios. Therefore, this study was initiated in early 2000 to determine the feasibility of blended CCBP and biosolids/biosolid products for use as growth media for horticultural ornamentals and turfgrass sod. Trace element concentrations in mixes used for both soil amendment and sod media were below USEPA regulatory limits. In the sod production component, growth media were uniformly spread to heights of 2, 3, and 4 cm on compacted subsoil, sprigged with bermudagrass [*Cynodon dactylon* (L.) Pers. x *C. transvaalensis* Burrt-Davy var. 'TifSport' (formerly Tift 94)], and maintained under ideal commercial sod field conditions. Following a maturation period of 99 d, sod was harvested and installed at the Georgia Experiment Station in Griffin. Remaining sod were destructively analyzed for determination of their physicochemical attributes. Field data collected from the ornamental beds showed yield and quality of flowers grown on CCBP-amended soil to outperform the commercially-amended soil under limited fertility conditions. Post-installation evaluations of sod made in April, 2001 did not reveal significant differences in rooting strength by mixture or sod thickness. All finished CCBP-containing sod retained significantly more volumetric water ($\Psi_m < -80$ kPa), while possessing 26-39% less gross (wet) weight than the control mix sod. The finished sod grown in selected combinations of bottom ash, fly ash, and biosolids possessed significantly greater biomass than the control sod mix, while requiring less supplemental fertilization. Utilization of the described CCBP-mixes as supplemental growth media in bermudagrass sod production was successful and may be a significant advantage when compared to some SE US soils. These experimental observations, in tandem with similar published results, indicate that utilization of CCBP in horticulture and turfgrass industries is technically feasible and environmentally-sound.

INTRODUCTION

Coal is the fossil fuel used most widely in US energy production. Current levels of carbon-based fuel reserves presuppose continued coal use for hundreds of years. Eighty percent of coal combustion by-products (CCBP) generated by the US electric utilities are comprised of fly ash, bottom ash, and boiler slag. Total annual production of fly ash (FA) in the US was 6 x 10^7 Mg, and bottom ash (BA) was approximately 1 x10^7 Mg in 1998.[1]

Fly ashes (FA) have been classified based on their elemental composition into two categories: class-C and class-F.[2] Class-C ashes are derived from sub-bituminous coals (mined in the western US) and are commonly low in S and high in base cations and alkalinity. This makes class-C fly ash more highly pozzolanic and therefore more valuable as a cement additive. These fly ashes (FA) also possess measurable calcium carbonate equivalencies (CCE) and are sometimes used as lime-substitutes. Class-F ashes are derived from bituminous coals and are high in Fe and relatively low in base cations. Some class-F fly ashes possess regulated metal concentrations which limit their potential use. It is important to remember that either class fly ash can be generated anywhere, depending only on the coal fuel source. Class-F fly ash (FA) is sometimes used in structural fill and other engineering applications. Unused FA is often impounded in lagoons or landfills.[1] Increasing regulatory pressure and elevated costs of CCBP disposal methods have stimulated discovery of innovative and practical CCBP applications.

Possibly the greatest limitation to incresed use of CCBP is its distinction for being chemically amorphous and spatially- and temporally-variable. Fly ash composition has been reported to vary with the parent coal from which it was derived, conditions during its combustion, efficacy of emission control devices, storage and handling of the byproducts and climate.[3] An investigation of ashes of differing geographic region reports wide value ranges for elemental concentrations and pH.[4] In a study of 29 FA samples collected from power plants throughout the SE US, results showed samples collected from a single power plant possessed a range of physicochemical attributes as varied as samples taken from different power plants.[5]

Physical characteristics of CCBP are also diverse. Sand-sized components of bottom ash (BA) (0.05-2.0 mm diam.) are relatively inert, possess comparatively low trace metal concentrations and have lighter densities than sand, potentially minimizing shipping and handling weight of consumer products containing them. Nearly all FA is silt-sized (2-50 µm diam.) with greater specific surface area and bulk densities than its coarser BA counterpart.

Numerous studies have evaluated FA as an alkalinity source/lime supplement. In a recent investigation, two acidic alfisols were amended with two alkaline FA at rates equivalent to 12.4, 49.4, and 123.6 Mg ha^{-1} (0.5-5.0 % v/v). These amendments resulted in significant increases in soil pH. At the highest application rate, soil pH of the less buffered soil increased 3 units and the pH of the more-clayey soil 2 units.[6] A less-recent greenhouse study examined the effect of 5 and 10 % (by mass) fly ash additions versus traditional liming of three acidic fine sands. Pre-mixed soil-filled pots were sowed with transplanted corn (*Zea mays* L.) seedlings. The higher rate of FA significantly raised soil pH above the control in all the fine sands as did the lower rate in 2 of 3 fine sands. The lime application made at 0.3% (by mass) also significantly raised all the soil pH levels above the control.[7] Positive correlations (r=0.92) between the CCE of fly ash and the final pH of soils receiving an 80 Mg ha^{-1} application have been reported.[8] Efficacy of acid neutralization relies on the addition of a 'high CCE' FA to an inherently acidic soil.

Coal combustion byproducts applied to soil to replace phosphorous and micronutrient fertilizers has been reported to correct nutritional deficiencies of P, Mg, Ca, S, Mo and B.[8-13] Delivery of certain plant essential elements (such as B) is generally more dependable than for others. For example, an examination involving sizable additions of pulverized FA containing 0.2% P did not result in a significant increase in the inorganic P fraction of treated soils.[14] In a field study of clover (*Trifolium subterraneum* var. Dalkeith), fly ash (FA) (0.5% P and 0.3% K) applications equivalent to 0, 5, 10, 20, 50 and 100 Mg ha^{-1} were combined with complete fertilizer, fertilizer minus P, fertilizer minus K, and no fertilizer split-plot treatments. Main-plot effects showed the 100 Mg ha^{-1} application of FA significantly increased dry matter production by an average of 56% at all sites. Results showed very little K was taken up into plant tissue unless provided in the complete fertilizer application, while tissue P showed a significant increase with increasing levels of FA.[15]

Mixtures of CCBP and organic waste products have been more reliable in consistently providing the primary nutrients needed to support agronomic crops. Sajwan et al. found sewage sludge/FA mixtures in ratios of 4:1, 2:1, 4:3, or 1:1, applied at rates of 124 to 248 Mg ha[-1] to be beneficial to the growth and nutritional status of cereals.[16] Results from a greenhouse study of corn indicated FA mixed with poultry manure sometimes produced more dry matter than FA and sewage sludge. Plant tissue analysis indicated the comparatively elevated levels of K and N in the poultry manure were responsible for the observed biomass increase.[8]

CCBP additions to soil have also been made for the purpose of improving soil physical properties. Because many CCBP possess silt to sand-sized particle diameters, they have been used for improving capillary porosity in sandy soils, and for increasing air-filled porosity in fine-textured soils. Studies investigating these attributes of coal ash date back to at least 1967.[17] A recent field experiment tested tillage of a class-F FA into an acidic, excessively drained sand. Rates of 316, 632, and 950 Mg ha[-1] FA significantly raised the water-holding capacity in the plow depth of the treated soil. Infiltration rates decreased dramatically (to approximately 80% of control plots) on those same soils.[18] A similar study reported sizable increases in capillary porosity of both fine and coarse sands to which FA had been pre-mixed. However, this slight gain was at the expense of non-capillary pores and saturated hydraulic conductivity.[19] Field and laboratory studies have shown modification of turfgrass soils with sintered fly ash to assist in drainage and infiltration rates of fine-textured turfgrass soils. The high-temperature sintering process pelletized fine fly ash into coarser, porous aggregates (similar to bottom ash). Sintered fly ash additions to the top 15 cm of soil at rates of 33% (v/v) resulted in reduced capillary porosity, but increased infiltration rate.[20]

Ostensibly, CCBP have practical agronomic value under specific conditions. Mixtures of CCBP and biosolids appear to result in a synergistic tandem of slow-release primary nutrients and exchangeable micronutrients. Unfortunately, these materials are considered low-analysis fertilizers and are expensive to apply. Additionally, trace metals associated with both societal waste products can accumulate in soils and potentially contaminate groundwater. The Clean Water Act limits cumulative applications of materials possessing measurable regulated metal concentrations. This restriction favors utilization of CCBP/biosolid/biosolid products in agronomic or horticultural applications where the soil media is concomitantly exported with the final product. This would include the 'soil' growth media of either container-grown horticultural plants or turfgrass sod. Additionally, morphology and growth habit of particular turfgrasses make them well-suited candidates for stabilization (vegetation) of CCBP land applications. Turves rarely amass high quantities of B in leaf tissue because B accumulates apically, and leaf tips are frequently removed by mowing practices. Bermudagrass (*Cynodon dactylon* L.) is an aggressive turf and adapted to a wide range of soil types, generally performing best on fine-textured soils with high fertility and available moisture. The salt tolerance of bermudagrass is considered good.[21]

An extensive investigation of an acidic, class-F fly ash application to a sod production field was conducted in the SE US.[22] Fly ash was applied to a Congaree silt loam (fine-loamy, mixed, active, nonacid, thermic Udifluvents) at rates equivalent to 0, 280, 560, and 1120 Mg ha[-1]. The FA material was tilled into the soil and seeded with centipedegrass [*Eremochloa ophiuroides* (Munro) Hack.] 2 d following. This planting was later abandoned because of germination failure in the treated plots. Soil tests indicated elevated soluble salts, particularly boron. The plots were re-tilled 8 months following and seeded again. In the following establishment period, diminished growth and vigor continued to be recorded as long as 57 days after planting on the plots treated with the two highest FA rates. The foremost-observed benefit of the FA applications was increased water retention at the 10 kPa tension level. Soil moisture at tensions >75 kPa were not measured. The authors asserted the phytotoxic effects of 560 and 1120 Mg ha[-1] application rates diminished 1 or more years following the initial application, and concluded turf farms were a viable utilization venue for fly ashes.[22]

Turfgrass sod production is a common agricultural enterprise in the SE US. A combination of elevated housing starts in the south (up 9.7% annually since 1997)[23] and increasing interest in traditional row-crop alternatives has instigated a consistent annual expansion in sod production hectarage. Most southeastern states currently have from 5 to 9 x 10³ ha in sod production, while Florida has nearly 3 x 10⁴ ha. Agricultural productivity on southern Piedmont soils is notoriously limited by edaphic constraints. Upland Piedmont soils are typically characterized by an ochric A horizon and a deep acidic B horizon with low available water holding capacity. Recently established sod fields, especially seeded fields, require adequate soil moisture and receive frequent irrigation applications. Following establishment of a root system by sod, favorable plant-available water-holding capacity in the upper soil profile can minimize irrigation requirements throughout the sod maturation period (5-23 months, depending on grass species).[24] This potential reduction in irrigation requirement could minimize variable cost contributions.

Adequate soil fertility is paramount in timely establishment and maturation of turfgrass.[25] Sod producers are advised to submit soil for testing annually and to maintain optimum fertility and pH levels on production fields. Any nutrient deficiency in farm soil, particularly a primary nutrient, could postpone sod maturity and bungle an opportune entry to market. Adequate soil phosphorous is particularly crucial in the preparation of seedbeds.[21] Fertilizer rate recommendations in the SE US vary by turfgrass species and soil properties. Assuming adequate soil P, an application of 100 kg N and 50 kg K ha^{-1} (K based on soil levels) is commonly made following the first mowing, regardless of species. Bimonthly N and K applications are sometimes extended until complete ground cover is achieved in production of bermudagrass and zoysiagrass (*Zoysia* spp.) sod. Iron is sometimes applied in a soluble foliar application antecedent to harvest.[24]

Thus, considering current levels of CCBP accumulation, productivity limitations of Piedmont soils, proven effectiveness of CCBP for physical and chemical soil conditioning, appropriateness of CCBP use in non-food chain applications, and the suitability of turfgrasses for vegetation of land applications of CCBP, our experimental objectives were to evaluate the suitability of fly ash, bottom ash, and biosolid mixtures as (1) soil amendments in ornamental bed or garden preparation, and (2) supplemental soil media for generation of a bermudagrass sod product in a typical commercial method.

MATERIALS AND METHODS

Ornamental Beds

Experimental growth media blending bottom ash, fly ash, and municipal waste were evaluated as soil amendments in establishment of pansy (*Viola tricolor* L.) in raised flowerbeds. The experimental growth media/soil amendment was comprised of 10 parts municipal compost, 5 parts bottom ash (BA), and 2 parts fly ash (FA)(v/v). A commercially-available mix of pine bark, peat moss, and perlite, was used as a control (Fafard 3B, Fafard, Inc. Anderson, SC). All mixes were easily handled and generally amiable. All mixture components were nitric or HF/aqua regia digested for elemental composition. The empirically derived physicochemical characteristics of the media mixtures are specified in Table 1. In November 2000, two application rates of experimental and control soil amendments were made at 500 and 1000 m³ ha^{-1} (5 and 10 cm application height, respectively) to four 1x2 m experimental plots established on a Cecil soil series (clayey, kaolinitic, thermic family of the typic Kanhapludults) in full sun. All applications were then roto-tilled to a depth of 15 cm. Three d following, pansies were transplanted into the prepared beds and established on 15 cm centers. The established flower beds were then mulched with a 4-cm layer of pine straw and bark to minimize soil evaporation. Fertilizer applications were withheld to intensify plant nutrient availability of any particular blend. Soils were collected and analyzed for extractable nutrients (Mehlich 3)[26] and elemental composition (HF/aqua regia digest) following removal of the pansies in April, 2001.

Table 1. Preliminary physicochemical properties of amendment materials used in 2000 horticultural and sod experiments

Treatment	†Solids	LOI	N	P	K	Ca	Mg	As	Cd	Cr	Cu	Ni	Pb	Se	Zn	pH	EC
			%					mg kg^{-1}								(1:5)	dS m^{-1}
Fly ash (FA)	99.9	7.1	<0.1	0.1	1.7	1.6	0.6	46.2	0.2	22.1	20.0	15.2	17.2	4.2	48.8	11.6	1.2
Bottom ash (BA)	79.0	9.0	<0.1	<0.1	0.0	0.1	0.0	1.4	0.1	18.0	13.0	17.0	1.5	0.7	71.0	6.4	0.1
Athens sewage sludge (SS)	20.5	21.1	3.3	2.3	0.2	1.6	0.2	3.2	7.0	71.0	270.0	43.0	110.0	1.6	867.0	7.2	6.3
Athens compost (CP)	70.0	34.3	1.0	1.1	0.2	0.7	0.1	3.4	1.2	36.0	127.0	28.0	35.0	1.7	536.0	4.6	5.2
Sod Production Experiment (mixes computed on a volumetric basis)																	
Control (2 Sand:1CP)	96.0	7.1	0.13	0.14	0.03	0.09	<0.1	0.4	0.2	4.7	16.5	3.6	4.6	0.2	69.7	5.2	0.6
1BA + 1CP (1:1)	74.0	17.7	0.35	0.38	0.07	0.24	<0.1	2.1	0.5	24.2	52.3	20.8	13.1	1.0	231.4	5.2	2.2
2BA + 1CP (2:1)	77.0	14.2	0.21	0.23	0.04	0.14	<0.1	1.8	0.3	21.7	36.5	19.3	8.4	0.9	166.8	5.3	1.6
4BA + 1FA + 1SS (4:1:1)	78.0	8.2	0.47	0.33	0.03	0.23	<0.1	8.5	1.1	27.8	51.9	21.9	19.5	1.4	187.7	7.2	0.8
Horticultural Experiment (mixes computed on a volumetric basis)																	
5BA + 10CP + 2FA	-	33.0	0.61	0.7	0.5	-	-	17.0	0.8	-	118.0	43.0	29.0	-	340.0	5.2	2.8
Commercial mix (Fafard 3B)	-	68.0	0.60	0.1	1.0	-	-	0.4	1.1	-	16.0	45.0	1.5	-	67.0	5.8	0.5
EPA-I								75	85	3000	4300	420	840	100	7500		
EPA-II								41	39	3000	1500	420	840	100	2800		

FA-Fly ash; BA-Bottom ash; SS-Athens anerobically digested sewage sludge; CP-Athens municipal compost

†Nutrient and metal concentrations calculated on this dry mass basis, - = data not available

EPA-I - maximum permissible concentration; EPA-II - maximum concentration in an exceptional quality material[27]

Concentrations of N, P, K, Ca, and Mg in experimental mixes derived only from organic matter contributions, not CCBP

Concentrations of As, Cd, Cr, Cu, Ni, Pb, Se and Zn in experimental mixes derived from both organic matter and CCBP contributions

Sod Production

All experimental soil amendments and sod growth media were considered exceptional quality sludges under USEPA regulations and unrestricted for horticultural or land applications not exceeding annual load limits[27] (Table 1). Based on preliminary greenhouse tests, three mixes of CCBP and biosolids/biosolid products were selected for the sod production study. The treatment mixes were formulated on a volume basis: 2 parts BA to 1 part municipal compost (CP) (2:1), 1 part BA to one part CP (1:1), and 4 parts BA to one part sewage sludge to one part FA (4:1:1). The control growth media consisted of two parts sand (fine to medium, felsic) mixed with one part compost (2 Sand:1 municipal compost (CP). Mix physicochemical properties are described in Table 1. The sod production field experiment was initiated in May, 2000. Mixes were uniformly spread to heights of 2, 3, and 4 cm (200, 300, 400 m^3 ha^{-1}) over a compacted Cecil subsoil (clayey, kaolinitic, thermic family of the typic Kanhapludults) in full sun. The experimental growth media were not tilled into the soil. Experimental design of the sod experiment was a two-way (growth media formulation by application height) strip-plot design with four replications (n=48). Mature 'TifSport' bermudagrass sod (*Cynodon dactylon* (L.) Pers. X *C. transvaalensis* Burrt-Davy [formerly Tift 94])[28] was obtained from a foundation stock field at the Georgia Crop Improvement Association, Inc., Athens, GA. The 48 m^2 of experimental growth media were established with viable sprigs at an equivalent rate of 60 kg moist sprigs per ha. A broadcast application of commercial fertilizer provided 12.2, 48.8, and 10.1 kg ha^{-1} of actual N, P, and K, respectively, on 9 June, one day after planting (DAP). Following ~50% vegetative cover of all plots [36 days after planting (DAP)], the field was mowed every 4 d with a motorized reel-mower at a height of 3 cm and the plots irrigated equally to prevent wilt. Leaf clippings of the fifth mowing (55 DAP) were collected, triple-rinsed with de-ionized (DI) H_2O, and measured for N content by dry combustion (CNS-2000, LECO Inc., St. Joseph, MI). These procedures were repeated every 8 d until sod harvest (99 DAP). Supplemental fertilizer sprays (24.5 kg ha^{-1} ammonium nitrate) were applied to all plots of a specific growth medium type when average leaf clipping N concentration fell below 80% of the well-fertilized bermudagrass reference (maintained on-site), every 8-d period. Application events were recorded (Figure 1).

Additional plant tissue was harvested at the sixth mowing (60 DAP), triple-rinsed with DI water, and nitric acid digested for trace elemental concentrations. Elemental composition of all plant tissue and growth media were analysed using inductively coupled plasma-mass spectroscopy (ICP-MS)(Elan-6000, PerkinElmer, Boston). Physicochemical properties (pH, EC, etc.) of growth media were analysed by standard methods.[29-30] Following the 99 d maturation period, bermudagrass sod was harvested with a sod-cutting device (Ryan Sod Cutter [45.7-cm width], Textron Golf, Inc., Augusta, GA). Sod grown on varying growth media volumes were all harvested at their corresponding depths, e.g., sod grown on the 4 cm application media volume were cut to include all 4 cm of the growth media. Accordingly, the height of the cutting blade was adjusted three times. The following day, 1.3 m^2 of sod grown on each type of media at each application volume (15.6 m^2 total) were installed both conventionally and on high-strength steel grids in a maintained turfgrass area of the Georgia Experiment Station in Griffin (Figure 1).

The remaining sod was destructively analysed immediately. Growth media were separated from shoot biomass and analysed for soil moisture retention and bulk density. Soil water retention was measured using pressure plates (Soil Moisture Equipment Co., Santa Barbara).[29] Growth media samples were tightly packed into steel cores of 2-cm radius and 0.9-cm height and saturated with 0.1 M $CaCl_2$ solution at atmospheric pressure. Four replications of every treatment were simultaneously measured at each level of matric potential (1.8, 5.9, 10, 80, 250 and 1450 kPa). Plant available water data reflects volumetric soil water available under tensions of 80-1450 kPa. Bulk density measurements were replicated 24 times. Sod biomass, referring to shoots and roots present in sod, were collected in triplicate subsamples and oven-dried (95 °C) before weighing. The sum of the media bulk density value and dry shoot biomass (on an area basis) comprised the sod dry weight. Soil moisture retention data was used to calculate soil water, and the sums of gravimetric soil water ($\Psi_m < -80$ kPa) and dry weight of the finished sod used to formulate total sod weight for comparing handling and installation ease.

Figure 1. (above) High-strength steel grids installed below transplanted sod material (September, 2000); (below) Grids and sod being pulled for rooting strength determination (April, 2001).

In April, 2001, the steel grids installed under the experimental and control sod were relocated, edged, and fastened by four hooks to a motorized winch. Tensile strength of sod fastening, or rooting strength, was measured by a load sensor (MSI 7200 Dyna-link, Measurement Systems Int., Seattle) upon force application and subsequent root failure (Figure 1). The experimental design of the horticultural experiment was a randomized complete block (RCB) and the sod experiment was a two-way strip-plot. Sod experiment soil moisture data, bulk density, LOI, and CEC measurements were conducted on growth media collected from only the 4 cm application volume sod and thus analysed as randomized complete blocks. The PROC GLM subroutine of SAS (SAS Institute, Inc., Cary, NC) facilitated analysis of all data.

RESULTS AND DISCUSSION

Ornamental Beds

CCBP-biosolid blends, as a flower garden soil amendment, resulted in enhanced plant growth of pansies compared to the control amended-soil. Total plant biomass and average flower

Table 2. Soil nutrient and chemical properties of amended flowerbeds following harvest, by type and application volume

Treatment	EC	pH	Exchangeable[26]							HF/aqua regia digested								
			Ca	K	Mg	Mn	N	P	Zn	As	Cd	Cr	Cu	Mo	Ni	Pb	Se	Zn
	dS m⁻¹ (1:5)		mg kg⁻¹															
500 m³ ha⁻¹ application																		
Commercial Mix	0.19	5.6	1122	330	187	141	7	67	14	2.1	0.06	25	15	0.3	10	9	0.1	64
5BA + 10CP + 2FA	0.18	6.0	2176	259	123	197	9	488	54	4.7	0.21	29	31	0.7	14	13	0.6	126
1000 m³ ha⁻¹ application																		
Commercial Mix	0.23	5.5	1136	335	245	125	6	43	11	2.7	0.07	30	16	0.4	14	12	0.2	78
5BA + 10CP + 2FA	0.23	6.0	2532	239	119	224	11	595	62	5	0.25	27	40	0.6	12	12	0.6	150
Common Soil Levels[31]										7.1	0.34	61	21	1.5	25	14	0.44	94

FA-Class-F fly ash; BA-Bottom ash; CP-Athens municipal compost

Table 3. Physicochemical media properties and sod growth parameters following harvest, by mixture type and application volume

Mix	Sod Biomass	Sod Rooting Strength	Bulk Density	†PAW	LOI
	g 42 cm⁻²	force of detachment (kPa)	g cm⁻³	cm cm⁻³	%mass
Control (2 Sand:1CP)	2.53 b	6.56	1.508 a	0.056 d	2.5 c
1BA + 1CP (1:1)	2.52 b	6.59	0.771 d	0.139 b	12.4 a
2BA + 1CP (2:1)	2.23 b	6.45	0.878 c	0.098 c	7.5 b
4BA + 1FA + 1SS (4:1:1)	3.13 a	6.56	1.039 b	0.215 a	6.5 b
LSD (α=0.05)	0.344	NS	0.075 (α=0.01)	0.033	2.3

Application volume (height)	Sod Biomass	Sod Rooting Strength
	g 42 cm⁻²	force of detachment (kPa)
2 cm	2.15 b	5.99
3 cm	2.40 b	6.26
4 cm	3.27 a	7.37
LSD (α=0.05)	0.643	NS

FA-Fly ash, BA-Bottom ash, SS-Athens anerobically digested sewage sludge, CP-Athens municipal compost,
†PAW-plant-available water from 10-1450 kPa tension, LOI-loss on ignition (440 °C)

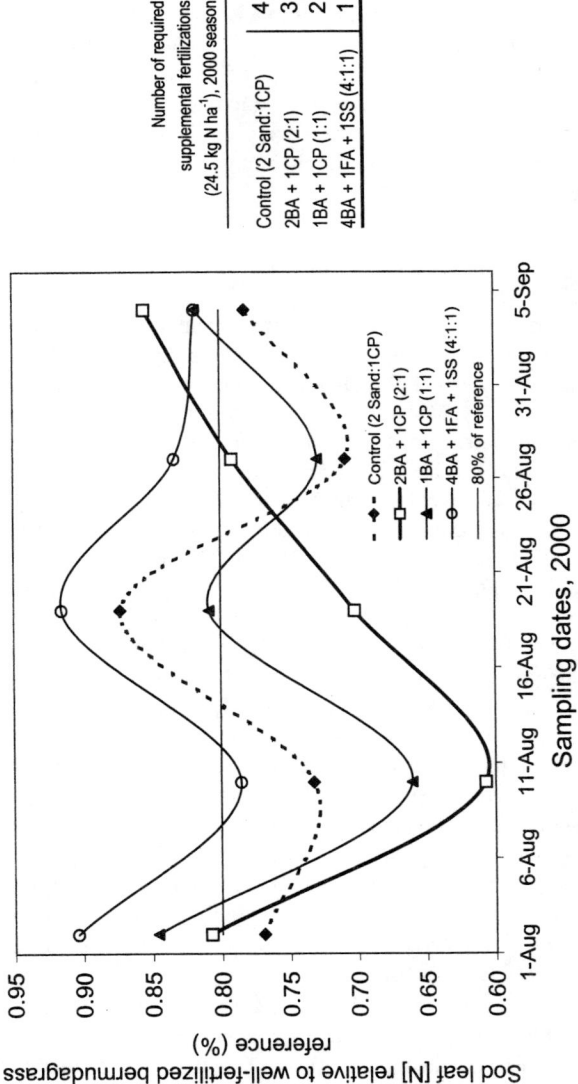

Figure 2. Leaf clipping N concentration of experimental bermudagrass sod relative to well-fertilized, on-site reference plants (4.7-5.6 % range, mean N conc.=5.2%).

mass of pansies grown in the CCBP-amended flower beds were significantly greater than pansies grown in control plots. Additionally, there was a significant effect of application volume. At the 5 cm application volume, the CCBP-based amendment increased yield of both plant biomass and flower mass by a factor of 6.8 compared to a 5 cm application of the control amendment. The 10 cm application increased yield of both plant biomass and flower mass by factors of 9.4 and 9.3, respectively, compared to the equivalent control amendment application (data not shown).

The observed increases in plant growth can be attributed, in part, to elevated exchangeable nutrient levels in the CCBP-treated soil versus the control soil. Available, Ca and P levels were nearly 2 and 9 times greater than those levels observed in the control plots, respectively. Mn and Zn levels were also elevated, but to a lesser extent (Table 2). Soil pH, electrical conductivity (EC) and exchangeable N levels were similar for the experimental and control beds. Total concentrations of regulated trace elements in the potting mix/soil amendment experiments were well below regulatory limits and similar to background levels in untreated soils (Table 2).[31] The results of the potting mix experiment were similar to those reported for poinsettia (*Euphorbia pulcherrima* Willd. ex Klotzsch) and peperomia (*Peperomia viridis* L.) grown in bottom ash mixes.[32]

Sod Production

In the bermudagrass sod production study, one experimental block was destroyed by surface runoff during a rain deluge in mid-June 2000. The remaining 3 replications were unaffected. Under irrigated, full-sun, and optimal-temperature conditions bermudagrass vigour is often directly related to N sufficiency. Mineralized soil N from fertilizer (1 DAP) and organic N in the compost and sewage sludge inclusions adequately supplied N during the first 2 months of production, as average leaf tissue N at the 2 August mowing (54 DAP) was $\geq 4\%$ by weight for bermudagrass grown on all mixes (range 3.81-4.86 N%, by mass). These organic N applications applied concomitantly with the growth media were sizable. For both the 2:1 and 2 Sand:1CP growth media, organic N was added at rates of 661, 496, and 331 kg N ha^{-1} for the 4, 3, and 2 cm height applications, respectively. For the 1:1 growth media, organic N was added at rates of 1000, 750, and 500 N ha^{-1} for the 4, 3, and 2 cm height applications, respectively. In the application of the 4:1:1 growth media, organic N was added at rates of 1687, 1265, and 844 kg N ha^{-1} for the 4, 3, and 2 cm height applications, respectively. Wong et al., showed sewage sludge additions to a sandy soil (0.25-3.5%) to result in elevated NH_4^+ concentrations as long as 70 days after application.[33] N mineralization from compost is also well substantiated. A recent growth chamber study indicates N from biosolid compost accounted for 33% of mineral N requirements in tall fescue (*Festuca arundinacea* Schreb.) culture.[34] A field study showed reduced N fertilizer requirements to be expected for the maintenance of turves when composted wastes are incorporated into a soil before the establishment of turfgrass.[35] Leaf tissue of 4BA+1SS+1FA grown sod maintained a nearly constant N concentration and required only one supplemental fertilization during the season (68 DAP). The control sod required 4 fertilizations over the same period (DAP 59-90; Figure 2). Warm-season sod establishment and biomass production have been accelerated by increasing N fertility rate.[36]

Mixture composition and application volume (height) significantly affected sod biomass production (Table 3), but application height and media type effects did not interact. The 4 cm height (for all mixtures combined) resulted in significantly greater biomass than the 2 and 3 cm application heights. Considering all experimental growth media were nutrient-rich and possessed favorable growth conditions when compared to the compacted subsoil counterpart, increased sod biomass was a predictable response to the greatest application volume (4 cm or 400 m^3 ha^{-1}). Because there were no application height by media type interactions, it can be inferred that there were no phytotoxic levels of nutrients or salts accumulated at the CCBP application rates implemented in this study. Additionally, leaf clippings of CCBP-grown bermudagrass sod collected 60 DAP did not possess concentrations of regulated metals significantly different from bermudagrass grown on control mixtures (data not shown). Moreover, leaf tissue concentrations of As, Cr and Se in all replications of every growth media were below ICP-MS detection limits (data not shown).

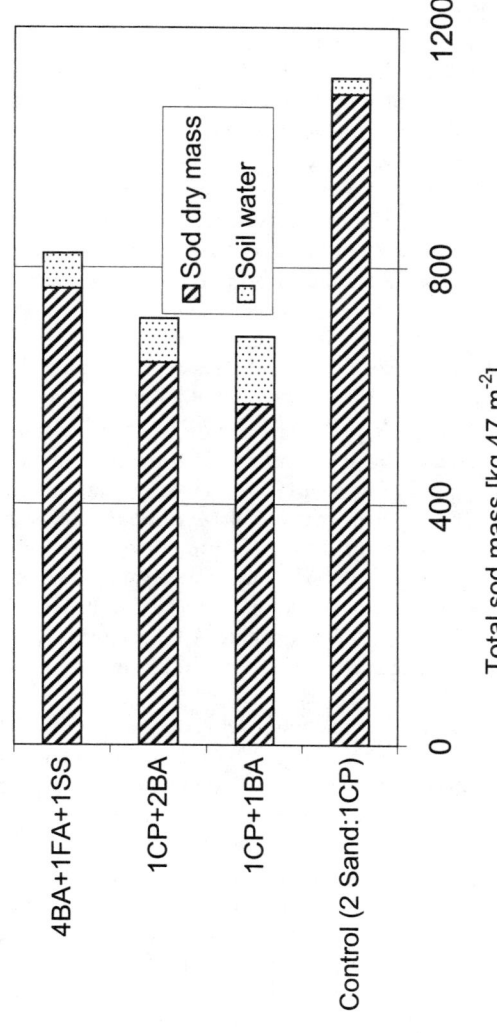

Figure 3. Total sod pallet mass by dry weight (bulk density + sod biomass) and soil water (gravimetric H_2O at $\Psi_m < -80$ kPa), based on oven-dry sod biomass (70°C) and 1.5 cm of growth media present.

The 4BA+1FA+1SS (4:1:1) experimental mix possessed significantly greater sod biomass than any other growth media, for all application heights combined (Table 3). This significant treatment effect was observed because of two likely reasons. The first reason is that although the organic matter addition in the 4:1:1 mix was only16.7% sewage sludge by volume, the sewage sludge possessed a greater density than the municipal compost and had significantly greater N and P content by mass (Table 1). Second, the chemistry of the actual organic N differed greatly between the two organic materials. Fresh manures reportedly provide greater quantities of mineral N than the same exact manure following composting.[37] Municipal compost used in this experiment was formulated from 33% yard waste and 67% sewage sludge by volume and was likely more resistant to short-term decomposition than fresh manure (sewage sludge).

Following sod harvest, laboratory tests of the experimental sod mixtures revealed significantly different values of bulk density and soil water retention (Table 3). Due to the high rate of composted organic matter inclusion and its inherent low mass, bulk densities of experimental sod growth mixes were low. Volumetric soil water near saturation (1.8 kPa tension, data not shown) of the CCBP-mixes was significantly greater than levels observed of the control mix. Additionally, both gravimetric and volumetric soil water retained by the CCBP-containing mixes at $\Psi_m < -80$ kPa were significantly greater than water retained in the control sod. Finished sod is comprised of dry weight (soil + vegetative mass) and water weight. However, the number of pallets which can be delivered by truck is limited by the weight of the finished product. Commercial sod weight can range from 0.8 to 1.6 Mg per pallet.[24] All CCBP-grown sod in this experiment possessed total weight between 0.7–0.8 Mg per pallet (Figure 3), while retaining significantly greater capillary water than the control sod. This clear benefit should be realized by the end-users at installation (handling weight) and throughout grow-in (irrigation requirements).

Plant available water (PAW) in sod proceeding harvest is highly desirable during periods of measurable evapotranspiration rates. The portion of capillary water designated as plant available is directly related to sod quality (even survivability) under transport and water-stressed conditions. Plant available water (PAW) appears to correlate with carbon content (LOI) by weight, with the exception of the 4:1:1 mixture (Table 3). Volumetric water retention of the 4:1:1 mixture greatly exceeds that of the compost-amended mixes near saturation. Conversely, at tensions greater than 80 kPa the volumetric water-holding capacity of the compost amended-CCBP mixtures greatly exceed the sewage sludge-amended 4:1:1 mixture. The particle size distribution and soil structure of the 4:1:1 mixture appears to result in proportionally greater macro-porosity when compared to the other mixes.

Fall of 2000 was warm, with average maximum temperatures in Griffin, GA exceeding 27, 25, and 16°C for Sept., Oct., and Nov., respectively.[38] All installed sod broke dormancy and flourished in April, 2001. Seven months following installation, there were no significant differences in rooting strength (Table 3) or turf quality (data not shown) among treatment mixes or application heights.

Container plants and sod are horticultural products associated with sizable retail markets. Implementing CCBP and biosolids as growth media in production of these items appear to be agronomically, environmentally and economically sound methods of societal waste disposal. All soil and plant materials associated with our experimental productions possessed trace element concentrations well below regulated levels. Under no conditions were the plants grown in the CCBP/biosolid experimental mixes outperformed by plants grown in traditional potting mixes or conventional soils. On the contrary, the experimental mix containing fly ash and sewage sludge fostered significantly greater bermudagrass biomass following the experimental period than the control mix, indicating an agronomic advantage in CCBP/biosolid use.

REFERENCES

1.   Miller, D. M., Miller, W. P., Dudka, S., and Sumner, M. E., Characterization of industrial by-products, in *Land Application of Agricultural, Industrial, and Municipal By-Products*, Soil Sci. Soc. Am., Madison, WI, 2000, 107.

2. Mattigod, S. V., Rai, D., Eary, L. E., and Ainsworth, C. C., Geochemical factors controlling the mobilization of inorganic constituents from fossil fuel combustion residues: I. Review of major elements, *J. Environ. Qual.*, 19,188, 1990.

3. Adriano, D. C., Page, A. L., Elseewi, A. A., Chang, A. C., and Straughan, I., Utilization and disposal of fly ash and other coal residues in terrestrial ecosystems: a review, *J. Environ. Qual.*, 9, 333, 1980.

4. Furr, A. K., Parkinson, T. F., Hinrichs, R. A., Van Campen, D. R., Bache, C. A., Gutenmann, W. H., St. John, L. E., Pakkala, I. S., and Lisk, D., National survey of elements and radioactivity in fly ashes: adsorption of elements by cabbage grown in fly ash-soil mixture, *J. Environ. Sci. Tech.*, 11, 1194, 1977.

5. Schumann, A. W., *Plant nutrient supply from ash-biosolid mixtures*, Ph.D. thesis, The University of Georgia, Athens, GA, 1997.

6. Matsi, T., and Keramidas, V. Z., Fly ash application on two acid soils and its effect on soil salinity, pH, B, P, and on ryegrass growth and composition, *Environ. Poll.*, 104, 107, 1999.

7. Molliner, A. M., and Street, J., Effect fly ash and lime on growth and composition of corn (*Zea mays* L.) on acid sandy soils, *J. Proc. Soil Sci. Soc. Florida*, 41, 217, 1982.

8. Schumann, A. W., and Sumner, M. E., Plant nutrient availability from mixtures of fly ashes and biosolids, *J. Env. Qual.*, 28, 1651, 1999.

9. Hill, M. J., and Lamp, C. A., Use of pulverized fuel ash from victorian brown coal as a source of nutrients for a pasture species, *Aust. J. Exp. Agric. Anim. Husb.*, 20, 377, 1980.

10. Elseewi, A. A., Straughan, I. R., and Page, A. L., Sequential cropping of fly-ash amended soils: effects on soil chemical properties and yield and elemental composition of plants, *Sci. Total Environ.*, 15, 247, 1980.

11. Elseewi, A. A., Bingham, F. T., and Page, A. L., Availability of sulfur in fly ash to plants, *J. Environ. Qual.*, 7, 69, 1978.

12. Elseewi, A. A., Grimm, S. R., Page, A. L., and Straughan, I., Boron enrichment of plants and soils treated with coal ash, *J. Plant Nutr.*, 3, 409, 1981.

13. Khan, M. R., and Khan, W. K., The effect of fly ash on plant growth and yield of tomato, *Environ. Poll.*, 92, 105, 1996.

14. Stout, W. L., Sharpley, A. N., and Pionke, H. B., Reducing soil phosphorous solubility with coal combustion by-products, *J. Environ. Qual.*, 27, 111, 1998.

15. Summers, R., Clarke, M., Pope, T., and O'Dea, T., Western Australian fly ash on sandy soils for clover production, *Commun. Soil Sci. Plant Anal.*, 29, 2757, 1998.

16. Sajwan, K. S., Ornes, W. H., and Youngblood, T. J., The effect of fly ash/sewage sludge mixtures and application rates on biomass production, *Environ. Sci. Health*, A30, 1327, 1995.

17. Salter, P. J., and Williams, J. B., Effects of pulverized fuel ash on the moisture characteristics of soils, *Nature*, 213, 1157, 1967.

18. Gangloff, W. J., Ghodrati, M., Sims, J. T., and Vasilas, B. L., Impact of fly ash amendment and incorporation method on hydraulic properties of a sandy soil, *Water Air Soil Poll.*, 119, 231, 2000.

19. Campbell, D. J., Fox, W. E., Aitken, R. L., and Bell, L. C., Physical characteristics of sands amended with fly ash, *Aust. J. Soil Res.*, 21, 147, 1983.

20. Patterson, J. C., and Henderlong, P. R., Turfgrass soil modification with sintered fly ash, in *Proc. 1st Int. Turfgrass Res. Conf.*, Harrogate, England, 1970, 161.

21. Beard, J. B., *Turfgrass: Science & Culture*, Prentice-Hall, Englewood Cliffs, NJ, 1973.

22. Adriano, D. C., and Weber, J. T., Influence of fly ash on soil physical properties and turfgrass establishment, *J. Environ. Qual.* 30, 596, 2001.

23. National Census Bureau, New privately owned housing units started in the South by purpose and design [Online]., available at: http://www.census.gov/const/www/quarterly_starts_completions.pdf, 2000, verified 7 Jan. 2002.

24. McCarty, B., Landry, Jr., G., Higgins, J., and Miller, L., *Sod Production in the Southern United States,* Clemson Univ. Coop. Extension Service, Bull. EC702, 1999.

25. Schlossberg, M. J., and Karnok, K. J., Root and shoot performance of three creeping bentgrass cultivars as affected by nitrogen fertility, *J. Plant Nutr.,* 24, 535, 2001.

26. Mehlich, A., Mehlich 3 soil test extractant: a modification of the Mehlich 2 extractant, *Commun. Soil Sci. Plant Anal.,* 5, 1409, 1984.

27. USEPA., *Part 503. Fed. Reg.,* 58, 9387, 1993.

28. Hanna, W. W., Carrow, R. N., and Powell, A. J., Registration of 'Tift 94' bermudagrass, *Crop Sci.,* 37, 1012, 1997.

29. Klute, A., Water retention. Laboratory methods, in *Physical and Mineralogical Methods,* Klute, A., Ed.,

Am. Soc. Agron., Madison, WI, 1986, 643.

30. Sparks, D.L., *Methods of Soil Analysis: Part 3—Chemical Methods,* Soil Sci. Soc. Am., Am. Soc. Agron., Madison, WI, 1993.

31. Helmke, P.A., The chemical composition of soils, in *The Handbook of Soil Science,* Sumner, M. E., Ed., CRC Press, Boca Raton, FL, 2000, chap. B1.

32. Bearce, B. C., Myers, S., Burch, M., Engstrom, B., and Smutna, L., Coal bottom ash as a growing medium for poinsettia, easter lily, and peperomia, in *Proceedings: 12th Int. Symposium on Coal Combustion By-Product (CCB) Management and Use,* Orlando, American Coal Ash Association, 1997, 1, chap. 4.

33. Wong, J. W. C., Lai, K. M., Fang, M., and Ma, K.K., Effect of sewage sludge amendment on soil microbial activity and nutrient mineralization, *Environ. Int.,* 24, 935, 1998.

34. Sikora, L. J., and Enkiri, N. K., Growth of tall fescue in compost/fertilizer blends, *Soil Sci.,* 164, 62, 1999.

35. Gentilucci, G., Murphy, J. A., and Zaurov, D. E., Nitrogen requirement for Kentucky bluegrass grown on compost amended soil, *Int. Turf. Res. J.,* 9, 382, 2001.

36. Ruemmele, B. A., Engelke, M. C., White, R. H., and Lehman, V., Alternative sod production method for zoysiagrass, *Int. Turf. Soc. Res. J.,* 9, 910, 2001.

37. Castellanos, J. Z., and Pratt, P. F., Mineralization of manure nitrogen-correlation with laboratory indexes, *Soil Sci. Soc. Am. J.,* 45, 354, 1981.

38. Georgia Automated Environmental Monitoring Network, Historical temperature data, Georgia Agriculture Experiment Station, Griffin, GA. Available at: http://www.griffin.peachnet.edu/cgi-bin/GAEMN.pl?site=GAGR&unit=Metric&report=hi, 2000, verified 25 Jan. 2002.

HEAVY METAL DISTRIBUTION AND BIOAVAILABILITY IN COAL ASH AND SLUDGE AMENDED ACID LATERITIC SOIL UNDER FIELD CONDITIONS

D. Chaudhuri, S. Tripathy, H. Veeresh[1], M. A. Powell, and B. R. Hart[2]

[1]Department of Geology & Geophysics
Indian Institute of Technology, Kharagpur, West Bengal, India 721 302
[2]Department of Earth Sciences
University of Western Ontario, London, Canada, N6A 5B7

1. ABSTRACT

Chemical fractionation of heavy metals in ash, sludge and amended soil was performed using sequential extraction. The metals were predominantly found to be present in the residual and carbonate fractions in the native soil and ash while amounts of Ni, Cd and Zn were significant in the exchangeable fractions of sludge. The distribution of metals in the extractable fractions differs according to the metal extracted, the treatment and the proportions of application. Among the metals, Ni and Zn and to a lesser extent Cd moved readily in the sludge amended soils due to lowering of pH and simultaneous increases in their most labile forms. Better yields of paddy and peanut with ash addition and reduction with sludge proportion were observed. The paddy straw and peanut shoot showed the maximum accumulation of metals indicating a physiological barrier in the transfer of metals from the shoot to the grain. Linear relationships were observed between the soil total metal concentrations with that in the crops. The study indicated that at the applied rate, the behavior of heavy metals in the amendments was primarily controlled by sludge.

2. INTRODUCTION

Coal is one of the most abundant fossil fuel used as a source of electrical energy throughout the world. Following the burning of coal for power generation, a large amount of fly ash is produced which needs to be disposed with negligible environmental impact. High ash content (30-50%) of Indian coals is contributing to approximately 110 million tonnes of annual ash generation with consequent management problems. Coal fly ash contains several plant essential macro and micronutrients, and therefore has been used as a soil amendment to alleviate the micronutrient deficiencies in agricultural soil demonstrating favourable response to crop growth and productivity.[1-5] The alkaline pH of ash has significant beneficial effects in addressing soil acidity problems and in turn reducing the availability of metals and lessening phytotoxocity[6-8] while the presence of predominantly silt size particles improves the soil physical properties.[9] However, environmental risk is also associated due to the potential buildup of element concentrations in soil, plant and ecosystem beyond toxic limits following ash utilization.[10,11]

Biosolids/sewage sludge from the treatment plants contain relatively high concentrations of trace metals both essential and non-essential for soil and plant and their indiscriminate application can cause contamination and phytotoxicity.[12, 13] Presence of high organic matter and substantial N and P in biosolids suggest its use preferentially as a fertilizer in agriculture or as a regenerator of soil.[14] Long-term field observations have shown that while sludge-applied metals can remain sufficiently available even in non-acid soils to harm sensitive crops,[15] the reduction in yield and retardation in plant growth are generally induced by high concentrations of metals in soils or in sludges,[16] high application rates of sludge[17] or by high cumulative sewage sludge loading rates.[18]

Fly ash alone is a poor source of the macronutrients N and P[9] mainly due to volatilization of nitrogen during coal combustion and most of the P being in relatively unavailable forms.[19] Application of sludge tends to increase the acidity in soils as a result of proton release from organic matter decomposition and mineralization of ammoniacal N. On the contrary, the combination of fly ash and biosolids offers a viable distributed recycling alternative to landfill and has the potential for agronomic use as substitutes for lime, fertilizer or simply as a soil amendment.[20] Evaluation of the bioavailability of metals and crop response of fly ash plus organic waste have earlier been studied.[20-22] Determination of specific chemical forms of metals and their mode of binding in soil is useful for predicting the bioavailability, mobility and transformation between chemical forms in agricultural or polluted soil.[23, 24] In recent years, various sequential extraction techniques have been used to fractionate metals in soil into a variety of operationally defined geochemical pools[25-27] and the advantages and limitations of the various extractants and techniques have been reviewed.[28]

The objective of this study was to determine the different geochemical forms of Cd, Cr, Cu, Ni, Pb and Zn and the changes in their distribution in an acid lateritic soil as a function of their speciation in sludge or sub-bituminous coal ash. The responses of a cereal and a leguminous crop to the amendment and the subsequent transport and accumulation of metals in them have also been investigated.

3. METHODOLOGY

3.1 Properties of soil and amendment

Soil collected up to a depth of 15 cm was air-dried, mixed and screened through 2-mm sieve prior to analysis. The pH was measured in deionized water at 1:2.5 (w/v) ratio soil/solution,[29] electrical conductivity (EC) in deionized water at 1:5 (w/v) ratio soil/solution,[29] organic carbon (OC) by Walkley-Black procedure,[30] cation exchange capacity (CEC) by ammonium acetate method,[29] available nitrogen by alkaline permanganate method[31] and total nitrogen by micro-kjeldahl technique[32] and is given in Table 1. Total element concentrations were determined on aqua regia digests of soil, ash and sludge[33, 34] samples ground to <150 μm and plant available element content was determined on a Mehlich 3 extraction[35] followed by measurement employing an inductively coupled plasma atomic emission spectrometer (ICP OES), Perkin-Elmer PE-3300DV. Selective major elements in soil (Table 1) were analyzed for total element content by using Philips PW 2404 X-ray Fluorescence spectrometer.[36] Coal fly ash (FA) from Kolaghat Thermal Power Plant and sewage sludge (SD) from Howrah Sewage Treatment Plant in India were collected, processed and analyzed for various parameters as in the case of soil and reported in Table 1 & 2.

3.2 Field Experiment

One time application of ash and sludge at 52 Mg/ha on dry weight basis either alone or in mixture proportions was made into the upper 15 cm layer of the selected soil in a field

Table 1. Selected properties of soil, sludge and fly ash used in this study

Parameter	Soil	SD	FA
Texture	Sandy Loam		
pH (1:2.5)	5.2	4.3	7.5
EC (dS/m)	0.04	2.99	0.13
CEC (Cmol/kg)	5.43	19.73	4.50
Organic C (g/kg)	5.4	144	8.6
Total N (%)	0.05	1.03	0.017
Total P (%)	0.02*	0.91	0.15
Total K (%)	0.05	0.26	0.07
Total Al (%)	2.98*	1.84	0.68
Total Ca (%)	0.12*	2.06	0.44
Total Fe (%)	1.27*	6.27	2.14
Total Mg (%)	0.039	0.38	0.087
Total S (%)	0.003	2.8	0.02
Available N (mg/kg)	76.86	2625	0
Mehlich III P (mg/kg)	13.1	96	360
Mehlich III K (mg/kg)	78.3	150	120
Mehlich III S(mg/kg)	19.6	9000	114

SD: Sewage sludge

FA:Fly ash

* Analysis done using WD-XRF

experiment laid out in a randomized block design with seven treatments and three replications. The treatments consisted of control, NPK (in crop specific recommended doses), SD00/FA 100, SD25/FA75, SD50/FA50, SD75/FA25 and SD100/FA00. Rice (Oryza *sativa* L.) variety IR36 was transplanted in July 1997 and harvested in October 1997. The land was kept fallow for a period of three months till sowing of peanut (Arachis *hypogaea* L.) variety AK-12-24 that was harvested in May 1998. Representative plant samples were collected from each treatment for analysis.

3.3 Plant Analysis

The plant samples were washed with Milli Q water and oven dried at 65^0 C. The dried plant parts were ground using Pulverisette (P-14, Fritsch) with zirconium blade to pass 20-mesh. Kernel samples were ground using an agate mortar and pestle to pass 2-mesh. Concentrations of Cd, Cr, Cu, Ni, Pb and Zn was determined in these samples through dry ashing followed by acid dissolution.[37] The digested samples were analyzed employing Perkin Elmer AAnalyst 300 Atomic Absorption Spectrometer (AA) with HGA 850 graphite furnace.

3.4 Heavy Metal Fractionation

Chemical forms of metals present in paddy soil are likely to be affected by both oxidizing and reducing conditions.[38-40] During this study, post harvest soil samples were collected under oxidizing conditions using a stainless steel agar from each plot. Chemical fractionation[41] of heavy metals was carried out on ash, sludge and both native and post harvest soils. Soil equivalent to 2.0 g on dry weight basis was sequentially extracted with 0.5 M KNO_3 for 16 hours (exchangeable fraction), with deionized water for 2 hours, (extracted two times and combined for adsorbed fraction), with 0.5M NaOH for 16 hours (organic fraction), with 0.05M Na_2-EDTA for 6 hours (carbonate fraction) and with 4M HNO_3 at 80^0C for 16 hours (sulphide/residual fraction). The concentrations of metals in each fraction were determined by employing an AA.

Table 2. Total and available contents (ppm) of heavy metal in soil and sludges used (Mean ± S.D., n=5)

Total[1]	Cd	Cr	Cu	Ni	Pb	Zn
Soil	0.62 ± 0.02	66 ± 1.44	11 ± 0.36	8 ± 2.07	18 ± 2.52	19 ± 0.16
FA	0.81 ± 0.032	20.84 ± 0.37	28.92 ± 0.77	14.23 ± 3.05	51.90 ± 4.60	97.84 ± 1.39
SD	3.54 ± 0.33	116 ± 1.54	328 ± 1.23	39 ± 2.2	226 ± 4.68	3137 ± 70.96
Available[2]						
Soil	0.20 ± 0.02 (32)	0.50 ± 0.08 (0.7)	2.02 ± 0.20 (18)	1.83 ± 0.63 (23)	5.52 ± 1.35 (31)	2.52 ± 0.04 (13)
FA	0.29 ± 0.04 (36)	0.85 ± 0.11 (4.1)	8.06 ± 0.39 (28)	2.14 ± 0.82 (15)	24.38 ± 1.69 (47)	51.06 ± 1.05 (52)
SD	0.96 ± 0.09 (27)	0.53 ± 0.08 (0.4)	44 ± 0.63 (13)	14 ± 1.06 (36)	3.33 ± 0.11 (1.5)	892 ± 14.73 (28)

1 Aquaregia extractable
2 Mehlich III extractable
3 Values in parentheses are relative availability(%) of metals calculated as the ratio of Aquaregia and Mehlich III extractable concentration.

SD Sewage sludge
FA Fly ash

Table 3. Effect of sludge, fly ash and mixture application on the distribution of heavy metals among sequentially extracted fractions in soil collected after harvest of paddy

mg/kg

SD:FA[1]	EXC	ADS	ORG	CARB	RES	Sum	Total[2]
Cd							
0/0	0.013	0.005	0.113	0.150	0.193	0.473	0.571
NPK	0.025	0.005	0.125	0.138	0.210	0.500	0.566
0/100	0.038	0.025	0.128	0.150	0.210	0.533	0.586
25/75	0.125	0.050	0.150	0.188	0.245	0.758	0.831
50/50	0.138	0.075	0.163	0.213	0.280	0.868	0.925
75/25	0.163	0.100	0.213	0.225	0.333	1.0	1.2
100/0	0.175	0.125	0.225	0.250	0.368	1.1	1.2
LSD (5%)	0.01	0.02	0.01	0.02	0.03	0.10	0.09
Cr							
0/0	0.438	0.100	6.6	2.71	39	49	57
NPK	0.450	0.100	6.6	2.78	41	51	63
0/100	0.525	0.125	3.1	6.74	44	55	71
25/75	1.4	0.175	6.9	3.15	50	61	76
50/50	1.5	0.225	7.2	3.39	52	64	81
75/25	1.5	0.225	7.4	3.69	57	70	83
100/0	1.6	0.250	7.6	3.66	61	74	86
LSD (5%)	NS	NS	0.02	NS	5.51	5.05	5.23
Cu							
0/0	0.675	0.250	3.5	0.763	1.3	6.5	8.2
NPK	0.713	0.225	3.5	0.800	1.4	6.6	8.2
0/100	0.825	0.375	4.4	0.913	1.5	8.0	9.5
25/75	0.925	0.450	4.6	0.988	1.6	8.6	10
50/50	1.0	0.575	4.9	1.1	1.8	9.3	11
75/25	1.1	0.675	5.1	1.2	1.9	9.9	12
100/0	1.2	0.725	5.3	1.4	2.0	11	14
LSD (5%)	0.02	NS	0.09	NS	0.03	1.32	1.87

SD:FA[1]	EXC	ADS	ORG	CARB	RES	Sum	Total[2]
Ni							
0/0	0.950	0.075	1.2	1.0	3.3	6.5	8.2
NPK	0.988	0.075	1.2	1.1	3.4	6.7	9.0
0/100	1.1	0.100	1.3	1.2	3.4	7.1	9.3
25/75	1.7	0.175	1.4	1.2	4.9	9.4	13
50/50	2.1	0.200	1.4	1.3	6.3	11	13
75/25	2.2	0.400	1.6	1.4	6.8	12	16
100/0	2.4	0.475	1.7	1.4	7.4	13	18
LSD (5%)	0.02	NS	0.02	0.02	0.23	2.36	2.61
Pb							
0/0	1.1	0.450	1.9	5.2	4.0	13	15
NPK	1.1	0.525	1.9	5.1	4.1	13	17
0/100	1.4	0.575	2.2	5.8	4.6	15	18
25/75	1.7	0.675	2.3	4.6	8.0	17	19
50/50	1.9	0.700	2.4	4.7	9.4	19	20
75/25	2.0	0.750	2.6	4.8	9.9	20	21
100/0	2.1	0.750	2.7	4.7	11	21	24
LSD (5%)	NS	NS	NS	0.02	1.18	1.58	1.87
Zn							
0/0	0.96	0.13	1.43	2.88	11	16	18
NPK	1.0	0.13	1.39	2.98	11	16	18
0/100	4.3	2.5	6.51	17	59	89	102
25/75	28	3.1	6.65	17	60	114	116
50/50	33	3.4	7.10	17	75	136	139
75/25	35	3.6	7.26	18	85	148	154
100/0	41	4.0	7.51	18	91	162	176
LSD (5%)	1.19	0.16	0.81	1.21	5.4	13	12

[1] SD Sludge
FA Fly ash
[2] Total AQR digests

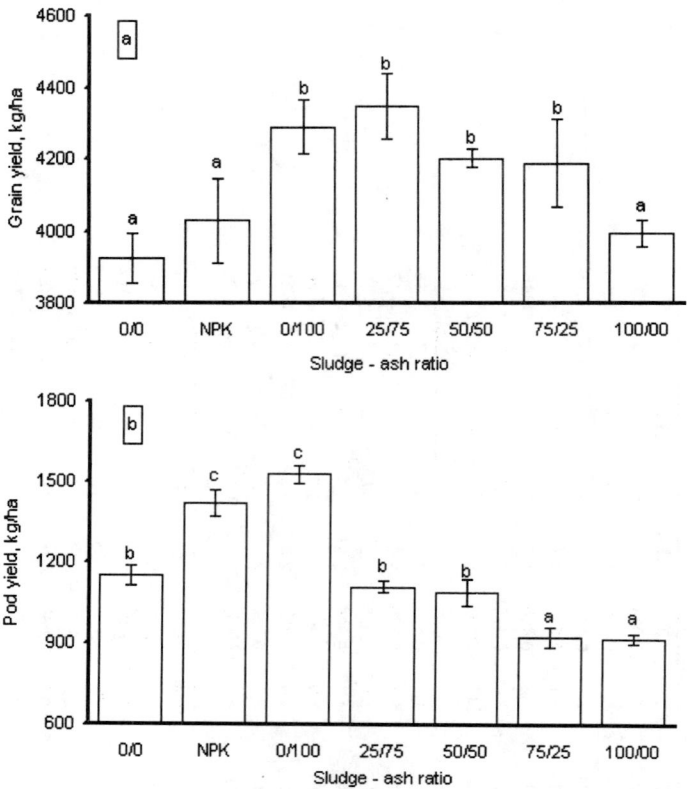

Figure 1. Yield response of paddy and peanut in different treatments.

4. RESULTS AND DISCUSSION

The soil besides being acidic has low OC, EC, CEC and available nitrogen (Table 1). Aluminium and Fe are found to be dominant in soil while it is deficient in Ca and Mg. Relative to the soil, the sludge was acidic with high OC, nitrogen, P, Fe and CEC with lesser concentrations of Al and the fly ash was alkaline with low CEC, OC, N, Al, Fe, but with sufficient plant available P and K. The concentrations of heavy metals in the sludge were below the ceiling limits of EPA, [42] where Cd = 85 mg/kg, Cr = 3,000 mg/kg, Cu = 4,300 mg/kg, Ni = 420 mg/kg, Pb = 840 mg/kg and Zn = 7,500 mg/kg.

4.1 Crop Responses and Yield

The response of both paddy and peanut yields to FA, SD and mixture addition to the soil is shown in Figure1a&b. The paddy yield (Fig. 1a) increased by 9% with application of ash at 52 Mg/ha relative to control and the highest was in SD25/FA75, but, with further increase in the sludge proportion it reduced. The NPK treatment showed higher yields than both control and 100% sludge. Peanut pod yield (Fig. 1b) showed systematic decrease with increase in sludge proportion. Application of sludge at 52 Mg/ha decreased the yield by 20%, while it increased by 33% with ash application at similar rate over the control. Ash addition had a positive impact on yield of both the crops, the magnitude of increase over control being higher in case of peanut. Application of 100% sludge to the soil decreased the yield in case of

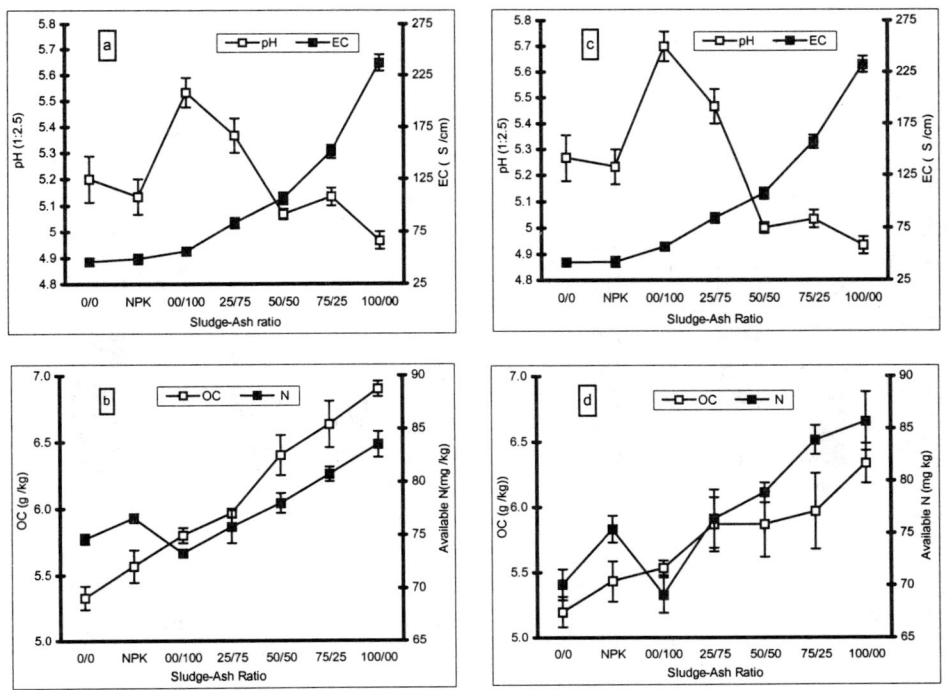

Figure 2. Changes in the pH, Electrical Conductivity (EC), Organic Carbon (OC), and Available Nitrogen in treatments after the harvest of paddy (a,b) and peanut (c,d.).

both crops and only in case of peanut the yield performance was lower than the control. In general, yield performances in the mixture amended soil were primarily controlled by sludge proportion.

Legumes grown on sludge treated soil often experience marked reduction in yield[43, 44] due to toxicity of metals directly affecting the plant[43] or indirectly by harming the microbial activity.[44-47] However, factors such as soil type, quality and quantity of sludge, crop variety etc. influence the performance of yield in one or the other way.

4.2 Post Harvest Soil Properties

The changes of pH, EC, OC and available N in the treatments after harvest of paddy and peanut are shown in Figure 2. The increase in the ratio of sludge in the amendments decreased the pH in both post harvest soils. The lowering of pH is attributed to the nature of the sludge, [48, 49] the high organic carbon content (144g/kg) and the high total (2.8%) and available (9000 mg/kg) sulphur content in the sludge. Addition of 100% ash increased the pH in post paddy (5.2 to 5.5) and post peanut (5.3 to 5.7) soils relative to control. Marginal increase in pH was noted in control, NPK, SD00/FA100 and SD25/FA75 of post peanut compared to post paddy soils. The electrical conductivity increased moderately with ash, sludge or mixture addition; the sludge amended ones showing comparatively higher values. The post peanut soils showed higher EC values over post paddy soil except for the control and NPK treatments. General increase in soil salinity levels following sludge addition are widely reported.[50, 51] Gradual increases in both OC and available N content in the post paddy soil were accompanied with sludge addition; the increase over control is sharp at 100% sludge addition, while the other treatments showed moderate increases. Organic carbon decreased in all the treatment in post peanut soils. Significant increases in organic carbon[52, 53]

295

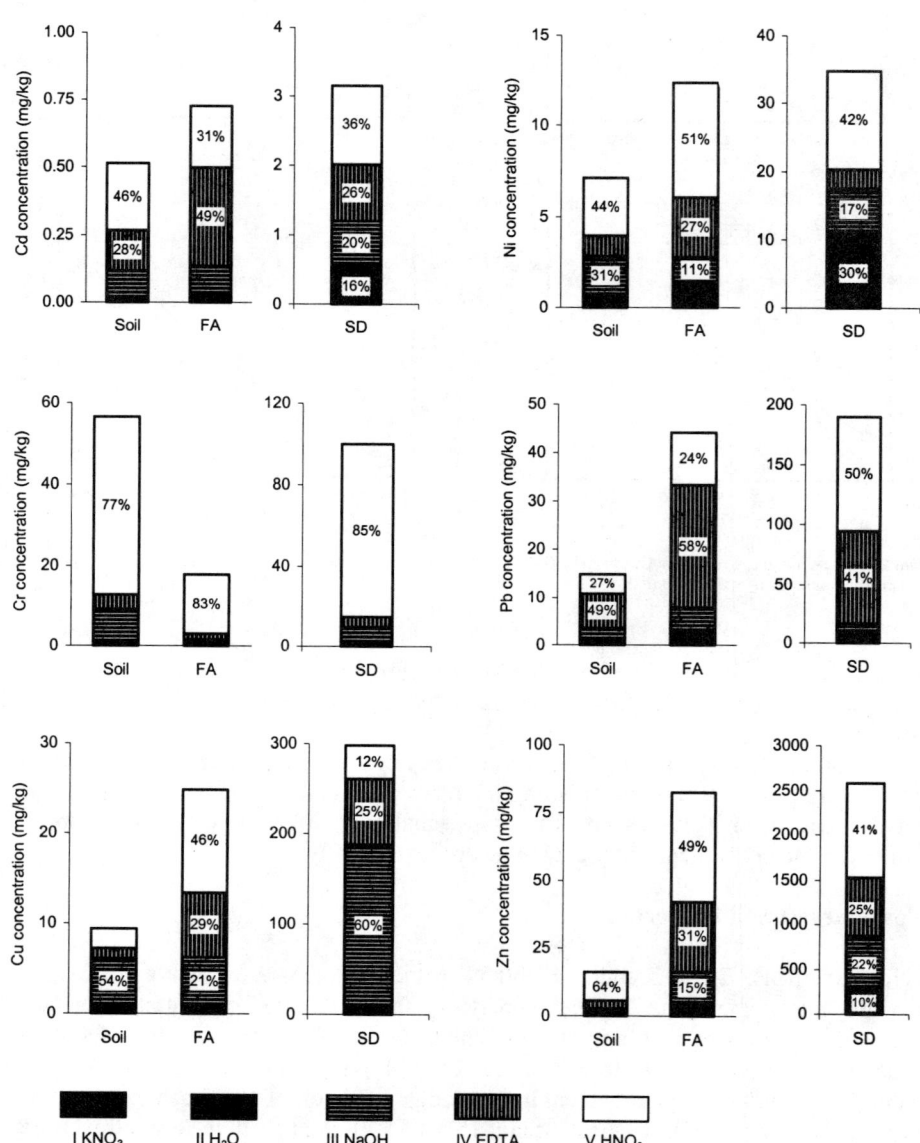

Figure 3. Chemical fractionation of heavy metals in soil, fly ash and sewage sludge.

Table 4. Effect of sludge, fly ash and mixture application on the distribution of heavy metals among sequentially extracted fractions in soil collected after harvest of peanut

mg/kg

Cd

SD:FA [1]	EXC	ADS	ORG	CARB	RES	Sum	Total
0/0	0.011	0.005	0.088	0.163	0.140	0.441	0.497
NPK	0.005	0.018	0.075	0.175	0.123	0.388	0.483
0/100	0.025	0.025	0.063	0.200	0.158	0.498	0.511
25/75	0.150	0.075	0.113	0.238	0.175	0.728	0.771
50/50	0.163	0.100	0.150	0.250	0.193	0.863	1.0
75/25	0.200	0.100	0.175	0.325	0.228	1.0	1.1
100/0	0.188	0.125	0.175	0.400	0.263	1.1	1.2
LSD (5%)	0.01	0.01	0.02	0.02	0.02	0.09	0.09

Cr

SD:FA [1]	EXC	ADS	ORG	CARB	RES	Sum	Total
0/0	0.400	0.075	6.5	2.7	39	49	55
NPK	0.413	0.100	6.6	2.8	41	51	61
0/100	0.488	0.075	3.2	7.0	47	57	70
25/75	1.4	0.125	7.2	3.1	51	63	81
50/50	1.4	0.175	7.4	3.4	58	71	85
75/25	1.5	0.175	7.5	3.6	66	78	93
100/0	1.5	0.200	7.6	3.7	72	85	97
LSD (5%)	NS	NS	0.05	NS	5.51	5.52	5.74

Cu

SD:FA [1]	EXC	ADS	ORG	CARB	RES	Sum	Total
0/0	0.650	0.175	3.2	0.800	1.5	6.3	8.0
NPK	0.688	0.275	3.3	0.863	1.6	6.7	8.8
0/100	0.850	0.325	4.2	1.1	1.8	8.2	12
25/75	0.988	0.400	5.7	1.2	1.9	10.1	12
50/50	1.1	0.525	6.2	1.2	2.1	11.1	14
75/25	1.3	0.550	6.5	1.4	2.2	11.9	15
100/0	1.3	0.600	6.7	1.4	2.3	12.4	16
LSD (5%)	NS	NS	0.08	0.02	NS	1.58	1.74

Ni

SD:FA [1]	EXC	ADS	ORG	CARB	RES	Sum	Total
0/0	0.763	0.050	1.3	0.975	3.2	6.3	7.8
NPK	0.825	0.050	1.3	1.0	3.3	6.6	8.0
0/100	0.975	0.075	1.8	1.3	4.2	8.4	10
25/75	2.7	0.400	2.1	1.4	4.8	11	14
50/50	3.3	0.425	2.4	1.7	6.2	14	17
75/25	3.7	0.475	2.5	1.7	6.7	15	18
100/0	3.9	0.600	2.6	2.0	7.3	16	20
LSD (5%)	0.27	NS	0.02	0.01	0.82	2.74	2.60

Pb

SD:FA [1]	EXC	ADS	ORG	CARB	RES	Sum	Total
0/0	0.975	0.375	1.9	4.9	3.8	12	15
NPK	1.0	0.425	2.0	4.9	3.9	12	15
0/100	1.5	0.475	2.4	6.4	4.4	15	18
25/75	1.7	0.525	2.6	6.5	7.4	19	21
50/50	1.9	0.550	2.8	7.7	9.0	22	24
75/25	2.0	0.675	2.8	7.9	9.7	23	24
100/0	2.2	0.700	3.0	8.4	10	24	26
LSD (5%)	NS	NS	NS	0.97	1.92	2.12	1.49

Zn

SD:FA [1]	EXC	ADS	ORG	CARB	RES	Sum	Total
0/0	0.83	0.10	1.3	2.6	10	15	15
NPK	0.86	0.15	1.3	2.7	10	15	16
0/100	-4.2	2.4	7.1	18	60	92	98
25/75	31	3.5	7.7	15	56	113	129
50/50	36	3.9	8.6	16	68	132	144
75/25	39	4.0	9.1	17	76	145	169
100/0	43	4.2	9.2	17	85	159	184
LSD (5%)	1.05	0.18	0.97	1.35	6.1	11	12

[1] SD Sludge
FA Fly ash
[2] Total AQR digests

in sludge-amended soils have been reported from earlier studies. Although the increase in OC and N in soils after mixture application would be beneficial for their fertility, it may be noted that the decline in pH may result in higher metal mobility.

4.3 Fractionation of Metals

Figure 3 shows the distribution of chemical forms of Cd, Cr, Cu, Ni, Pb and Zn in native soil, ash and sludge and the contribution of major fractions to the "total" of all fractions. On comparison of the sum of the metals extracted from each fraction with the total concentration (aqua regia digest) of each metal in soil, the recoveries were found to be within 20%.

Metals in the exchangeable (EXC) phase have been considered to be nonspecifically adsorbed that can easily be replaced by competing cations; metals in association with organic matter by complexation, adsorption and chelation as organic fraction (ORG); metals precipitated or coprecipitated as carbonates represent the carbonate fraction (CARB) and metal fractions in resistant silicate material represent residual (RES) and are likely to be available only after digestion at elevated temperatures.[54]

4.4 Distribution of Heavy Metals in Soil, Ash and Sludge

The EXC and the water-soluble fractions are found to be very low in soil, ash and sludge except for Cd (16%), Ni (30%), Zn (10%) in sludge and Ni (10%) in ash. In soil and sludge, the RES and to a lesser extent the CARB fractions are the two major chemical forms in which Cd is distributed, while the distribution is reverse in the case of ash. Earlier studies have reported Cd in sludge to be associated with CARB followed by ORG and RES fractions[55] CARB, ORG and RES[56] or CARB, RES and the EXC fractions.[57] The total Cr in sludge (85%), ash (83%) and soil (77%) is present in RES phase indicating very low mobility and also have been reported earlier.[57] Cu is mostly organic bound in soil and sludge whereas is most predominant in the RES fraction in ash. Presence of Cu in ORG followed by CARB and RES fraction[55, 56] and mainly with RES fraction[57] has been reported. Highest occurrence of Ni was in the RES form in all the three materials while the second major form was the ORG bound Ni in the case of soil and sludge and the CARB form in case of ash. Ni has been found to be equally present in the CARB, ORG and RES fraction in sludge,[56] mainly in the RES, ORG and the CARB fractions[57] or mainly in the EXC and ADS fractions.[14] The soil and ash have majority of the Pb held in the CARB form followed by the insoluble RES fraction. The sludge has half of the total Pb in the RES fraction followed by 41% in the CARB form. Previous studies have reported Pb to be present mostly in the ORG fraction[58, 59] or in the RES and CARB fractions.[57] Zn was predominantly found to be in RES followed by CARB and ORG fractions in soil, ash and sludge. On the contrary, Zn in sludge has earlier been reported to be associated with carbonate and organic fractions[56, 60] and carbonate, ORG and RES fractions[57] or mainly in the EXC and ADS fractions.[14] Heavy metals in the soil and ash are primarily found to be present in the RES and CARB fractions and hence are poorly mobile. Conversely, significant amounts of Ni, Cd and Zn being present in the EXC fractions in sludge are more labile and easily mobile.

4.5 Heavy Metal Distribution in Post Harvest Soil

The concentrations of metals in various chemical fractions in post harvest soil after each crop are given in Table 4 and 5. The predominant fractions recovered in the control after harvests are: Cd – RES (40-41%), Cr – RES (80%), Cu – ORG (51-54%), Ni – RES (51%), Pb – CARB (41%), and Zn – RES (66-67%). Similar extractabilities of metals have also been reported.[55, 61]

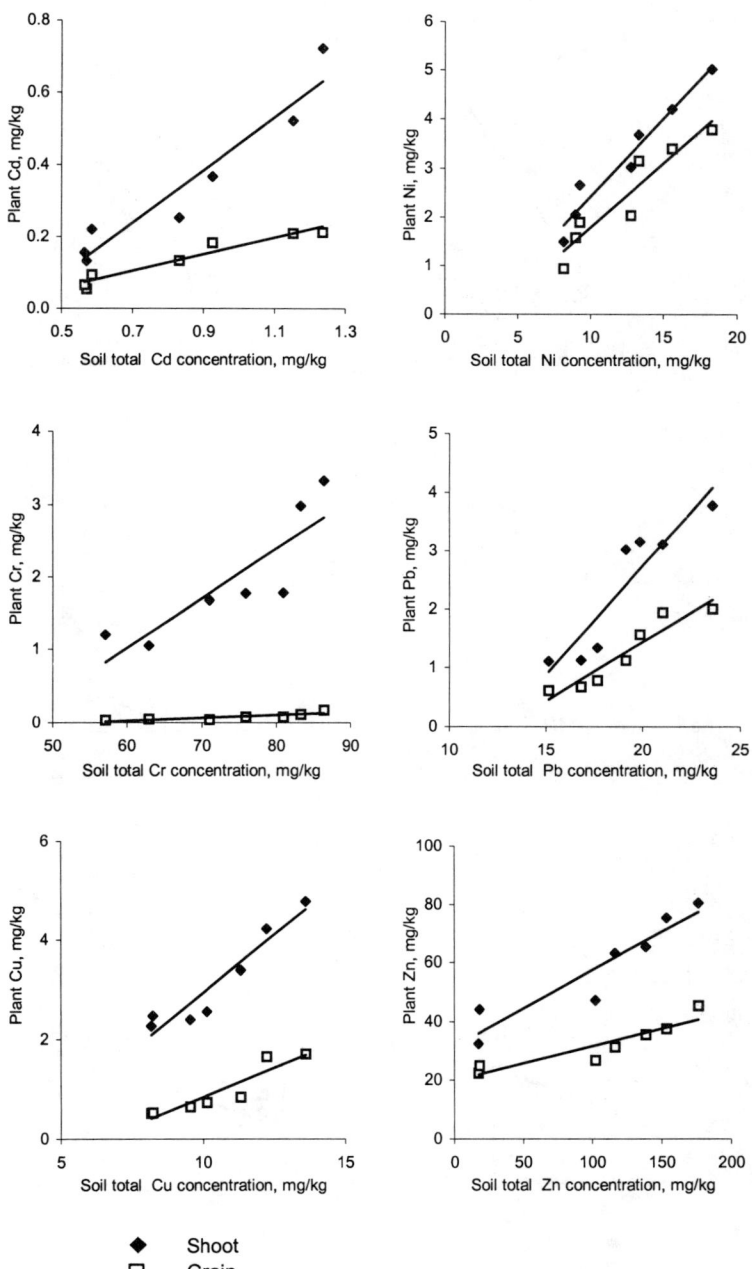

◆   Shoot
☐   Grain

Figure 4a. Relationships between plant concentrations and total soil Ni, Pb and Zn concentrations in Paddy.

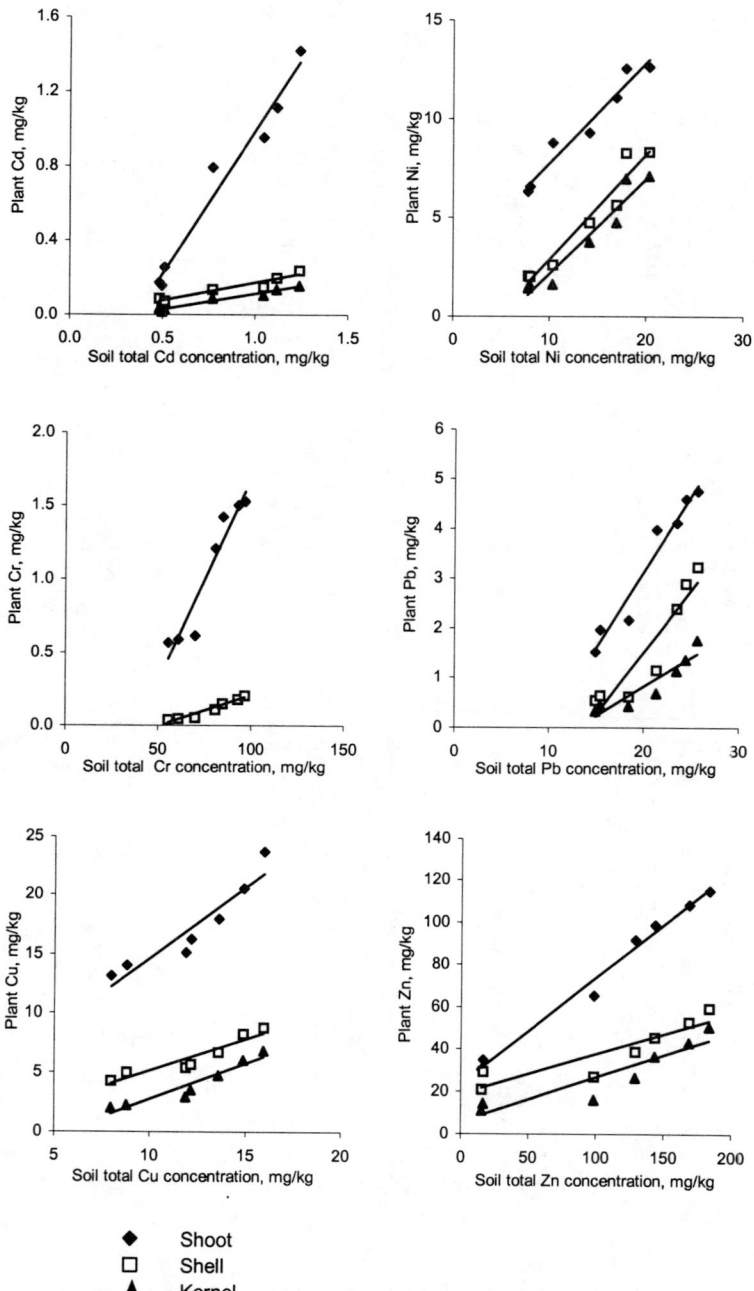

Figure 4b. Relationships between plant concentrations and total soil Ni, Pb and Zn concentrations in Peanut.

Table 5. Range of plant heavy metal concentration (mg/kg) and linear regression parameters between plant and total soil metal (AQR) concentration.

Crop	Tissue	Cd Range	Cd Intercept	Cd Slope	Cd R²	Cr Range	Cr Intercept	Cr Slope	Cr R²	Cu Range	Cu Intercept	Cu Slope	Cu R²
Paddy	Straw	0.13-0.72	-0.28	0.73	0.92**	1.2-3.3	-3.11	0.069	0.75*	2.3-3.7	-1.72	0.47	0.90**
	Grain	0.06-0.21	-0.06	0.23	0.94**	0.03-0.18	-0.20	0.004	0.69	0.51-2.0	-1.49	0.23	0.86**
Peanut	Shoot	0.16-1.4	-0.55	1.54	0.97**	0.57-1.6	-1.07	0.028	0.81*	13.1-23.7	2.72	1.20	0.879**
	Shell	0.07-0.23	-0.12	0.19	0.93**	0.04-0.20	-0.20	0.004	0.70	4.3-8.8	-0.24	0.60	0.891**
	Kernel	0.02-0.16	-0.06	0.17	0.96**	ND	-	-	-	2.0-6.8	-3.22	0.24	0.91**

Crop	Tissue	Ni Range	Ni Intercept	Ni Slope	Ni R²	Pb Range	Pb Intercept	Pb Slope	Pb R²	Zn Range	Zn Intercept	Zn Slope	Zn R²
Paddy	Straw	1.5-5.0	-0.77	0.32	0.95**	1.1-3.8	-5.80	0.42	0.86**	33-81	31.47	0.26	0.88**
	Grain	0.94-3.8	-0.86	0.26	0.89**	0.61-2.1	-4.58	0.37	0.79*	22-46	19.84	0.12	0.84**
Peanut	Shoot	6.3-13	2.68	0.51	0.89**	1.5-4.7	-3.01	0.31	0.75*	30-115	23.84	0.50	0.86**
	Shell	2.1-8.3	-2.41	0.53	0.89**	0.5-3.1	-3.53	0.25	0.77*	21-60	18.82	0.39	0.95**
	Kernel	1.4-7.1	-2.63	0.48	0.87**	0.3-1.8	-1.52	0.12	0.80*	11-51	5.96	0.28	0.89**

* Significant at 5%
** Significant at 1%

Application of amendments to soil increased Cd concentration in all the fractions. The distribution of Cd in both control and amended soil collected after paddy harvest was RES>CARB>ORG>EXC>ADS. While CARB$_{Cd}$ predominated over the RES$_{Cd}$ in post harvest peanut control, NPK and 100% ash amendments, EXC$_{Cd}$ increased over ORG$_{Cd}$ in the mixture and 100% sludge amended soil. Increase in Cd concentrations in EXC (3-4%) and CARB (5-16%) with simultaneous decrease in ORG (4-5%) and RES (7-10%) fractions in the amended soil after peanut harvest suggest a redistribution of Cd with time to more mobile fractions. The distribution of Cd in soils is influenced by pH, redox potential, and total Cd concentration.[62-64] Presence of 15-17% EXC$_{Cd}$ in post harvest sludge amended soils and the increase by 3-4% in it after peanut harvest can be attributed to the resultant low pH of sludge amended soils and the presence of 16% EXC$_{Cd}$ in sludge itself. 100% ash applied soils had only 7% EXC$_{Cd}$ as compared to the 3% in the native soil and marginally decreased (2%) in the post peanut harvest soil. Cd in ash was mainly in the CARB form (49%) and the increase in the CARB$_{Cd}$ in 100% ash amendment suggests its reduced solubility relative to the sludge and mixture treatments.

In this study, very small percentages of Cr were detected in the EXC and ADS fractions in all the treatments. A trend towards increase in its insoluble forms was noted indicating the contributions by FA, SD or both mainly resided in the less available fractions. Previous studies[57] have reported more than 80% of Cr to be present in residual fraction in sludge treated soil. The distribution pattern of Cr (RES>ORG>CARB>EXC>ADS) remained unchanged in all post-harvest amended soils except in case of 100% ash amendments where ORG and CARB fractions changed positions. A marginal increase in ORG and RES forms and a decrease in CARB forms of Cr were noted after the harvest of peanut relative to post harvest paddy soil.

Copper recovery was highest in the ORG fraction in all the treatments and the distribution was in the order of ORG>RES>CARB>EXC>ADS. Similar observations have earlier been made on Cu to be predominantly present in the ORG fraction[41, 55] or mainly in inorganic precipitated form[65] in sludge-treated soils. Marginal increase in ORG$_{Cu}$ (2-4%) and CARB$_{Cu}$ (7-8%) was noted in sludge applied soil only after peanut harvest. Insignificant changes in EXC, ADS and RES concentrations of Cu were noted in the post harvest amended soils of either crop. Distribution of Cu in native soils was not significantly affected by any amendment. The solubility and the bioavailability of Cu appears to be controlled mainly by the organic matter from sludge since it acts as a source of Cu and also serves as its major adsorbent and to a limited degree by the pH.[59]

Amendment contributed Ni increased its concentration significantly in all the chemical fractions and had greatest effect on the EXC and the RES fractions. Changes in distribution patterns associated with increase in Ni concentrations in EXC (5-7%) and ORG (3-4%) fractions were noted after peanut harvest in the sludge amended soils with subsequent decrease in the RES fraction (10-11%). Sludge amendment to acidic soils decreases Ni in the RES fraction with associated increase in the ORG fraction.[57] Conversely, the control, NPK and only ash amended soils showed decrease in the EXC (3-4%) fraction with subsequent increase in the RES (2%) fraction. Ni distribution in control, NPK and 100% ash amended soil was: RES>ORG>CARB>EXC>ADS and changed to RES>EXC>ORG>CARB>ADS in sludge amended soils. Relative to EXC$_{Ni}$ (12%) in post peanut control, EXC$_{Ni}$ in 100% sludge treatment increased to 24% and did not change at all with similar levels of ash application. Increased mobility with the increase in EXC$_{Ni}$ in sludge treated soils was observed while and no such indication for the ash amendments was noted.

No significant changes were observed in EXC, ADS and ORG fractions of Pb in either of the post harvest native or amended soils. Decrease in the CARB$_{Pb}$ associated with increase in the RES fraction was observed in sludge treated soil relative to the control in both the post harvest soils. The redistribution of Pb from the CARB in native soils to the RES in the sludge treated soils can be attributed to its chalcophilic properties[62] and the likely formation of PbS as the available S content in sludge was high (Table 1). Pb contributed by ash addition increased the $CARB_{Pb}$ concentration in 100% ash amended soil, since 58% of Pb was present in the CARB fraction of ash. The distribution of Pb in native soil was CARB>RES>ORG>EXC>ADS and with sludge amendment the concentrations in the RES fraction increased over the CARB fraction. Previous studies[63, 66-68] on contaminated soils also reported low exchangeable Pb. Pb in sludge treated soils have been reported to be present mainly in the CARB fraction.[41, 57]

Major increases in the EXC fraction of Zn was observed in sludge treated soils. 100% sludge application increased the Zn by 19% and 21%, respectively in the post paddy and peanut soil relative to the control. Between the similar treatments, Zn concentration in the EXC fraction increased by 3% from post paddy to post peanut soil. A simultaneous decrease in the RES form was also noted in 100% sludge amendment by 10% and 13%, respectively in post paddy and post peanut soil relative to the control. Only ash application showed no significant change in EXC, ADS fractions; a marginal increase in ORG, CARB and RES fractions in both the post harvest soils. The sludge proportion in the mixtures appears to be controlling the distribution pattern of Zn. The order of extraction for Zn was RES>CARB>ORG>EXC>ADS in control, NPK and 100% ash application that changed to RES>EXC>CARB>ORG>ADS in sludge and mixture treated soils. Zn in sludge treated soils mainly occurred in the CARB and RES forms.[41, 57] Prolonged application of sludge to soil increases Zn in the CARB, [57, 60] or ORG and EXC fractions with decrease in the RES fraction.[57]

4.6 Relationship between Total Soil and Plant Metal Concentrations

The relationships between the concentration of metals in paddy, peanut and that of the total concentration in soil are shown in Fig 4a&b. Even though the metal concentrations differed between the crop, between the vegetative tissues and the storage organs; it is clear that most of the relationships are linear and there is no evidence of plateau within the ranges of soil concentration of metals established in this study.

Table 5 shows the concentration ranges in plant parts and the parameters of linear regression for individual metals. The ranges of concentration of metals in the vegetative tissues are higher than the storage organs. The slope of regression indicates the transfer efficiency of metals from soil to the plant.[69] It is seen that the vegetative tissues have greater slopes than the storage tissues (grain or kernel) indicating a physiological barrier in the transfer of heavy metals from the vegetative parts to the storage organs. The slopes of Cr in vegetative and storage tissues were less in paddy and peanut indicating low transfer efficiency from the soil to the plants. Cr concentration in peanut kernels was below analytical detection limits. The fractionation of Cr in the post harvest soils also indicates it to be mainly present in the resistant form thereby reducing its bioavailability. The transfer efficiency in the paddy straw was Cd>Cu>Pb>Ni>Zn>Cr, while the order for peanut shoot was Ni>Cd>Cu>Zn>Pb>Cr. The orders indicate high uptake of Cd in the vegetative tissue of both crop, and Ni in the shoot of peanut only. The increase of Ni and Cd in EXC fractions in post peanut soils increased the bioavailability and thereby increasing their accumulation in plants. The slopes of Cd, Cu, Ni, and Pb were very similar in grain tissues in paddy indicating not much difference in the transfer of these metals from the soils to the crop. Compared to paddy grain, peanut kernels had higher transfer efficiency for Cu, Ni and Zn. The transfer efficiency of Ni was highest in peanut kernel, while that of Cu and Zn were similar. The transfer efficiency of Zn was found to be low even though its range of concentration in plant tissue was the highest among the metals studied. This may be attributed to substantially high concentrations of total Zn in amended soil reflected in the low slope for all tissues and the regression being performed considering all the treatments. Earlier studies [59, 70] had also shown that Zn uptake by plants is higher in acid than in alkaline soil.

5. CONCLUSION

Heavy metals in the soil and ash were primarily present in the residual and carbonate fractions, while significant amounts of Ni, Cd and Zn were present in the exchangeable fraction of the sludge. Changes in the distribution pattern with increase in the exchangeable fraction of Ni, Zn, and, Cd to a lesser extent was noted in post peanut soils indicating an increase in the soluble and mobile forms of these metals in the soil over time. Ash application to the soil did not increase the soluble forms of any of the metals studied. Sludge application did not change the distribution of Cu and Cr, while Pb was redistributed into its residual form indicating that Cu and Pb mobility was not pH dependent. The changes in the metal distribution in the mixture treated soils were primarily controlled by the sludge. Paddy and peanut yields increased over control with 100% ash addition, but decreased in sludge treated soils more so in the case of peanut. The total and plant available concentrations of all heavy metals increased in soil with sludge, ash and their mixture additions and related linearly to their concentrations in both vegetative and storage tissues. The vegetative tissues showed higher transfer efficiency for all metals than the storage organs indicating a physiological barrier. Despite the migration of Zn in the more mobile forms in sludge treated soils resulting in higher range of concentration in plant parts, the transfer efficiency was low. The incorporation of ash did not show any metal accumulation in the soil and also there was no evidence of increase in their mobile forms. This short-term study represents a small fraction of residence time of metals in sludge amended soils and hence the role of organic matter in metal immobilization cannot be regarded as permanent. Potential relevant risks could arise from mineralization of organic matter and consequently from the release of metals into their more soluble forms over time.

ACKNOWLEDGEMENTS

This work forms a part of sponsored projects, Land Restoration Through Waste Management, Multipurpose Waste Recycling Project and Fly Ash Management in India. The authors thank India Canada Environment Facility, New Delhi, India, Canadian International Development Agency and International Development Research Center, Canada for their funding.

REFERENCES

1. Adriano, D.C., Page, A.L., Elseewi, A.A., Chang, A.C., Straughan, I.R., Utilization and disposal of fly ash and other coal residues in terrestrial ecosystems, *J. Environ. Qual.*, 9, 333, 1980.
2. El-Mogazi, D., Lisk, D.J., Weinstein, L.H., A review of physical, chemical and biological properties of fly ash and effects on agricultural ecosystems, *Sci. Total Environ.*, 74, 1, 1988.
3. Bilski, J. J., Alva, A.K, Sajwan, K. S., Fly ash, In *Soil Amendments and Environmental Quality*, Rechcigl, J. E., Ed., CRC Press, Inc., 1995, chapter 9.
4. Sikka, R., Kansal, B. D., Effect of fly ash application on yield and nutrient composition of rice, wheat and on pH and available nutrient status of soils, *Bioresource Tech.*, 51, 199, 1995.
5. Kalra, N., Jain, M. C., Joshi, H. C., Choudhary, R., Harit, R. C., Vatsa, B. K., Sharma, S. K., Kumar, V., Fly ash as a soil conditioner and fertilizer, *Bioresource Tech.*, 64, 163, 1998.
6. Wlliams, J.H., Zinc, copper and nickel-suggested safe limits in sewage sludge treated soils, In *Environmental Effects of Organic and Inorganic Contaminants in Sewage Sludge*, Davis, R. C., Hucker, G. and Hermite, P.L., Eds., D. Reidel Publishing company, Holland, 82-90.
7. Fang, M., Wong, J. W. C., Li, G. X., Changes in biological parameters during co-composting of sewage sludge and fly ash residues, *Bioresource Tech.*, 64, 55, 1998.

8. Wong, J. W. C., Jiang, R. F., Su, D. C., The accumulation of boron in Agropyron Elongatum grown in coal fly ash and sewage sludge mixture, *Water Air and Soil Pollut.*, 106, 137, 1998.

9. Carlson, C.L., Adriano, D.C., Environmental impact of coal combustion residues, *J. Environ. Qual.*, 22, 227, 1993.

10. Alva, A., Paramasivam, S., Prakash, O., Sajwan, K.S., Ornes, W.H., van Clief, D., Effects of fly ash and sewage sludge amendments on transport of metals in different soils, In *Biogeochemistry of trace elements in coal and coal combustion byproducts*, Sajwan, K.S., Alva, A.K. and Keefer, R.F., Eds., Kluwer Academic/Plenum Publ., 1999, 207.

11. Clark, R.B., Zeto, S.K., Baligar, V.C., Ritchey, K.D., Nickel, lead, cadmium, and chromium concentrations in shoots of maize grown in acidic soil amended with coal combustion byproducts, *Biogeochemistry of trace elements in coal and coal combustion byproducts*, Sajwan, K.S., Alva, A.K. and Keefer, R.F., Kluwer Academic/Plenum Publ., 1995, 59.

12. Sauerbeck, D.R., Plant element and soil properties governing uptake and availability of heavy metals derived from sewage sludge, *Water Air and Soil Pollut.*, 57-58, 227, 1999.

13. Heckman, J.R., Angle, J.S., Chaney, R.L., Residual effects of sewage sludge on soybean: I. Accumulation of heavy metals, *J. Environ. Qual.*, 16, 113, 1987.

14. Scancar, J., Milacic, R., Strazar, M., Burica, O., Total metal concentrations and partitioning of Cd, Cr, Cu, Fe, Ni and Zn in sewage sludge, *Sci. Total Environ.*, 250, 9, 2000.

15. Alloway, B., Jackson, A., The behavior of heavy metals in sewage sludge-amended soils, *Sci. Total Environ.*, 100, 151, 1991.

16. Moreno, J. L., Garcia, C., Hernandez, T., Ayuso, M., Application of composted sewage sludges contaminated with heavy metals to an agricultural soil, *Soil Sci. Plant Nutr.*, 43, 565, 1997.

17. Brallier, S., Harrison, R.B., Henry, C.L., Dongsen, X., Liming effects on Cd, Cu, Ni and Zn in a soil amended with sewage sludge 16 years previously, *Water Air and Soil Pollut.*, 86, 195, 1996.

18. Berti, W.R., Jacobs, L.W., Chemistry and phytotoxicity of soil trace elements from repeated sewage sludge applications, *J. Environ. Qual.*, 25, 1025, 1996.

19. Bradshaw, A.D., Chadwick, M.J., *The restoration of land*, University of California Press, Berkeley, 1980.

20. Schumann, A.W., Sumner, M.E., Plant nutrient availability from mixtures of fly ash and biosolids, *J. Environ. Qual.*, 28, 1651, 1999.

21. Menon, M.P., Ghuman, G.S., James, J., Chandra, K., Effects of coal fly ash amended composts on the yield and elemental uptake by plants, *J. Environ. Sci. Health*, 27, 1127, 1992.

22. Adriano, D.C., Page, A.L., Elseewi, A.A., Chang, A.C., Cadmium availability to Sudan grass grown on soils amended with sewage sludge and fly ash, *J. Environ. Qual.*, 11, 197, 1982.

23. Miller, W. P., Martins, D. C., Zelazny, L. W., Effect of sequence in extraction of trace metals from soils, *Soil Sci. Soc. Am. J.*, 50, 598, 1986.

24. Tsadilas, C.D., Mstsi, T., Barbayiannis, N., Dimoyiannis, D., Influence of sewage Sludge application on soil properties and on the distribution and availability of heavy metals fraction, *Commun. Soil Sci.. Plant Anal.*, 26, 2603, 1995.

25. Lake, D.L., Kirk, P.W.W., Lester, J.N., Fractionation, characterization and speciation of heavy metals in sewage sludge and sludge-amended soils, A review, *J. Environ. Qual.*, 13(2), 175, 1984,

26. Chang, A.C., Page, A.L., Warneke, J.E., Grgurevic, E., Sequential extraction of soil heavy metals following a sludge application, *J. Environ. Qual.*, 13, 33, 1984.

27. Shuman, L.M., Fractionation method for soil microelements, *Soil Sci.*, 140, 11, 1985.

28. Beckett, P. H. T., The use of extractants in studies on trace metals in soils, sewage sludges and sludge-treated soils, *Adv. Soil Sci.*, 9, 143,1989.

29. Jackson, M.L., *Soil Chemical Analysis*, Prentice-Hall of India, Bombay, 1967.

30. Nelson, D.W., Sommers, L.E., Total carbon, organic carbon and organic matter, In *Methods of Soil Analysis*, Part II, 2nd Edition, Page, A, L., Miller, R. H. and Keeney, D. R., Eds., American Society of Agronomy, Madison, Wisconsin, 1982,539.

31. Subbaiah, B.V., Asija, G.L., A rapid procedure for the estimation of available nitrogen in soils, *Current Sci.*, 25, 259, 1956.

32. Bremner, J. M., Mulvaney, C.S., Nitrogen -Total. In *Methods of Soil Analyses*, Part 2, 2nd Edition, Page, A.L., Miller, R.H. and Keeney, D.R., Eds., American Society of Agronomy, Madison, Wisconsin, 1982, 595.

33. Lynch, J., Provisional Elemental Values for Four New Geochemical Sil and Till Reference Materials, Till-1, Till-2, Till-3, and Till-4. Geological Survey of Canada contribution paper 1996, 701, 1996,277.

34. Maier, E.A., Griepink, B., Muntau, H., Vercoutere, K., Certification of the total contents of the Aqua Regia soluble contents of Cd, Co, Cu, Pb, Mn, Ni, and Zn in a sewage sludge, Commission of the European Communities, Community bureau of reference, 1993, 60.

35. Tran, T.S., Simard, R. R., *Soil sampling and Methods of Soil Analysis*. Lewis Publ., Boca Raton, FL, 1993, 43.

36. Jenkins, R., Gould, R.W., Dalegedcke, *Quantitative X-ray Spectroscopy Series*, 20, Marcel Dekker, Inc, New York, USA, 1998.

37. Jones, J. B., Case, V.W., Sampling, handling and analyzing plant tissue samples, In *Soil Testing and Plant Analysis*, 3rd Edition, Westerman, R. L., Eds., Soil Science Society of America, Madison, Wisconsin, 1990, 404.

38. Bingham, F. T., Page, A. L., Mahler, R. J., Ganje, T. J., Cadmium availability to rice in sludge-amended soil under " flood" and " nonflood" culture, *Soil Sci. Soc. Am. J.*, 40, 714. 1976.

39. Dutta, D., Mandal, B., Mandal, L. N., Decrease in availability of zinc and copper in acidic to near neutral soil on submergence, *Soil Sci.*, 147, 187, 1989.

40. Sajwan, K. S., Lindsay, W. L., Effects of redox on zinc deficiency in paddy rice, *Soil Sci. Soc. Am. J.*, 50, 1264, 1986.

41. Sims, J.T., Kline, J.S., Chemical fractionation and plant uptake of heavy metals in Soils amended with Co-composted sewage sludge, *J. Environ. Qual.*, 20, 387, 1991.

42. EPA, Part 503, Standards for the use or disposal of sewage sludge. 1995.

43. Giordano, M., Mortvedt, J. J., Mays, D. A., Effects of municipal wastes on crop yields and uptake of heavy metals, *J. Environ. Qual.*, 4, 394, 1975.

44. McGrath. S. P., Brookes P.C, Giller K.E., Effects of potentially toxic metals in soil derived from past application of sewage on nitrogen fixation by Trifolium repens L., *Soil Biol. Biochem.*, 20, 415,1988.

45. Giller, K.E., McGrath, S.P., Hirsch, P.R., Absence of nitrogen fixation in clover grown on soil subject to long-term contamination with heavy metals is due to survival of only ineffective Rhizobium, *Soil Biol. Biochem.*, 21, 841,1989.

46. Koomen, I., McGrath, S. P., Giller, K.E., Mycorrhizal infection of clover is delayed in soils contaminated with heavy metals from past sewage sludge applications, Soil *Biol. Biochem.* , 22, 871,1990.

47. Brookes, P. C., McGrath, S.P.,Heijnen, C., Metal residues in soils previously treated with sewage sludge and their effects on growth and nitrogen fixation by blue-green algae, *Soil Biol, Biochem.*, 18, 345, 1986.

48. Epstein, E., Taylor, J.M., Chaney, R.L., Effects of sewage sludge and sludge compost applied to soil on some soil physical and chemical properties, *J. Environ. Qual.*, 5, 422, 1976.

49. Crawford, D.M., Baker, T.G., Maheswaran, J., Changes in soil chemistry associated with changes in soil pH in Victorian pastures, *Aust. J. Soil Res.*, 33, 491, 1995.

50. Fleming, G. A., Davis, R. D., Contamination problems in relation to land use, In *Processing and Use of Organic and Liquid Agricultural Wastes*, Reidel, Dordrecht, 1985, 304.

51. Navas, A., Bermudez, F., Machin, J., Influence of sewage sludge application on physical and chemical properties of Gypsisols, *Geoderma, 87*, 123, 1998.

52. Stadelmann, F. X., Furrer, O. J., Long-term effects of sewage sludge and pig slurry applications on micro-biological and chemical soil properties in foiled experiment. In *Long-term Effects of Sewage Sludge and Farm Slurry Application*, Williams, J. H., Guidiand, G. and L'Hermite, P., Eds., Elsevier, London, 1985, 136.

53. Bernal, M.P., Roig, A., Garcia, D., Nutrient balances in calcareous soils after application of different rates of pig slurry, *Soil Use and Manage.*, 9, 9, 1993.

54. Yong, R. N., Galvez-Cloutier, R., Selective sequential extraction analysis of heavy-metal retention in soil, *Can. Geotech.J.*, 30, 834, 1993.

55. Sposito, g., Lund, J, Chang, A. C., Trace metal chemistry in arid-zone soils amended with sewage sludge:I. Fractionation of Ni, Cu, Zn, Cd and Pb in solid phases, *Soil Sci. Soc. Am. J.*, 46, 260, 1983.

56. Emmerich, W.E., Lund, L.J., Page, A.L., Chang, A.C., Solid phase form of heavy metals in sewage sludge-treated soil, *J. Environ. Qual.*, 11, 178, 1982.

57. McGrath, S.P., Cegarra, J., Chemical extractability of heavy metals during and after long time applications of sewage sludge to soil, *J. Soil Sci.*, 43, 313, 1992.

58. Taylor, R., Xiu, H., Mehadi, A., Shuford, J., Tadesse, W., Fractionation of residual cadmium, copper, nickel, lead and zinc Iin previously sewage sludge-amended soil, *Commum. Soil Sci. Plant Anal.*, 26, 2193, 1995.

59. Planquart, P., Bonin, G., Prone, A., Massiani, C., Distribution, movement and plant availability of trace metals in soils amended with sewage sludge composed; application to low metal loadings, *Sci. Total Environ.*, 241, 161, 1999.

60. Knudtsen, K., O'Connor, G. A., Characterization of iron and zinc in Albuquerque sewage sludge, *J. Enviorn. Qual.*, 16, 85, 1987.

61. Pitchel, J., Anderson, N., Trace metal bioavailability in municipal solid waste and sewage sludge composts, *Bioresource Tech.*, 60, 223, 1997.

62. Kabata-Pendias A, Pendias H., *Trace Elements in Soils and Plants*, CRC Press, Boca Raton, FL., 1992.

63. Chlopecka, A., Bacon, J. R., Wilson, M. J., Kay, J., Forms of cadmium, lead, and zinc in contaminated soils from southwest Poland, *J. Environ. Qual.*, 25, 69, 1996.

64. Ma, L. Q., Rao, G. N., Chemical fractionation of cadmium, copper, nickel, and zinc in contaminated soils, *J. Environ. Qual.,*26, 259, 1997.

65. Walter, I., Cuevas, G., Chemical fractionation of heavy metals in a soil amended with repeated sewage sludge application, *Sci. Total Environ.*, 226, 113, 1999.

66. Ramos, L., Hernandez, L. M., Gonzalez, M. J., Sequential fractionation of copper, lead, cadmium and zinc in soils from or near Donana National Park, *J. Environ. Qual.*, 23, 50, 1994.

67. Benschoten, V., Reed, J. E., Matsumoto, B. E., McGarvey, P.J., Metal removal by soil washing for an iron oxide coated sandy soil, *Water Environ. Res.*, 66, 168, 1994.

68. Heil, D., Hanson, A., Zohrab, S., The competitive binding of lead by EDTA in soils and implications for heap leaching remediation, *Radioactive Waste Manage., Environ. Rest.*, 20, 111, 1996.

69. McGrath, S. P., Zhao, F. J., Dunham, S. J., Crosland, A. R., Colemar, K., Long-term changes in the extractibiltiy and bioavailability of zinc and cadmium after sludge application, *J. Environ. Qual.*, 29, 875, 2000.

70. Smith. S., Effect of soil pH on availability to crops of metals in sewage sludge treated soils. I. Nickel, copper and zinc uptake and toxicity to ryegrass, *Environ. Pollut.*, 85, 321, 1995.

MINE SOIL REMEDIATION USING COAL ASH AND COMPOST MIXTURES

John J. Sloan[1] and Don Cawthon[2]

[1]Texas Agricultural Experiment Station
Dallas, TX 75252
U.S.A.
[2]Texas Agricultural Experiment Station and Tarlton State University
Stephenville, TX 76402
U.S.A.

ABSTRACT

Combinations of coal combustion ashes and composted animal manures may accelerate revegetation of drastically disturbed landscapes, such as surface mines. The objective of this study was to evaluate the effect of coal ash plus compost mixtures on soil chemistry and plant growth in acid mine soils. Scrubber sludge (flue gas desulfurization residue + fly ash) or bottom ash was mixed with three types of compost (dairy manure, poultry litter, and biosolids) at rates of 0, 33, 67 and 100 % (v/v). The coal ash + compost mixtures were blended with acid mine soil (pH 4.0) at rates of 15, 30, and 45% (v/v) (equivalent to 150, 300, and 450 dm^3 m^{-3}) and placed in pots in a greenhouse. Ryegrass seeds were planted in each pot and harvested after two months growth with no fertilization. Applying scrubber sludge residue alone at rates of 15, 30 and 45% (v/v) decreased ryegrass yield, but increased soil pH from 4.0 to 7.2, 7.1, and 7.6, respectively. The same rates of bottom ash increased soil pH to 5.2, 5.6, and 6.5, respectively, but had little effect on ryegrass yield. Composted dairy manure and biosolids increased ryegrass growth at rates up to 45% (v/v) when applied alone. Composted poultry litter increased ryegrass growth at a 15% rate, but decreased it at rates of 30 and 45% due to excess dissolved salts. Copper and Zn uptake were correlated to organic matter application rate. Leachate concentrations of P were increased by the addition of organic amendments and sulfur concentrations were increased by the addition of scrubber sludge. Bottom ash had no significant effect on heavy metal uptake or leachate composition. The results demonstrate that combinations of animal manure compost with coal combustion ashes can effectively stimulate biomass production in acidic surface mine soils.

INTRODUCTION

Many regions of the country and world are faced with the challenge of dealing with multiple waste byproducts. Eastern Texas is dominated by agriculture, surface mining and electrical generating activities (Fig. 1). Each of these activities generates waste byproducts or, in the case of surface mining, disrupts natural ecosystem functions. Since these activities occur in relative close proximity to each other, there is a unique opportunity to combine the waste products to create soil amendments that can significantly improve the re-vegetation of drastically disturbed surface-mine soils.

Surface mining destroys the upper layer of soil that typically has the best combination of biological, chemical and physical properties to support vegetative growth. A common mining practice is to replace the top soil with subsurface materials that do not contain acid-producing minerals (e.g., pyrite). Although these topsoil replacement materials do not generate acidic runoff,

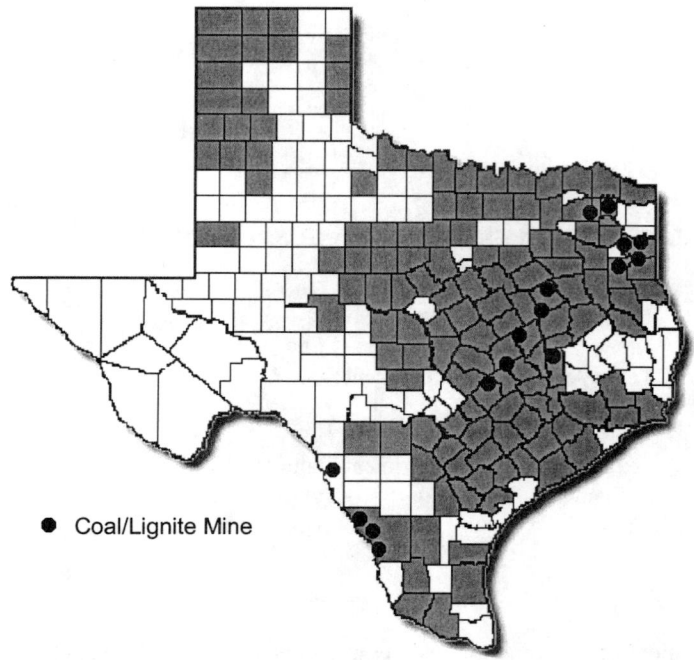

Figure 1. Location of coal and lignite mines in Texas. Shaded areas are counties that produced more than 275,000 metric tons of animal manure (beef, dairy, and/or poultry) during 2000 (courtesy of Foundation for Organic Resources Management).

they lack the proper combination of properties to support healthy plant growth. They tend to have acid pH values (<5.5) and are essentially devoid of organic matter. Typically, high rates of inorganic fertilizer are needed to reestablish vegetation, typically grasses. In fact, these areas usually require a high level of management for many years in order to reestablish a self-sustaining ecosystem.

Re-vegetation of land that suffers from topsoil depletion would benefit from addition of appropriate amendments that neutralize acidity and contain organic matter and essential plant nutrients. Such amendments would foster the re-establishment of soil microbiology[1], reduce the environmental and food-chain risks associated with heavy metals, and provide suitable physical qualities to promote rapid plant root and shoot growth.

The various ashes generated by burning coal in order to produce electricity have properties that could prove beneficial for acidic surface mine soils.[2] Fly ash is the fine particulate material filtered from the exhaust stream. Fly ash amendments have improved the physical properties of both heavy clay and sandy soils, and to a lesser extent, improved soil chemical properties to maximize plant growth.[3] Fly ash can supply essential plant nutrients, such as Ca, S, B, Mo, and Se. However, excessive loading of some trace elements, including B, Se, As, and Mo, and highly soluble salts are of concern when using coal combustion by-products as soil amendments.[4] Fly ash mixed with biosolids at the Bunker Hill Superfund Site corrected surface and subsoil acidity, increased field water holding capacity and electrical conductivity in proportion to the amount applied, up to 112 Mg ha^{-1}.[5]

The use of lime slurries to remove SO_2 from the combustion vapors creates a wet flue gas desulfurization (FGD) residue, also called scrubber sludge.[6] In some cases, the fly ash is combined with the FGD residue to create a material that is easier to handle. FGD residues were used effectively as a boron source on Hubbard loamy sand and Renova silt loam soils in Minnesota.[7, 8] The placement of the FGD residue at 0.15m depth was determined to inhibit soybean seedling emergence while a depth of 0.25m showed no significant impairment to seedlings.[7]

Table 1. Chemical properties of coal combustion scrubber sludge and bottom ash. Scrubber sludge is an 85:15 combination of fly ash:FGD residue (w:w).

Parameter	Scrubber sludge	Bottom ash
pH	10.7	7.2
Bicarbonate alkalinity as HCO3 (mg kg^{-1})	<20	920
Carbonate alkalinity as CO3 (mg kg^{-1})	1580	<20
TCLP extractable elements (mg L^{-1})		
As	0.08	0.02
Ba	0.29	2.27
Cd	<0.01	<0.01
Cr	0.09	<0.01
Pb	<0.05	<0.05
Hg	0.0051	<0.0002
Se	0.30	<0.02
Elemental Analysis (%)		
Silica, SiO2	50.1	59.8
Alumina, Al2O3	12.9	11.3
Titania, TiO2	0.8	0.9
Ferric Oxide, Fe2O3	10.9	16.4
Calcium Oxide, CaO	12.9	6.6
Magnesium Oxide, MgO	1.8	1.7
Potassium Oxide, K2O	1.2	0.6
Sodium Oxide, NaO	0.3	0.3
Sulfur Trioxide, SO2	6.2	0.3
Phosphorus Pentoxide, P2O5	0.1	0.01
Strontium Oxide, SrO	0.1	0.15
Barium Oxide, BaO	0.2	<0.01
Manganese Oxide, Mn3O4	0.1	0.03

Although fly ash and FGD residue have received some attention as soil amendments, little information is available on the use of bottom ash for similar purposes. Drum-type boilers that burn pulverized coal produce a course-textured cinder-like bottom ash. This ash is usually hydraulically sluiced from the boiler to an onsite lagoon for long term storage. Previous research found that plant growth was not inhibited by the inclusion of bottom ash in a potting medium.[8-12] However, after studying microbial activity at sluiced coal ash disposal lagoons, Klubek et al.[13] concluded that application of bottom ash to soil would require the addition of low C:N ratio residues to stimulate microbial community development. Properly composted and cured animal manures should have low C:N ratios, plus supply other essential nutrients for microbial and plant growth.

In addition to mining activities, the eastern half of Texas is also the setting for several animal production industries, including beef, dairy and poultry production. In most of the counties, the combined manure production from all animal operations exceeded 275,000 Mg during the year 2000 (Fig. 1). Many livestock and poultry operations in this area have reached the point where they can no longer apply manure to adjacent soils due to excessive nutrient loads. Composting animal manures prior to transportation stabilizes the organic matter and reduces the overall bulkiness. Therefore, composted manure can be transported more economically than raw manures, making it easier to remove the material from watersheds that are already excessively nutrient rich.

Utilization of coal combustion byproducts is an increasing national and international concern as annual production of these materials in the U.S. reached 98.2 million Mg in the year 2000.[14] At the same time, proper utilization of animal manures and biosolids is currently receiving national and international attention due to non-point source loading of nutrients and pathogens into surface and underground water resources. Due to the cost of transportation, most electrical generating plants in East Texas store their various coal ashes onsite. In many cases, the electrical plant is located adjacent to the surface-mine where the coal is extracted. Large dairies and

311

confined animal feeding operations are beginning to compost their animal manure in order to reduce its bulkiness and prepare it for possible transportation out of the nutrient-rich watershed. Thus, the situation in Eastern Texas is conducive to the use of coal ashes and composted animal manures for remediating surface mine soils.

Since the production of coal combustion by-products and animal manures continues to exceed the rate of use, there is a need to develop strategies that utilize the beneficial properties of these waste materials while protecting soil and water quality. Utilizing coal ashes and animal manures for the purpose of revegetating surface mine soils may have a positive effect on the environment by preventing excessive buildup of nutrients in soils adjacent to animal operations and by reducing the erosion of sediments from disturbed mine soils through the rapid reestablishment of vegetation. The objective of this research was to evaluate the ability of coal ash and animal manure compost mixtures to improve vegetative growth in an acidic surface mine soil.

MATERIALS AND METHODS

Coal ashes: Two types of coal combustion ashes were used in the study. Both ashes were collected from an electrical generating power plant near Hallsville, Texas that produces 303,000 Mg fly ash, 204,000 Mg flue gas desulfurization (FGD) residue, and 86,700 Mg bottom ash per year. At the power plant, most of the fly ash is mixed with dewatered FGD residue to produce a more manageable product that we call scrubber sludge. The scrubber sludge is transferred by conveyor to a storage pit for temporary storage before removal to an onsite disposal area. For this study, we used scrubber sludge that was a 85:15 mixture of fly ash:FGD residue (w/w, dry wt.). The scrubber sludge was collected from the temporary storage pit. Bottom ashes are hydraulically sluiced from the boiler into storage lagoons. Later the lagoon is drained and a bulldozer pushes the bottom ashes into large mounds in order to reduce storage space and reutilize the disposal lagoon. The bottom ash used in this study was collected from one of the large storage mounds. Coal ashes were air-dried, ground with a hammer mill grinder, and passed through a 2 mm sieve. Chemical properties of the scrubber sludge and bottom ash are shown in Table 1.

Composts: Poultry litter compost was produced from 1-year old bedding from a broiler production facility in Eastern Texas. Water was added to the poultry litter to raise the moisture content to 40% before composting. Dairy manure came from a freestall confined dairy operation in Como, Texas. Manure was hydraulically flushed from the stalls to a holding tank, from which a submersible pump moved the slurry into a solid/liquid separator. Solid waste was deposited on a concrete slab while liquid effluent was returned to the lagoon for recycling as flush water or for irrigation. Solid manure was left on the concrete slab until the moisture content dropped from about 75% to 65%. Both poultry and dairy manures were thermophylically composted (>54°C for 3 days) in a rotating drum for approximately 10 days. No additional bulking agents were added to the manures prior to composting. Biosolids were collected from the wastewater treatment plant in Sulphur Springs, Texas and blended 50:50 with soft wood shavings, followed by composting in the same way as the poultry litter and dairy manure. The composted organic materials were screened through a 10 mm sieve to remove any large particles.

Soil: Rather than stockpiling the original topsoil for later reuse, the mine operators identified non-acid producing subsurface overburden for use as a topsoil replacement after extraction of coal. We collected overburden material from the upper 30 cm of an area that had been leveled and prepared for revegetation. Soil was air-dried, ground in a hammer mill grinder and then passed through a 2 mm sieve to make particle sizes more uniform.

Mixtures and application rates: Scrubber sludge or bottom ash was mixed with each form of compost at rates of 0, 33, 67, and 100% (v/v), resulting in a total of 12 mixtures. Each of the mixtures was added to mine soil at rates of 150, 300 and 450 $dm^3 m^{-3}$. For scrubber sludge applied alone, the rates were equivalent to 165, 324, and 478 g kg^{-1} mix and for the bottom applied alone,

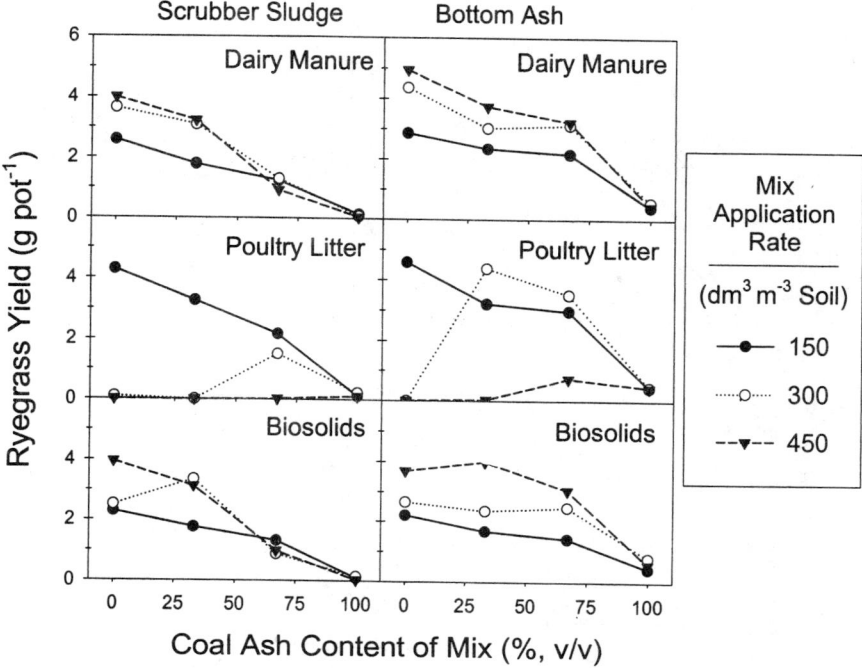

Figure 2. Effect of scrubber sludge (left column) or bottom ash (right column) addition to composted dairy manure, poultry litter, or biosolids on ryegrass growth when mixes were applied to an acidic mine soil (pH 4.0) at rates of 150, 300, and 450 dm^3 m^{-3} (15, 30, or 45% (v/v)). Ryegrass yield for the control (no soil amendments) was 0.38±0.02 g pot^{-1}.

the rates were equivalent to 160, 316, and 468 g kg^{-1} mix. Each treatment was replicated 3 times. An unamended mine soil control was also included. Twenty ryegrass seeds were planted in each pot. Plants were harvested after two months growth and analyzed for yield and heavy metal concentrations.

Leachate chemistry: In a separate procedure, a portion of each of the amended soils was leached with water to determine the potential for groundwater contamination from the coal ash/compost mixtures. Amended soil was placed in a 7.6-cm diameter by 15.2-cm long polyvinyl chloride (PVC) leaching column and leached with 1 pore volume of deionized water. Leachate water was filtered through Whatman No. 1 qualitative filter paper and analyzed for inorganic P, S, and heavy metals.

RESULTS AND DISCUSSION

Ryegrass yield

Ryegrass yields varied widely for the various mixtures and application rates (Fig.2). In general, ryegrass yield decreased as the amount of coal ash in the composted manure increased for every application rate. This was primarily due to the dilution of nutrients in the composted manure by the coal ash. However, as scrubber sludge became the dominant ingredient in the mix, the decrease in ryegrass yield was greater than could be explained by nutrient dilution. Due to a pH>10 and a high carbonate content (Table 1), scrubber sludge quickly increased soil pH to above 7 (Fig. 3b). It is likely that a large and rapid increase in soil pH and salt content had an adverse effect on plant growth, microbial activity, and nutrient availability (Carlson and Adriano, 1992).[4]

313

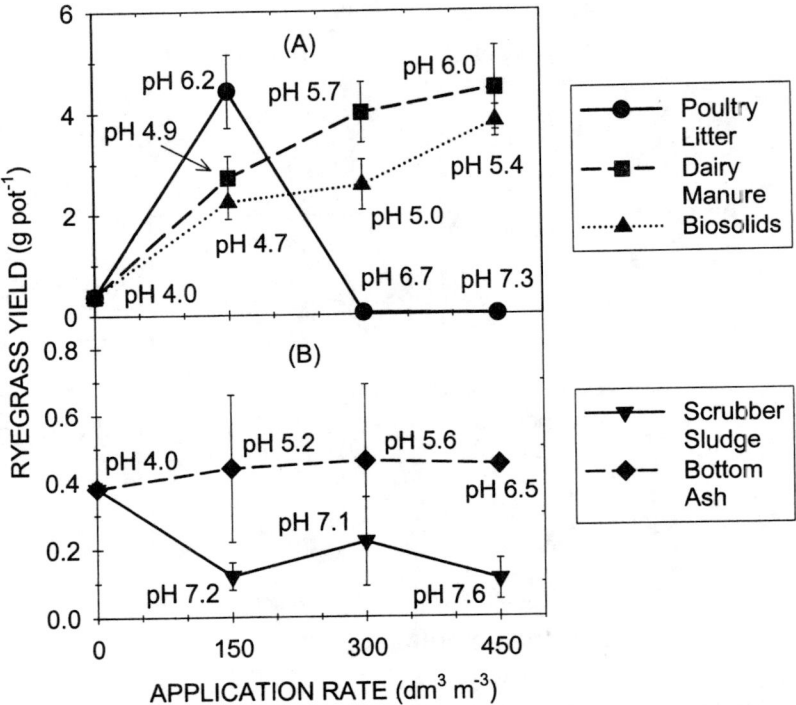

Figure 3. Effect of (A) composted organic amendments and (B) coal combustion ashes on ryegrass yield when applied as individual ingredients to an acidic surface-mine soil (pH=4). Values next to each point are the corresponding soil pH value for that treatment.

In the case of composted dairy manure and composted biosolids, there was very little interaction between the bottom ash content of the mixture and the mixture application rate. Bottom ash clearly had no effect on the composted manures other than to decrease ryegrass yield by diluting the amount of available nutrients.

Ryegrass yield responses were more easily explained by evaluating responses to individual ingredients. Ryegrass yields increased with dairy manure and biosolids application rates up to 450 dm^3 m^{-3} when applied without coal ashes (Fig. 3A). Both of these amendments also increased soil pH from 4.0 to above 5.0 for application rates above 300 dm^3 m^{-3}. Aluminum and Mn can be present at toxic concentrations at pH 4, but are quickly converted to non-toxic forms at pH > 5.0.[15,16] Sloan and Basta[17] showed that anaerobically-digested biosolids applied without lime can increase the pH of soil solution and remove toxic Al^{3+} Our results suggests that in regards to revegetation of acidic mine soils, composted dairy manure and biosolids should be applied to mine soils at relatively high rates (450 dm^3 m^{-3}). Although actual application rates would depend on the chemical properties of the specific composted manures, benefits from using high application rates would be expected because of similar management practices at dairy operations and municipal wastewater treatment plants.

Composted poultry litter produced a different response than the other composted manures when applied to the acidic mine soil. The 150 dm^3 m^{-3} rate of composted poultry litter increased ryegrass yield equivalent to the 450 dm^3 m^{-3} rates of composted dairy manure and biosolids, but 300 and 450 dm^3 m^{-3} rates prevented ryegrass growth (Fig. 3A). Composted poultry litter was also more effective at increasing soil pH than composted dairy manure and biosolids. Composted chicken litter increased soil pH from 4.0 to 6.2, 6.7, and 7.6 for the 150, 300, and 450 dm^3 m^{-3} application rates, respectively. Negative responses above the 150 dm^3 m^{-3} rate were due to excess salts present in the compost (data not shown). Results suggest that care must be taken when using

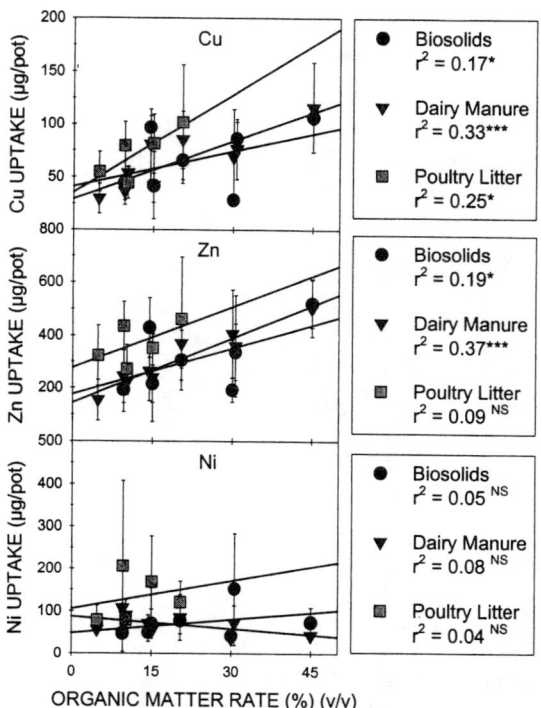

Figure 4. Effect of composted manures on ryegrass uptake of Ni, Zn, and Cu from an acidic surface mine soil.

composted chicken litter. More generally, all organic matter sources should be evaluated prior to application to soil to avoid harmful concentrations of salts or other elements.

Ryegrass yield did not significantly respond to bottom ash applied alone, but there was a very slight trend for increased growth with application rate. The bottom ash, for the most part, was a relatively inert material and supplied very few plant nutrients. However, the bottom as did contain small amounts of bicarbonates, which was probably responsible for the increase in soil pH from 4.0 to 5.2, 5.6, and 6.5 for the 150, 300, and 450 $dm^3 m^{-3}$ application rates, respectively. Although the increased pH was probably beneficial for plant growth, there was little response in ryegrass yield due to limited soil nutrients. From a disposal perspective, bottom ash can be applied to soil at high rates with no adverse effects to plant growth. Bottom ash, which is a coarse-textured material, may improve the physical properties of a clay-textured soil. Coarser-textured soils have better internal drainage, which generally provides a better environment for root respiration and microbial activity.

Ryegrass growth responded negatively to scrubber sludge at the lowest application rate of 150 $dm^3 m^{-3}$ (165 g kg^{-1}) when it was applied alone without organic matter composts (Fig. 3B). However, ryegrass responded favorably when scrubber sludge was mixed with the composted manures and applied at rates resulting in scrubber sludge additions greater than 165 g kg^{-1} (Fig. 2). Soil pH was increase from 4.0 to between 7 and 8 for all scrubber sludge application rates. Although increased soil pH is generally favorable for plant growth in an acid soil, the excess salts added with large applications of scrubber sludge were detrimental to plant growth. Although not measured in this study, high application rates of scrubber sludge and flue gas desulfurization residues can result in boron toxicity.[7] Stehouwer et al.[18, 19] found that FGD residue rates 120 g kg^{-1} increased ryegrass growth in acidic mine overburden soils, while increasing soil pH from 3.95 to >7. Our results suggest that only small amounts (150 $dm^3 m^{-3}$) of scrubber sludge should be applied directly to acidic mine soil.

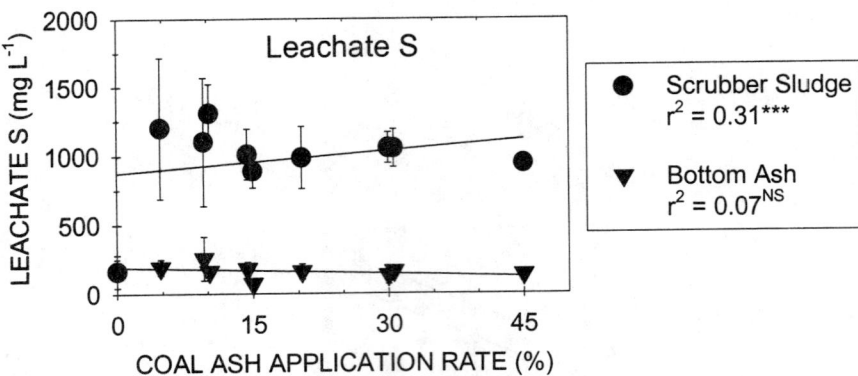

Figure 5. Effect of coal ash applications on leaching of sulfur from an acidic surface mine soil.

Heavy metal uptake

In general, heavy metal uptake by ryegrass was very low (Fig. 4) and could be attributed to composted manure amendments rather than coal ashes. Copper and Zn uptake, but not Ni, were positively correlated to application rates of composted dairy manure, poultry litter, and biosolids. Copper and Zn uptake from soil that received composted poultry litter was slightly higher than from soil amended with composted dairy manure or biosolids. However, heavy metal uptakes from all three organic matter sources were similar in magnitude. Pant tissue concentrations of Cd, Cr, and Pb were below detection limits of 0.15, 0.5, and 0.5 mg kg^{-1}, respectively.

There were probably two reasons heavy metal uptake was not correlated to coal ash applications. First, heavy metal concentrations were generally low in the coal ash materials (Table 1), so the amount added to soil from those sources was insignificant. Second, both coal ash materials increased soil pH to levels that would decrease heavy metal solubility, and thus their availability for plant uptake.[17]

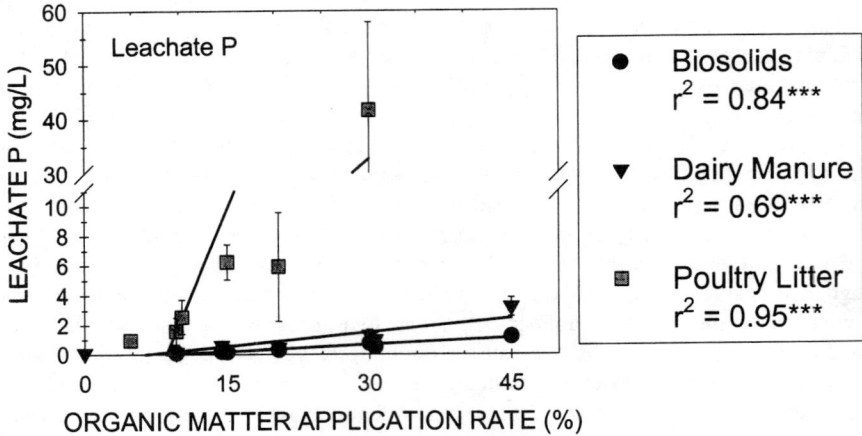

Figure 6. Effect of composted organic amendments of leaching of inorganic phosphorus from an acidic surface-mine soil.

316

Leachate

The FGD residue, which was a major component of the scrubber sludge, contained a large amount of sulfur (Table 1). Consequently, S concentrations in leachate water increased significantly with even small additions of FGD residue (Fig. 5). Bottom ash, which contained very little sulfur, had no significant effect on leachate S concentrations (Fig. 5). The high S content of FGD residue can increase soil solution SO_4-S for up to two years after application.[8]

Organic matter additions significantly increased the concentrations of inorganic P in leachate from the mine soil (Fig. 6). Composted poultry litter had much greater quantities of soluble P than either composted dairy manure or biosolids. Part of the detrimental affect to ryegrass that resulted from applications of 300 dm^3 m^{-3} composted poultry litter might be due to the antagonistic effect of high P levels on the availability of micronutrients. None of the organic matter amendments or coal ashes affected Cu, Ni, or Zn concentrations in mine soil leachate (data not shown).

CONCLUSIONS

Composted animal manures and biosolids successfully promoted ryegrass growth in acidic surface mine soil when applied at rates that prevented salt injury to plants. Composted dairy manure and biosolids were beneficial to ryegrass growth at application rates up to 45% (v/v), but composted poultry litter was detrimental to growth when applied at rates above 15% (v/v). Scrubber sludge, a combination of fly ash and flue gas desulfurization residue, increased the pH of the acidic surface mine soil, but was detrimental to plant growth when applied at rates above 150 dm^3 m^{-3} (165 g kg^{-1}). Bottom ash had small positive benefits on plant growth and soil pH at rates up to 450 dm^3 m^{-3} (468 g kg^{-1}), probably due to its modification of soil physical conditions. Overall results from the study support the idea of combining animal manures and coal ashes to create amendments that promote revegetation of acidic mine soils in regions that produce excess quantities of these materials.

ACKNOWLEDGMENTS

This research was supported by the Texas Agricultural Experiment Station and Texas A&M University at Commerce. Coal ashes, mine soil and some laboratory analyses were provided by Central and South West Services, Inc (now American Electric Power).

REFERENCES

1. Schutter, M.E. and J.J. Fuhrmann. 1999. Microbial responses to coal fly ash under field conditions. J. Environ. Qual. 28:648-652.

2. Kalra, N., M.C. Jain, H.C. Joshi, R. Choudhary, R.C. Harit, B.K. Vatsa, S.K. Sharma, and V. Kumar. 1998. Fly ash as a soil conditioner and fertilizer. Bioresource Technology. 64:163-167.

3. Korcak, R.F. 1995. Utilization of coal combustion by-products in agriculture and horticulture. Agricultural Utilization of Urban and Industrial By-products. ASA Special Publication no. 58:107-130.

4. Carlson, C.L. and D.C. Adriano. 1993. Environmental impacts of coal combustion residues. J. Environ. Qual. 22:227-247.

5. Henry, C., and S. Brown. 1997. Restoring a superfund site with biosolids and fly ash. Biocycle (Nov) 79-83.

6. U.S. Environmental Protection Agency. 1988. Wastes from the combustion of coal by electric utility power plants. EPA/530-SW-88-002. U.S. Govt. Print. Office, Washington D.C.

7. Ransome, L.S., and R.H. Dowdy. 1987. Soybean growth and boron distribution in a sandy soil amended with scrubber sludge. J. Environ. Qual. 16:171-175.

8. Sloan, J.J., R.H. Dowdy, M.S. Dolan, and G.W. Rehm. 1999. Plant and soil responses to field-applied flue gas desulfurization residue. Fuel. 78:169-174.

9. Butler, S.H. and B.C. Bearce. 1995. Greenhouse rose production in media containing coal bottom ash. J. Environ. Hortic. 13 (4):160-164.

10. Cary, E.E., M. Gilbert, C.A. Bache, and W.H. Gutenmann. 1983. Elemental composition of potted vegetables and millet grown on hard coal bottom ash-amended soil. Bull. Environ. Contam.Toxic. 31:418-423.

11. Chen,Y, A. Gottesman, T. Aviad, and Y. Inbar. 1991. The use of bottom-ash coal-cinder amended with compost as a container medium in horticulture. Second symposium on horticultural substrates and their analysis, Guernsey, UK, 10-14 Sep. 1990; Acta-Horticulturae. 294:173-181

12. Woodard, M.A., B.C. Bearce, S. Cluskey. and E.C. Townsend. 1993. Coal bottom ash and pine wood peelings as root substrates in a circulating nutriculture system. HortScience. 28(6):636-638.

13. Klubek, B., C.L. Carlson, J. Oliver, and D.C. Adriano. 1992. Characterization of microbial abundance and activity from three coal ash basins. Soil Biol. Biochem. 24(11):1119-1125.

14. American Coal Ash Association. 2002. Coal Combustion Products (CCP) Production and Use Survey – 2000. American Coal Ash Assoc. 6940 South Kings Hwy., Suite 207, Alexandria, VA 22310-3344

15. Foy, C.D. Acid soil tolerances of two wheat cultivars related to soil pH, KCl-Extractable aluminum and degree of aluminum saturation. J. Plant Nutr. 10:609-623, 1987

16. Pavan, M.A., F.T. Bingham, and P.F. Pratt. 1984. Redistribution of exchangeable calcium, magnesium, and aluminum following lime or gypsum applications to a Brazilian oxisol. Soil Sci. Soc. Am. J. 48:33-38.

17. Sloan, J.J. and N.T. Basta. 1995. Remediation of acid soils by using alkaline biosolids. J. Environ. Qual. 24:1097-1103.

18. Stehouwer, R.C. P. Sutton, R.K. Fowler, and W.A. Dick. 1995a. Minespoil amendment with dry flue gas desulfurization by-products: Element solubility and mobility. J. Environ. Qual. 24:861-869.

19. Stehouwer, R.C. P. Sutton, and W.A. Dick. 1995b. Minespoil amendment with dry flue gas desulfurization by-products: Plant Growth. J. Environ. Qual. 24:861-869.

SOIL AMENDMENTS PROMOTE VEGETATION ESTABLISHMENT AND CONTROL ACIDITY IN COAL COMBUSTION WASTE

R.M. Danker[1*], D.C. Adriano[1], Bon-Jun Koo[1], C.D. Barton[2], and T. Punshon[1,3]

[1] Advanced Analytical Center for Environmental Sciences, The University of Georgia, Savannah River Ecology Laboratory, Drawer E, Aiken, SC 29802 USA
[2] USDA Forest Service, Center for Forested Wetlands Research c/o Savannah River, Ecology Laboratory, Drawer E, Aiken, SC 29802, USA
[3] Rutgers University, Division of Life Sciences, 604 Allison Road Piscataway, NJ 08854, USA

1. ABSTRACT

The effects of adding various soil amendments and a pyrite oxidation inhibitor to aid in the establishment of vegetation and to reduce acid drainage (AD) from coal fly ash and coal reject (FA + CR*) were assessed in an outdoor mesocosm study. Preliminary greenhouse experiments and field observations at the U.S. Department of Energy's Savannah River Site (SRS) indicated that plants would not survive in this material without altering its physical and chemical characteristics. Samples of mixed FA + CR were obtained from a field site at the SRS. The following treatments were used: Biosolid only (Treatment A), Biosolid + Surfactant (Treatment B), Topsoil + Surfactant (Treatment C), and Biosolid + Topsoil + Surfactant (Treatment D). Leaching was induced due to inadequate rainfall. Loblolly pine seedlings (*Pinus taeda*) inoculated with ectomycorrhizal fungi - *Pisolithus tinctorius* (Pt) and *Scleroderma cepa* (Sc) - were transplanted into each mesocosm tank. Soil solution samplers were installed in each unit at 15 and 41 cm depths. Samples were taken periodically and measured for pH, EC, and other parameters.

The results indicate that the addition of amendments can aid in the revegetation of a FA + CR landfill and control AD. Pine seedlings growing in treatments with biosolid application were significantly taller than the treatment without it; however, there were no significant differences concerning diameter, biomass, and plant tissue concentrations of Al, Fe, and Mn for the pines. Biosolid addition also appears to be effective for mitigating proton generation. Sodium lauryl sulfate (SLS) and topsoil addition were not as important to plant survival and growth as biosolid addition; nonetheless, SLS and topsoil addition did not appear to be disadvantageous to growth in the treatment with biosolid addition (Treatment D). Based on leachate data, the topsoil + surfactant treatment had a much lower initial pH (pH ~ 3 or below) than the other treatments, and Al concentrations were correspondingly high. Electrical conductivity, in general, has been decreasing since the inseption of the study and appears to indicate that the addition of biosolid + surfactant

---

*CR = coal reject – refers to raw coal discarded due to its low combustion quality.

Chemistry of Trace Elements in Fly Ash, edited by Sajwan *et al.*
Kluwer Academic/Plenum Publishers, 2003

(Treatment B) is the most effective treatment for inducing the lowest sulfate and metal concentrations. Preliminary results indicate that the use of amendments is essential for plant growth and establishment in pyrite enriched coal waste sites.

2. INTRODUCTION

Worldwide, coal-fired power plants burn gigatonnes of coal annually, thus producing enormous amounts of coal combustion products (CCPs). In the United States, 860 million tonnes of coal were burned during 2000, generating 98 million tonnes of CCPs[1]. The general types of CCPs are fly ash (FA), bottom ash (BA), boiler slag (BS), and flue gas desulfurization residue (FGD or synthetic gypsum)[1]. Fly ash is the primary CCP produced of which little (<30%) has been reutilized for beneficial purposes in the United States[2]. Due to a lack of utilization, the accumulating FA becomes a waste disposal problem[2]. Currently, the most widely accepted disposal methods for FA are landfilling, stockpiling, and storage in settling ponds[3].

The mineralogical, physical, and chemical properties of FA are extremely variable and depend on the nature of the parent coal, combustion conditions, emission control device efficiency, storage and handling methods, and climatic conditions[2, 3]. Certain elements (As, B, Mo, and Se) present in FA can bioaccumulate and could become critical in the food chain[4, 5]. The pH is an important factor in determining the bioavailability of FA derived metals and can vary from 4.5-12.0, being primarily dependent on the S content of the parent coal and amount of lime (i.e. for desulfurization) added to the material[6].

High levels of metals/metalloids in animals exposed to FA and CR left over from combustion activities had been reported at the U.S. Department of Energy's Savannah River Site (SRS)[7]. For example, studies at the SRS indicate that Se could cause morphological deformities in both teeth and spinal columns in bullfrog tadpoles[7]. Research concerning the problems related to FA and CR storage and disposal merits attention. By-products arising from worldwide reliance on coal as a major energy source will continue to pose serious ecological problems.

Located at the SRS, the 488 D-Area Ash Basin is an unlined, earthen basin approximately 8.5 ha in size that contains approximately 1 million tonnes of dry ash and CR[8] (Fig. 1). The CR is pyritic in nature resulting in the generation of acid drainage (AD) that has contributed to a deterioration in groundwater quality and poses a threat to the biota in down gradient wetlands[8]. Pyrite (FeS_2) is commonly associated with coal as well as metal ore deposits (including Zn, Cu, U, Au, and Ag) (e.g. from old mining sites). The exposure of pyrite and other iron sulfides to air and water oxidizes the sulfides resulting in AD. This process is complex due to the involvement of chemical, biological, and electrochemical reactions that are sensitive to various environmental conditions[9]. The general stoichiometry can be described by the reaction:

$$FeS_{2(S)} + 3.75O_2 + 3.5H_2O \quad Fe(OH)_{3(S)} + 2H_2SO_4 \qquad [Eq. 1]$$

where iron sulfide and other mixed-metal sulfides decompose upon exposure to the atmosphere, producing sulfuric acid and insoluble ferric iron hydroxide from hydrolysis[10]

The AD is highly acidic (pH can be <2) and is often enriched with Fe, Mn, Al, SO_4^{2-} and other trace elements[9]. The kinetics of pyrite oxidation depend on oxygen availability, abundance of iron-oxidizing bacteria, surface area of the exposed pyrite, and chemical characteristics of the influent water[11]. *Thiobacillus ferrooxidans* is the primary iron-oxidizing bacteria involved in pyrite oxidation and catalyses the reaction. Anionic

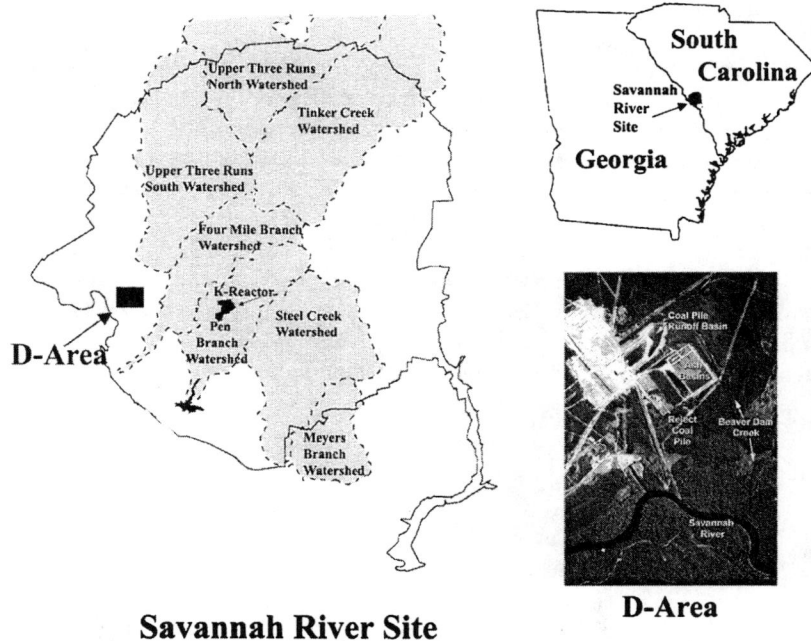

 — note: (two map/photo labels appear within the figure)

South
Carolina

Savannah
River
Site

Georgia

D-Area

Savannah River Site

D-Area

Figure 1. Location of the fly ash–coal reject landfill site (D-Area Ash Basin, DOE Savannah River Site, SC).

surfactants (e.g. sodium lauryl sulfate - SLS), applied at concentrations greater than 25 mg L^{-1} can reduce this bacterial activity consequently slowing acid production[12]. In previous studies, SLS was found to inhibit *Thiobacillus ferrooxidans* activity at lower concentrations than other generally available surfactants[11]. In essence, surfactants wash away the protective slime coating of the bacteria, breaking the surface tension of the cell wall and causing the cell to lyse[11]. However, anionic surfactants are very soluble and thus are susceptible to leaching. Accordingly, surfactant solutions may have a short period of effectiveness, generally only 2 or 3 months, but such a time frame of effectiveness may be sufficient to promote vegetative establishment on problem sites[11].

Use of a vegetative cover to control AD via enhanced evapotranspiration has been hypothesized by Barton et al.[13]. Not only would a vegetative cover influence the redox conditions and AD generation of a waste site, but enhanced buffering capacity due to organic metabolites from root exudates and plant decay could help to break the acid production cycle[11]. A healthy root system can compete with acid-producing bacteria for both oxygen and water, and organic acids can be formed by beneficial heterotrophic soil bacteria and fungi creating an unfavorable environment for *Thiobacillus ferrooxidans*[11]. Biosolids, such as municipal sewage sludge and animal waste, are an important group of soil amendments that are increasingly being used in agro-forestry and reclaimed lands[2]. As well as supplying plant nutrients, the organic matter (OM) in biosolids enhances aeration, porosity, tilth, and water retention capacity of soils[2].

The main objective of this research was to examine the use of various amendments to facilitate the establishment of vegetation and inhibit AD generation in coal combustion waste (in this case combined coal FA and CR). The study was aimed to elucidate these effects in a large–scale mesocosm study using dry FA and CR from the 488 D-Area Ash Basin on the SRS.

3. MATERIAL AND METHODS

An outdoor mesocosm study was initiated on 13 June 2000. Field samples of mixed FA + CR from the D-Area Ash Basin were used for the experiment. The material was contained in 61 x 244 cm galvanized steel cattle tanks fitted with leachate ports. About 30 cm of the mixed FA + CR was deposited in the bottom of the tanks with the amendment material (15 cm) and the topsoil added (7 cm) on the surface: Treatment A – Biosolid only, Treatment B - Biosolid + Surfactant, Treatment C - Topsoil + Surfactant, and Treatment D – Biosolid + Topsoil + Surfactant. The biosolid material consisted of sanitary sewage sludge, poultry waste, and wood chips composted for 90 days. Sodium lauryl sulfate was applied in the amount of approximately 0.24 L of a 0.6 % SLS solution to the mixed FA + CR to inhibit acid-producing bacteria. The soil type of the topsoil was Dothan sand (Fine-loamy, siliceous, thermic Plinthic Paleudults) and was collected on the SRS.

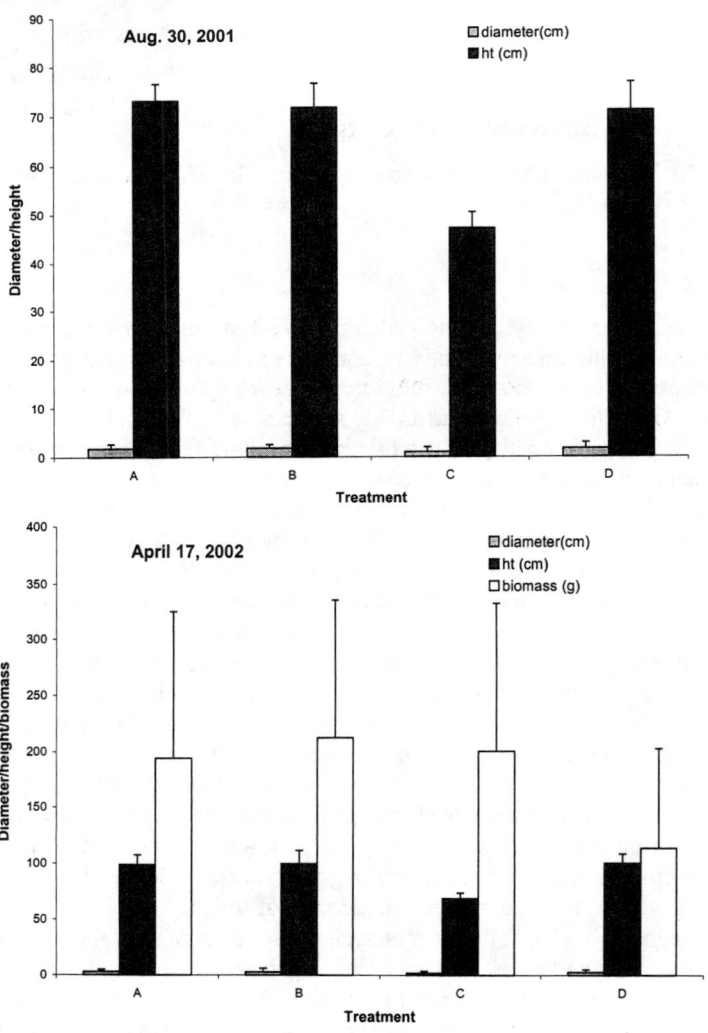

Figure 2. Height, diameter, and biomass data for pines (mean and standard deviation where n=3 for height and diameter; mean and standard deviation where n=6 for biomass)

Table 1. Chemical characteristics of substrate materials used in the mesocosm study.* [a]

Parameter	Topsoil	Biosolid/compost	FA-CR mix
pH (1:1)	5.07 (0.62)[a]	6.83 (1.03)	1.72 (0.75)
EC (1:5) (mS cm^{-1})	0.07 (0.08)	2.66 (0.53)	5.92 (1.86)
OM (%)		31.9	
NO_3-N	10.3 (3.0)	31.0 (14.7)	BDL
P[†]	6.8 (2.9)	BDL	BDL
K[†]	17.2 (10.0)	79.6 (19.2)	2.3 (1.5)
Mg[†]	48.7 (34.3)	146.0 (107.6)	174.3 (82.2)
Ca[†]	208.2 (118.3)	766.3 (64.5)	526.0 (47.0)
Al[‡]	4193.8 (146.3)	430	1767.9 (495)
Fe[‡]	3510.5 (195.5)	351	20476.4 (13827.1)
Mn[‡]	222.3 (8.89)	2.33	7.18 (12.3)
Zn[‡]	14.3 (11.1)	4.95	1.96 (6.00)
Cd[‡]	BDL	0.13	0.1 (0.58)
Pb[‡]	8.15 (1.1)		28.4 (12.4)
As[‡]	9.88 (0.72)	BDL	64.7 (43.0)
Se[‡]	BDL	BDL	8.88 (5.80)

*mg kg^{-1} except where noted otherwise.
[†]Mehlich-1 method (HNO_3-H_2SO_4).
[‡]U.S. EPA method 200.2 (HNO_3-HCl).
BDL = below detection limit
[a]Values in parentheses represent standard deviation.

These tanks were set up in a restricted access area at the Savannah River Ecology Laboratory, having a randomized order with three replicates per treatment. The treated materials were moistened and then equilibrated before planting. Loblolly pine seedlings (*Pinus taeda* L.) (9 seedlings/tank) inoculated with *Pisolithus tinctorius* (Pt) and *Scleroderma cepa* (Sc) were transplanted in mid-April 2001. An automatic sprinkler system was set up, and lysimeters (Soilmoisture Equipment Corp., Goleta, CA) were installed at two depths (15 cm and 41 cm) in mid-June 2001. After one year of growth, biomass samples of the pine seedlings were taken mid-April 2002; two seedlings were sampled, keeping disturbance of the soil to a minimum. Rainfall data taken by a weather station at the SRS showed the average rainfall per month from June 2001 to December 2001 to be 62 mm; however, there was very little rainfall for the months of October to December with the average being only 16 mm. For January 2002 to June 2002, the average was 38 mm. Rainfall averages compiled since 1952 showed an average rainfall per month of 100 mm for June through December, 76 mm for October to December, and 106 mm for January to June. Thus, drought-like conditions were indicated for much of the study period.

For periodic sampling of the soil solution, the tanks were slightly oversaturated by the sprinkler system and suction applied to the ceramic lysimeters after an overnight equilibration. The pH and EC were measured immediately after collection. Samples were taken on the following dates: 6/29/2001, 7/13/2001, 8/31/2001, 11/29/2001, 1/24/2002, 3/27/2002, 5/28/2002, and 7/31/2002. Solution samples were filtered using 0.45 μm nylon syringe filters, acidified by adding HNO_3 (1% of sample volume), and maintained in cold storage for Inductively Coupled Plasma Optical Emission Spectroscopy (ICP/OES) analysis. Ion Chromatography (IC) was used for anion analysis.

For plant biomass and analysis of plant tissue metal concentrations, plant material was oven-dried at 65°C until no further weigh loss occurred. The material was then separated into leaves, stems, and roots and weighed to determine dry biomass. Dried plant tissue was ground using a sample mill (Thomas, Arthur H., Wiley, 2-mm mesh, Philadelphia, PA) and digested in a 5 M HNO_3 + H_2O_2 by microwave (CEM Corp. MDS-2000, Matthews, NC) in pure Teflon PFA vessels.

Analysis of variance (ANOVA), general linear models (GLMs), and PROC Univariate were calculated using SAS (SAS, 1999) to determine significant differences within the data.

4. RESULTS AND DISCUSSION

4.1 Substrate Material

The FA + CR substrate material exhibited high acidity (pH = 1.72) and high EC (5.92 mS cm^{-1}). Low pH values (pH ~ 1 or lower) had been observed in sulfide-rich tailings from impoundments and weathered mine sites[14, 15]. Since the FA + CR mix is composed of alumino-silicate and sulfide minerals, high concentrations of Fe, Al, and other trace elements were evident. The low buffering capacity and low nutrient content of the FA + CR substrate are not conducive to plant establishment or growth. On the other hand, the biosolid exhibited a high OM content (~ 32%) and should provide ample buffering capacity for the substrate. The biosolid data also indicate the presence of primary and secondary plant nutrients that, although initially low, may serve as a slow release fertilizer in addition to serving as a buffering agent. The topsoil exhibited near circumneutral pH conditions and low quantities of nutrients.

Table 1. Chemical characteristics of substrate materials.* [a]

4.2. Effect on Pine Growth

Growth measurements for the pines (4 and 12 months after transplanting) indicated that there was no significant difference in diameter induced by the treatment; however, differences in height were apparently due to the biosolid application, with the pines in treatments containing biosolid (Treatments A, B, and D) being significantly ($P<0.0001$) taller than the topsoil + surfactant treatment (Treatment C) (Fig. 2).

Soils with pH values below 5.5 generally contain exchangeable Al at sufficiently high concentrations to be toxic to plants[16]. In addition, the oxidation of sulfur containing compounds may result in the formation of high acidity in the soils, generally noted by pH values below 4.0, which is highly unfavorable to the growth of most plant species[17]. For the shallow samples the pH did not drop below 3.5, and no plant mortality was noted.
Soil solution (i.e., leachate) data indicate that Treatment C had a much lower initial pH than the other treatments (pH ~ 3 or below), with Al concentrations being correspondingly high (Fig. 6). The potential for Al toxicity at these levels is high and may have contributed to the difference in plant height.

Although results indicated that the topsoil addition was not as important to plant survival and growth as biosolid addition, the topsoil addition did not appear to be

Table 2. Al, Fe, and Mn concentrations (mg kg^{-1}) in leaf tissue (mean and standard deviation where n=6).* [a]

Treatment	Al	Fe	Mn
A	251 (81)[a]	79 (17)	237 (100)
B	191 (85)	72 (10)	176 (120)
C	335 (108)	86 (38)	328 (139)
D	298 (114)	85 (26)	322 (178)

*Analysis of variance (ANOVA) indicates no significant difference.
[a]Values in parentheses represent standard deviation.

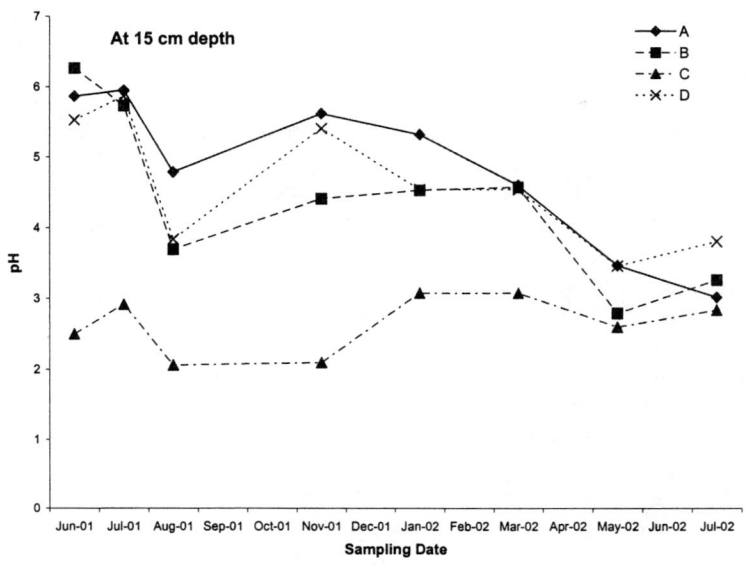

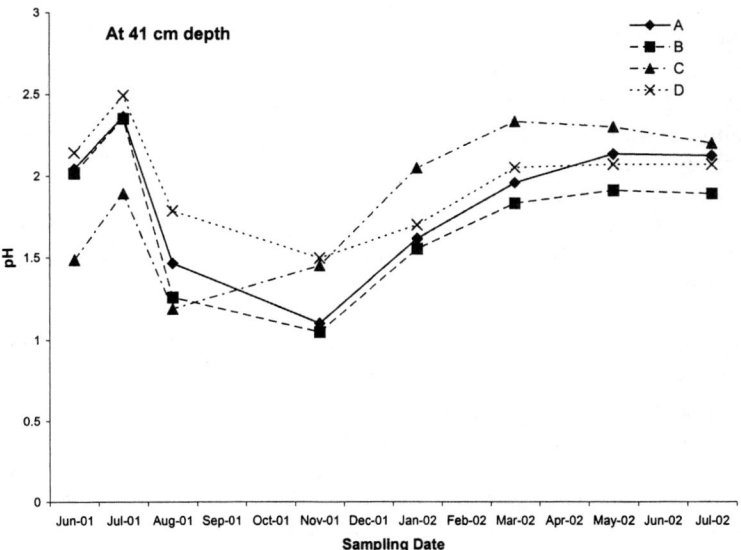

Figure 3. pH of soil solution/leachates from two sampling depths (15 and 41 cm) (mean and standard deviation where n=3)

detrimental to growth in the treatment with biosolid addition (Treatment D). In addition, height differences may be attributed to the OM content in the biosolids which supplied plant nutrients and enhanced physical soil properties[2]. Like the data for the diameter, biomass analysis showed no significant difference (P=0.4891, n=6) for the different treatments. There were also no significant differences in plant tissue concentrations of Al, Fe, and Mn for the pines (Table 2).

4.3. Effect of Treatments on pH and EC

In comparison with the initial data (Table 1) of the untreated FA + CR material, the pH was dramatically increased from 1.7 to an average pH of 5 (for the four treatments) at the shallow depth (15 cm) in the beginning of the study (6/29/2001) (Figs. 4 and 5). The pH lingered for about a month through the second sampling date (7/13/2001) and rapidly declined on the third sampling date (8/31/2001); conversely, the EC started low (5.92 mS cm^{-1}) at the inception of the study, stayed somewhat low through the second sampling date, and drastically increased on the third sampling date – exactly the opposite of pH. For the lysimeter soil solution samples collected at the 15 cm depth, pH values appear to have gradually decreased over time for Treatments A, B, and D, but not for Treatment C. Initially, the OM and/or humic compounds in the biosolid can complex Fe^{3+}, which otherwise at this state may serve as an electron acceptor that can exacerbate the oxidation of the pyrite. Over time, the OM undergoes oxidation and the buffering potential decreases eventually resulting in a lowering of the pH.

Based on the pH, the effect of the SLS on acid generation was negligible. These results may be due to the solubility and subsequent leaching and/or biodegradation of the surfactant. However, this may also suggest that the primary mechanism of the oxidation of pyrite in these materials is one of a physicochemical, rather than of a biological nature. A study by Barton et al. indicated a similar phenomenon for these substrate materials where Fe^{3+} served as the primary oxidizing agent, i.e., electron acceptor[13].

The OM content of the biosolid likely inhibited oxidation by serving as an electron donor (i.e., lowering oxidation) and complexing of Fe^{3+}. The EC of the leachates followed an inverse correlation to that of the pH. When pH values were low, EC concentrations were high. The IC data indicate that sulfate was the most dominant anion; only low concentrations of chlorides and nitrates were observed. The dominance of the sulfates was apparently due to the FeS_2 oxidation; the sulfates may serve as the main ion pair for the metals. Sulfate concentrations somewhat correspond with the EC (Fig. 4). The EC as well as the sulfate concentrations appear to be decreasing over time.

From the sulfate data (Fig. 5), the biosolid + surfactant (Treatment B) appears to have induced the lowest sulfate concentrations for the 41 cm depth samples, while the biosolid only (Treatment A) induced the lowest sulfate concentrations for the 15 cm depth samples. The peaks for the 41 cm samples occurred around the 19 November 2001 sampling date, with the leachate pH at its lowest (Fig. 3). The correspondence between the EC and sulfate concentrations was expected due to exorbitant generation of SO_4^{2-} from pyrite oxidation of the reject coal, becoming the dominant electrolyte in the soil solution/leachate. The dominating mitigating effect of the biosolid on acidity (i.e., low pH) as well as total dissolved solids (TDS), i.e., as indicated by the EC (TDS can be estimated by multiplying EC (mS cm^{-1}) by 640 for soils with EC values between 0.1 and 5.0 mS cm^{-1} and 800 for those with EC values > 5.0 mS cm^{-1}), is very obvious throughout virtually the duration of the study but especially during the initial 5 months[18]. This might be due to the time needed to equilibrate the various amendments with the substrate and to decompose the biosolid.

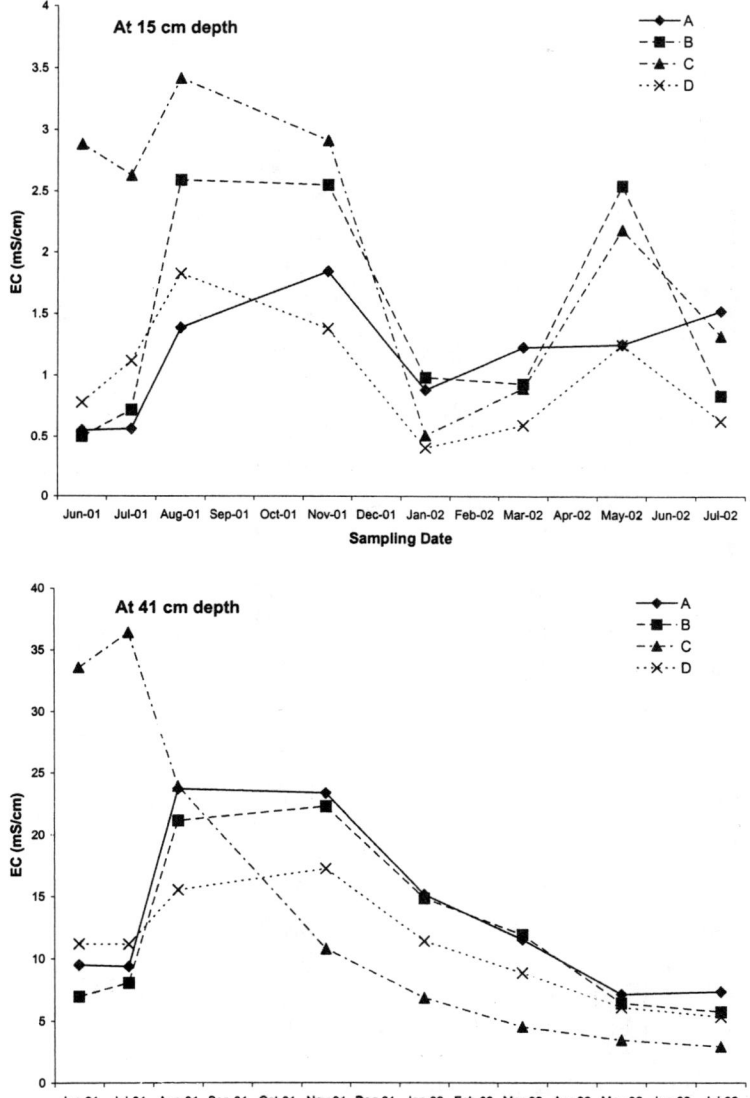

Figure 4. EC (mS cm^{-1}) of soil solution/leachates from two sampling depths (15 and 41 cm) (mean and standard deviation where n=3)

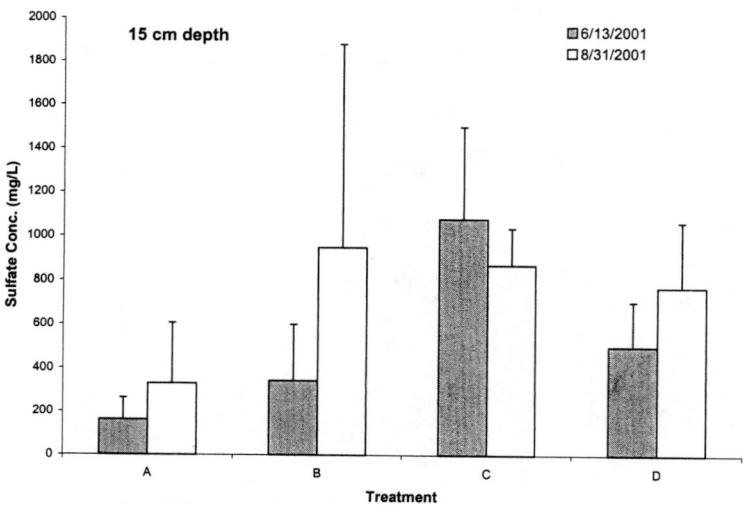

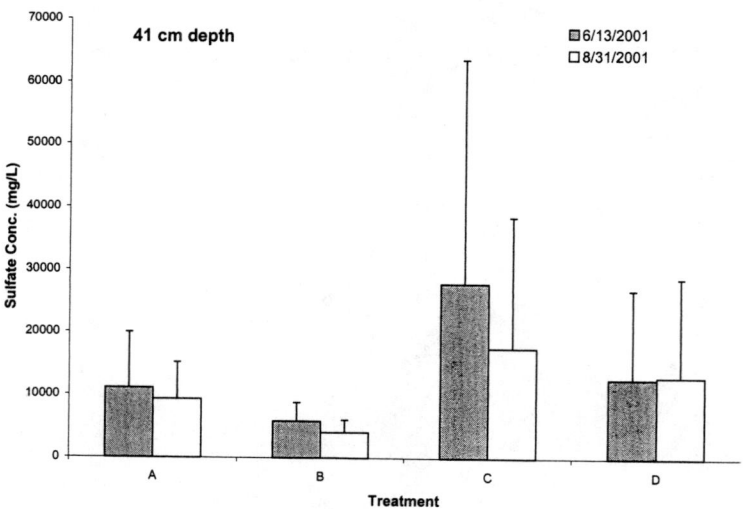

Figure 5. Sulfate concentration (mg L^{-1}) of soil solution/leachates from two sampling depths (15 and 41 cm) (mean and standard deviation where n=3)

4.4. Effect on Metal Concentration

As expected, Fe was the most dominant metal in the leachate. From these data, it appears that the biosolid + surfactant (Treatment B) was the most effective at reducing the concentrations of metals in the lower depth. This occurred even though the pH values of the treatments containing biosolid were similar.

Iron and aluminum concentrations for the 15 cm depth samples peaked around November 2001 coinciding with the fall of pH. For the 41 cm depth samples, the biosolid only (Treatment A) was consistently higher in both Fe and Al concentrations than the biosolid + surfactant (Treatment B); this may be due surfactant addition in Treatment B.

Based on initial chemical characteristics of the substrate material (Table 1), Treatments C and D were expected to have the highest Al concentration. Leachate data indicated that Treatment C had a much lower initial pH than the other treatments (pH ~ 3 or below); Al concentrations were correspondingly high (Fig. 6). While Fe could have been derived primarily from the dissolution of the pyrite in the CR, the Al could have been dissolved by such low pH from the clay minerals, the alumino silicates in the FA and Al oxyhydroxides.

Manganese concentrations for the 15 cm depth samples corresponded with the pH and Fe/Al data peaking around November 2001. Manganese concentrations for the 41 cm depth samples followed the same general trend peaking at around the same date. For the 41 cm sampling depth, samples from Treatment A had consistently higher Mn concentrations than those from Treatment B; once again this may be due to surfactant addition in Treatment B. The topsoil had a significantly higher concentration of Mn than the CR material (Table 1), and there were instances at the initial sampling dates in which treatments containing the topsoil (Treatments C and D) had higher concentration of Mn than treatments that did not contain it (Treatments A and B). At a pH above 4, Mn is not as readily available to plants, which may explain why it is not problematic. For the 15 cm sampling depth, the pH of the leachates has been above 4 for all of the treatments except Treatment C for most of the experiment.

5. CONCLUSIONS

The results indicate that revegetation of a FA + CR landfill is feasible for controlling AD with the addition of proper amendments. Differences in height of the pine seedlings appear to be due to biosolid application, with those growing in the treatments containing biosolid being significantly taller than the treatment without it; however, there were no significant differences concerning diameter, biomass, and plant tissue concentrations of Al, Fe, and Mn for the pines. Biosolid addition appears to be effective for mitigating proton generation, supplying plant nutrients, and enhancing physical soil properties. Surfactant and topsoil addition were not as important to plant survival and growth as biosolid addition; nonetheless, SLS and topsoil addition did not appear to be detrimental to growth in the treatment with biosolid addition (Treatment D). Based on leachate data, the topsoil + surfactant treatment had a much lower initial pH (pH ~ 3 or below) than the other treatments and Al concentrations were correspondingly high.

As such, differences in plant growth were observed. Based on these 1-year data, the most critical factor limiting plant establishment in pyritic coal waste (i.e., that has some resemblance to pyritic mine waste), i.e., extreme acidity, can be mitigated by adding amendments (especially biosolid and top soil) in order to limit the diffusive flux of O_2 from the atmosphere, to enhance the buffering capacity of the substrate, and to provide a more favorable rooting environment including initially supplying plant nutrients.

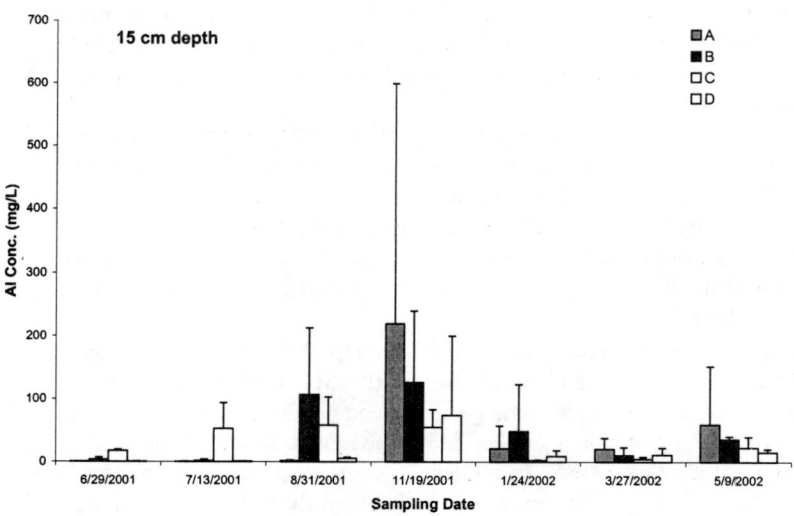

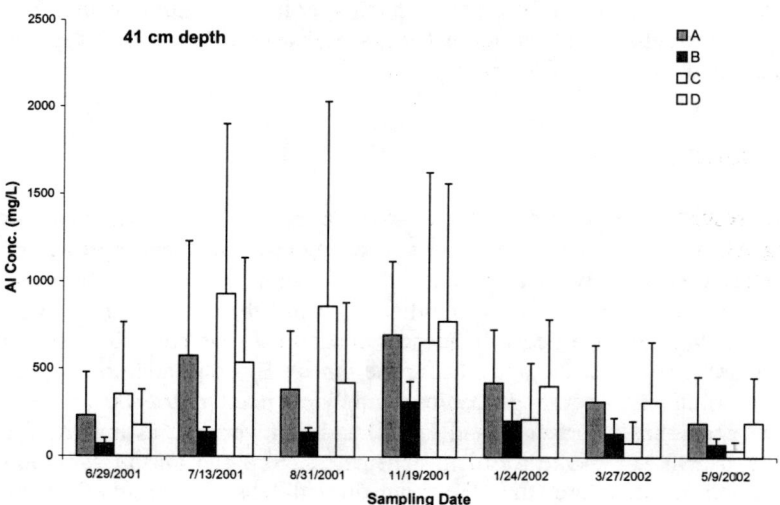

Figure 6. Fe and Al concentration (mg L^{-1}) of soil solution/leachates from two sampling depths (15 and 41 cm) (mean and standard deviation where n=3)

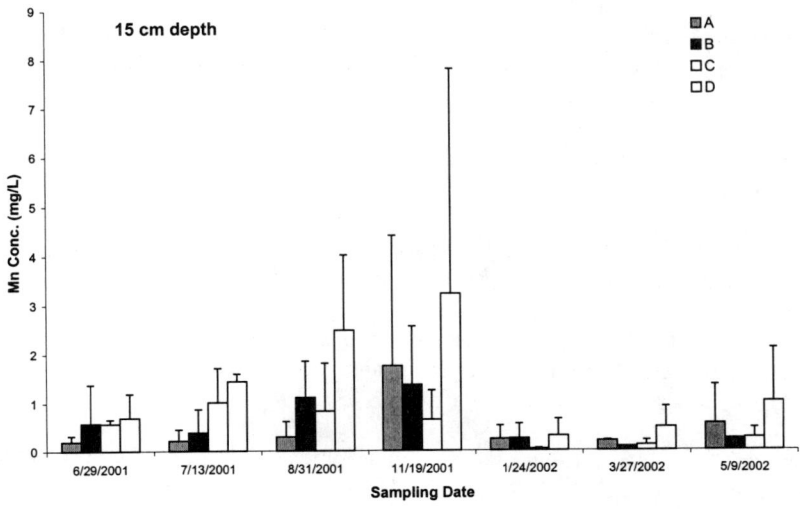

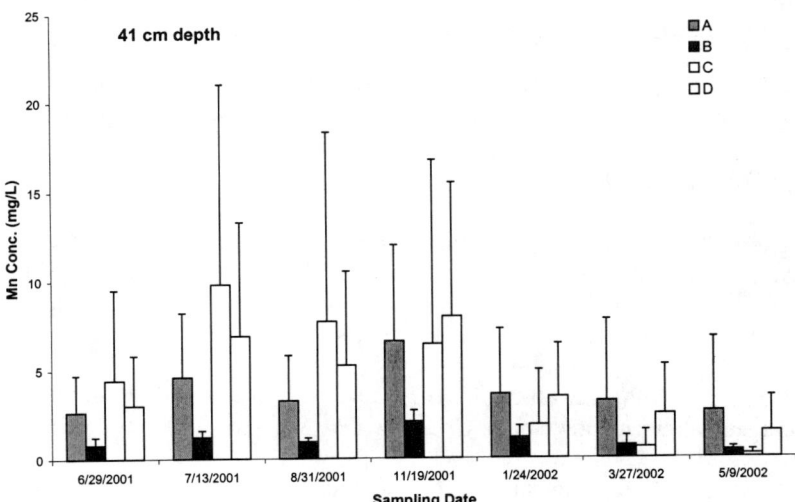

Figure 7. Mn concentration (mg L^{-1}) of soil solution /leachates from two sampling depths (15 and 41 cm) (mean and standard deviation where n=3)

ACKNOWLEDGEMENTS

This research was supported by Financial Assistance Award number DE FC09-96SR18546 from the DOE to the University of Georgia Research Foundation.

REFERENCES

1. Punshon, T., Seaman J. C., and Sajwan, K. S., Introduction: The production and use of coal combustion products, in *Chemistry of Trace Elements in Fly Ash*, Sajwan, K.S., Alva, A.K. and Keefer, R.F., Eds., Kluwer Academic Publishers, 2002.

2. Adriano, D. C., *Trace Elements in Terrestrial Environments*, Second Edition, Springer, New York, 2001, Chap. 3.

3. Adriano, D. C. and Weber, J. T., Influence of fly ash on soil physical properties and turfgrass establishment, *J. Environ. Qual.*, 30, 596, 2001.

4. Adriano, D. C., Page, A. L., Elseewi, A. A., Chang, A. C., and Straughan, I., Utilization and disposal of fly ash and other coal residues in terrestrial ecosystems: a review, *J. Environ. Qual.*, 9, 333, 1980.

5. Carlson, C. L. and Adriano, D. C., Environmental impacts of coal combustion residues, *J. Environ. Qual.*, 22(2), 227, 1993.

6. Keefer, R.F., Coal ashes - Industrial Wastes or Beneficial Byproducts? in *Trace Elements in Coal and Coal Combustion Residues*, Keefer, R. F. and. Sajwan, K. S., Eds., Lewis Publishers, Ann Arbor, 1993, 3-9.

7. Rowe, C. L., Hopkins, W. A., and Coffman, V.R., Failed recruitment of southern toads (*Bufo terrestris*) in a trace element-contaminated breeding habitat: direct and indirect effects that may lead to a local population sink, *Arch. Environ. Cont. Toxicol.*, 40, 399, 2001.

8. Barton, C. D., Marx, D. C., Adriano, D. C., and Bartley, H., Use of a vegetative cover to control acidic drainage from coal combustion waste, in Proceeding of the American Society For Mining Reclamation; 19th Annual National meeting, Barnheisel, R. I., Ed., ASSMR, Lexington, KY, 2002.

9. Evangelou, V. P., and Zhang, Y. L., A review: pyrite oxidation mechanisms and acid mine drainage prevention, *Crit. Rev. Env. Sci. Tec.*, 25(2), 141, 1995.

10. Nordstrom, D. K., Aqueous pyrite oxidation and the consequent formation of secondary iron minerals, in Acid sulfate weathering, Soil Science Society of America Special Publication 10, Kittrick, J.A., et al., Ed.,. Madison, WI, 1982.

11. Kleinman, R., L., P., and Rastogi, V., Reducing acid mine drainage liabilities using bactericides & other control technologies, in 13th Annual National Meeting American Society for Surface Mining and Reclamation Workshop #8, 1996.

12. Kleinman, R., L., P., and Erickson, P. M., Control of acid drainage from coal refuse using anionic surfactants, Bureau of Mines Report of Investigations 8847, 1983.

13.   Barton, C., Romanek, R., Seaman, J., and Paddock, L., Geochemistry of an abandoned landfill containing coal combustion waste: implications for remediation, in *Chemistry of Trace Elements in Fly Ash,* Sajwan, K.S., Alva, A.K. and Keefer, R.F., Eds., Kluwer Academic Publishers, 2002.

14.   Blowes, D. W., Reardon, E. J., Jambor, J. L., and Cherry, J. A., The formulation and potential importance of cemented layers in inactive sulfide mine tailings, *Geochem. Cosmochim. Acta*, 55, 965, 1991.

15.   Nordstrom, D. K., Alpers, C. N., Ptacek, C. J., and Blowes, D W., Negative pH and extremely acidic mine waters from Iron Mountain, California, *Environ. Sci. Technol.*, 34, 254, 2000.

16.   Thomas, G. W., Problems encountered in soil testing methods, in *Soil Testing and Plant Analysis, Part 1*, SSSA Spec. Publ. 2, SSSA, Madison, WI. 1967, 37-54.

17.   Thomas, G. W., and Hargrove, W. H., The chemistry of soil acidity, in *Soil acidity and liming*, 2nd Ed. Agronomy Monographs 12, ASA, CSSA, and SSSA, Adams, F., Ed., Madison, WI, 1984, 3-56.

18.   Sparks, D. L., *Environmental Soil Chemistry*, CA Academic Press, San Diego, 1995, Chap. 10.

ABOUT THE EDITORS

Kenneth S. Sajwan is a Professor and the Director of the Environmental Science Program in the Department of Natural Sciences and Mathematics at Savannah State University. Dr. Sajwan earned a B.S. in Agriculture and Animal Husbandry, an M.S. in Agronomy from India, and Ph.D.'s in Science from the Indian Institute of Technology, Kharagpur, India, and in Soil Chemistry and Plant Nutrition from Colorado State University. After completing his Ph.D. from Colorado State University he devoted nine months as a Reader to teaching and research at the graduate level in Water Use Management at the Indian Institute of Technology-Roorkee (formerly University of Roorkee) in India. Dr. Sajwan joined Savannah State University during the Fall of 1992 as an Associate Professor and was promoted to full Professor in 1996. Prior to coming to Savannah State University, he had worked as an Assistant Professor at the University of Georgia's Savannah River Ecology Laboratory in Aiken, South Carolina. His previous work experience also includes a World Bank consultancy to Columbia, South America, and research associateships at the University of Wisconsin and the University of Kentucky. Dr. Sajwan also holds an Adjunct Professorship appointment at Alabama Agricultural & Mechanical University and the University of South Carolina at Aiken.

Dr. Sajwan has taught a number of courses during his career at Savannah State University. These courses include Introduction to Environmental Science, Limnology, Contaminant Hydrology, Environmental Ethics, Environmental Law, Environmental Chemistry and Analysis, Environmental Health Safety and Risk Assessment, Ecology and Evolutionary Biology, Hazardous Waste Management, Principles of Ecotoxicology, Special Topics, Seminar and Senior Research. Dr. Sajwan has been recognized as a devoted and talented teacher and his accomplishments are reflected in the success of his students and his uncanny ability to motivate, challenge and inspire his students to excel academically both in the classroom and beyond. Dr. Sajwan has received several awards for his outstanding contribution to teaching and research. Dr. Sajwan is a recipient of the Board of Regent's University System of Georgia Teaching Excellence Award for the year 2002, and the White House Initiative Millennium Award for Teaching and Research Excellence for the year 2001. In addition, he is the recipient of the Board of Regents' of the University System of Georgia's Distinguished Professor of Teaching and Learning Award for the 1998-99 academic year at Savannah State University. Dr. Sajwan is also the recipient of the 1999 International Award for Innovative Excellence in Teaching, Learning, and Technology for the year 1999.

Dr. Sajwan has edited two books, *Trace Elements in Coal and Coal Combustion Residues, and Biogeochemistry of Trace Elements in Coal and Coal Combustion Byproducts,* and two laboratory manuals, *General Biology Laboratory Manual and Environmental Science Laboratory Manual*, and has published over hundred (100) articles in peer reviewed journals, serials, conference proceedings, and symposia. Dr. Sajwan is very well known nationally and internationally in his area of research. His primary research includes biogeochemistry of trace elements, environmental chemistry, ecotoxicology, and chemical equilibrium in soils. Currently, Dr. Sajwan is investigating the potential benefits and environmental impact of the application of coal ash and organic waste mixtures to agricultural lands for crop production.

335

Ashok K. Alva is currently a Research Leader at the USDA-Forage and Vegetable Crops Unit Laboratory, Prosser, Washington. He received his Ph.D. degree in Soil Chemistry and Plant Nutrition from the Pennsylvania State University in 1983. He has a wide range of research expertise in Universities in the United States, Australia, Denmark, Thailand and India. In addition to several invitational visits to international universities and research institutes, he has been involved in cooperative research projects in Brazil, and China.

He has conducted extensive research on the use of coal combustion byproducts as agricultural soil amendments including fly ash and the gas desulfurization gypsum. His research expertise includes: reaction and transport of agricultural chemicals in soils; heavy metal toxicity to plants and techniques to alleviate such toxicities; soil solution chemistry-speciation of metals in soil solution using thermodynamic chemical speciation models; sequential fractionation of metals and relationship between the sequential fractionation and single extraction; chemistry of acid soils and factors affecting plant growth; application of tissue analysis for diagnosis and correction of nutritional status of plants. His current research also includes development of irrigation and nutrient best management practices in sandy soils to increase the nutrient uptake efficiency, maintain economically competitive agricultural production industry, and minimize non-point source pollution of groundwater.

He has authored and co-authored 19 invited book chapters; and over 208 papers in refereed journals, conference proceedings, and extension publications. Dr. Alva has made over 95 invited presentations at international, national, and local meetings. He has supervised 13 postdoctoral fellows, 7 visiting scientists, and several graduate students. Dr. Alva is a Fellow of the American Society of Agronomy, an Associate Editor of the Journal of Environmental Quality, a member of the Editorial board for the Journal of Crop Production, a Co-Editor for the Florida State Horticultural Society Proceedings (1995-98), and serves as a technical reviewer for 14 international journals and several competitive grant proposals. He received the University of Florida Meritorious Faculty Performance Awards in 1997 and 1998, and a Superior Research Leadership Award from the USDA-ARS - Pacific West Area Director, 2000 and 2001.

Robert F. Keefer is an Emeritus Professor from the Division of Plant and Soil Sciences, College of Agriculture and Forestry, at the West Virginia University. He was awarded a B.S. degree in General Agriculture from Cornell University and M.S. and Ph.D. degrees in Soil Science (Agronomy) from the Ohio State University. He worked two years as an organic chemist with Hercules Powder Company, and then began his academic and research career in 1965 at West Virginia University. For thirty years, he taught courses in Soil Fertility, Soil Conservation and Management, Advanced Soil Fertility, the Chemistry of Soil Organic Matter and was part of a team-taught course on Plant Disorders. He devoted a year to teaching, conducting research, and developing a graduate program in soil science at Makerere University in Uganda and assisted one of his graduate students in Togo (West Africa).

Dr. Keefer's research has been broad-beginning with soil fertility and soil test correlation, then branching into plant nutrition/soil chemistry relationships, especially dealing with micronutrients. Interest over the years shifted to use of agricultural manures, municipal wastes such as sewage sludge, and industrial byproducts such as coal ashes and sawdust, particularly with respect to environmental aspects of plant nutrition, toxicities, and heavy metal transport in soils, waters, plants, and animals.

The geographical positioning of West Virginia in the center of the eastern U.S. coalfield and the recent concern with maintaining or improving the quality of our environment led Dr. Keefer to develop a research program emphasizing constructive use of coal combustion by-products in reclaiming surface-mined land. His proficiency in this area was evidenced by the many calls he received on this topic; besides being asked to be a keynote speaker at the International Conference on Metals in Soils, Waters, Plants, and Animals held in Orlando, FL in April-May 1990. Dr. Keefer's sabbatical was devoted mainly to producing a digest of research on the chemical composition and leaching characteristics of coal combustion by-products by the Electric Power Research Institute (EPRI). He was a major contributor to a book section on arsenic mobilization and bioavailability in soils published in Advances in Environmental Science and Technology, entitled *Arsenic in the Environment*.

His latest accomplishment has been writing a book entitled, A Handbook of Soils for Landscape Architects, to be published by Oxford University Press this year (2000). This book was developed from a course in soils that Dr. Keefer taught at WVU for landscape architectural students. The handbook contains soil science information that can be readily understood and used by persons who do not have a scientific background, but wish to know more about soils. The many illustrations clarify the importance of soils information in growing plants for landscape applications.

INDEX

long-term leaching behavior of, 87
Infiltration, 277
Iron deficiency, 255
Iron sulfide, 136-137
Iron, 83, 254, 258, 324, 329, 330
Iron-oxidizing bacteria, 107, 320

Jarosite, 136-137
Jurbanite, 135-137

Kaolinite, 17, 33, 91, 94, 237-238
   Fe and Al coated, 237, 240-247
   physico-chemical characteristics of, 239

Land application, 251-261; *see also* Soil
      amendments, US EPA 503 regulation
   biosolids, 253
   CCBs, 266
   coal ash, 252
   FGD in agriculture, 265-267
   fly ash, 26
   implications of trace elements, 265-267, 271
Landfill Directive, 46
Landfills, 1, 4, 10, 13, 50, 64, 189-190, 204,
      252
   disposal of fly ash/coal reject materials, 321
   geochemical investigations, 105-140
   remediation of, 138
   use of dry cover, 105, 107-108
Langmuir isotherm, 184-185
Leachate, 29
   chemistry, 76
   salinity, 265, 271
Leaching behavior
   elemental transport/distribution in amended
      soils, 189-201
   cumulative effects, 197-199
   experimentation, 190-192
   impacts on groundwater, 22, 201, 258-260
   implications of landfilling, 75
   industrial waste, 87
   major nutrient elements, 194-197, 199-
      200
   metals, 83, 197-198
   of fly ash, 83, 329-331
   of textured soils, 189
   of trace elements
   procedures for assessment of, 75-76, 79-80
   risk assessment, 79-88
   significance of pH, 193-194
   stages of, 25-42
Lead, 189, 198-199, 219-220, 260, 290, 298-
      300, 302
   desorption/remobilization perspectives, 219-

234
Legume crops, 290, 295
Lepidocrocite, 135-137
Ligand exchange, 238, 241, 246
Lime supplement, 276
Limestone, 63-66
   agricultural uses of, 63-64
Liming agent, 92, 143
Lining of irrigation canals, 26
Loblolly pine seedlings, 319, 323
Lycopersicon esculentum, 251

Macroporosity, 219, 232
Maghemite, 136-137
Magnesium, 194, 196, 200, 266
Magnetite, 135-137
Maize, 255
Manganese, 45, 92, 255, 259, 324, 329, 331
Matrix potential, 91-92, 96-97
Maximum contaminant level (MCL), 260
Maximum contaminant level goal (MCLG), 260
Medicago sativa, 252
Mercury, 29, 45, 55
Mesocosm, 265-267
Metal leaching behavior, 83
   trends, 83
Metal mobility, 220
Metal speciation, 20
Metal standards
   for soil amendment products, 70-73
Metal-organic complexes, 230
Micas, 161
Microbial methylation, 205, 213
Microspheres, 2
Mine fills, 78
   use of CCBs in, 78-88
Mine soil remediation, 309-312
Mine Water Leaching Procedure (MWLP), 75-
      76, 79-80
   benefits of, 88
   comparison with TCLP, 87
   environmental risk evaluation by, 79-80
      case study on CCBs, 79-88
   predicting long-term leaching behavior of
      wastes, 87-88
Mineral solubility, 134
Mineralization, 33, 42
Molybdenum, 255, 257, 260, 265
   adsorption studies on coated kaolinite, 239,
      245-247
    effect of pH, 238, 245-247
   toxicity of, 258
Monomethylarsenate, 205
Montmorillonite, 91, 94